Common Goals for Sustainable Forest Management

Common Goals for Sustainable Forest Management

Divergence and Reconvergence of American and European Forestry

EDITED BY:
V. Alaric Sample and Steven Anderson

IN COLLABORATION WITH:
Franz Schmithüsen
Dominique Danguy des Deserts
Dennis C. Le Master
Patrice Harou

FOREST HISTORY SOCIETY
DURHAM, NORTH CAROLINA

PINCHOT INSTITUTE FOR CONSERVATION
WASHINGTON, DC

The Forest History Society is a nonprofit, educational institution dedicated to the advancement of historical understanding of human interaction with the forest environment. The Society was established in 1946. Interpretations and conclusions in FHS publications are those of the authors; the Society takes responsibility for the selection of topics, the competence of the authors, and their freedom of inquiry.

The Pinchot Institute for Conservation is a nonprofit natural resource policy, research, and education organization dedicated to leadership in conservation thought, policy, and action. The Institute was established in 1963.

This book is published with support from the U.S. Forest Service, the Pinchot Institute for Conservation, and the Lynn W. Day Endowment for Forest History Publications.

Printed in the United States of America

Forest History Society
701 William Vickers Avenue
Durham, North Carolina 27701
(919) 682-9319 / www.foresthistory.org

©2008 by the Forest History Society. All rights reserved. No part of this publication may be reproduced or transmitted in any form or by any means, electronic or mechanical, including photocopying, recording, or by any information storage and retrieval system, without permission in writing from the publisher.

First edition

Design by Zubigraphics, Inc.

Library of Congress Cataloging-in-Publication Data

Common goals for sustainable forest management divergence and reconvergence of American and European forestry / edited by V. Alaric Sample and Steven Anderson.
   p. cm.
   Summary: "This book contains presentations from two colloquia that were conducted in 2005 in recognition of the centennial of the U.S. Forest Service. Collectively, they explore the history of forest management and practice in the United States and Europe in relation to current and future challenges to sustainable forest management"--Provided by publisher.
   Includes bibliographical references.
   ISBN 978-0-89030-070-1 (pbk. : alk. paper)
   1. Sustainable forestry--United States--Congresses. 2. Sustainable forestry--Europe--Congresses. 3. Forest management--United States--Congresses. 4. Forest management--Europe--Congresses. 5. Forests and forestry--United States--History--Congresses. 6. Forests and forestry--Europe--History--Congresses. I. Sample, V. Alaric. II. Anderson, S. (Steven), 1956–
   SD143.C568 2008
   634.9'20973--dc22
                           2008001708

CONTENTS

Acknowledgments ix

Foreword
Dale Bosworth xiii
Jean-David Levitte xvii

Introduction
V. Alaric Sample 1

Common Roots
Chapter 1. European Forest Conditions Prior to 1805
David Adams 12

Chapter 2. European Silviculture and the Education
of Gifford Pinchot in Nancy
Marie-Jeanne Lionnet and Jean-Luc Peyron 30

Chapter 3. A Historical Perspective on
French-U.S. Forestry Cooperation
François Le Tacon, Jean Pinon, and Francis Martin 40

Chapter 4. Science and the Forest:
Achievements, Evolution, and Challenges
Yves Birot and François Houllier 54

Chapter 5. Forest Policy in America as a Developing Nation:
Jeffersonian Democracy, the Taming of the American
Wilderness, and the Rise of the Conservation Movement
Michael Williams 72

Divergence in European and American Forestry
Chapter 6. Le Coup d'oeil Forestier:
Shifting Views of Federal Forestry in America, 1870–1945
Char Miller 94

Chapter 7. Breaking Old Stereotypes:
John Muir, Gifford Pinchot, and American Forestry
John Perlin 114

Chapter 8. Back to the Future: The Rise, Decline, and Possible Return of the U.S. Forest Service as a Leading Voice for Conservation in America, 1900–2000
Paul Hirt 124

Chapter 9. Man, Nature, and Forest: The Great Debates of Ideas
Christian Barthod 154

Chapter 10. The Evolution of Forest Management in Europe
Heinrich Spiecker. 168

Reconvergence in Sustainable Forest Management

Chapter 11. The Emerging Consensus on Principles of Sustainable Forest Management: Common Goals for the Next Century of Conservation
V. Alaric Sample 190

Chapter 12. European Forests:
Heritage of the Past and Options for the Future
Franz Schmithüsen. 216

Chapter 13. The Future of Wood in Our Evolving Societies
Michel Vernois 250

Chapter 14. The National Forest Program and the National Strategy for Biodiversity in France
Cyrille Van Effenterre and Jean-Jacques Bénézit. 258

Chapter 15. New Challenges for Forestry in Germany
Konstantin von Teuffel. 264

Chapter 16. The Continuing Evolution in Social, Economic, and Political Values Related to Forestry in the U.S. and Europe
Dennis C. Le Master and Franz Schmithüsen. 278

Chapter 17. European and U.S. Influence on Forest Policy at the Food and Agriculture Organization of the United Nations
Jean-Paul Lanly 300

Chapter 18. The International Dialogue on Forests: Convergences and Divergences between Europe and the United States
Gérard Buttoud 320

Chapter 19. The Role of Forest Conservation in Meeting Global Challenges of the 21st century: The Necessity for International, Multisectoral Cooperation
Jeffery Burley 328

Chapter 20. The Evolution of Forestry Education in the United States and Europe: Meeting the Challenge of Sustainable Forestry
J. E. de Steiguer, Patrice Harou, and Terry L. Sharik 346

Epilogue 371

Appendix
Agenda
Toward Sustainable Forest Management: The Divergence and Reconvergence of European and American Forestry
A Colloquium in Commemoration of the Centennial of the United States Forest Service 375

Contributors 387

ACKNOWLEDGMENTS

This book is a testament to what can be accomplished through cooperation across national boundaries and among public and private organizations all dedicated to a common purpose—advancing the conservation and sustainable management of the world's forests. Based on contributions from numerous individuals, this anthology celebrates the centennial of the U.S. Forest Service. Forestry leaders from many countries came together in a year-long symposium that bridged two continents. It was an opportunity for everyone involved to reflect on the legacy from which this unique organization sprang, and the leadership responsibility it has now assumed.

The Pinchot Institute was privileged to lead the organization of this symposium, and I would like to take this opportunity to thank all who helped make it possible. First and foremost, I thank U.S. Forest Service Chief Dale Bosworth for his leadership, support, and energetic participation throughout this effort, with the able assistance of Jerilyn Levi, Stana Federighi, Denise Ingram, Gloria Manning, Richard Paterson, and Carla Hawley. Ed Brannon and the Forest Service staff at Grey Towers National Historic Site not only provided much of the original inspiration for recognizing the role of European forebears in influencing the education of Gifford Pinchot and the development of the U.S. Forest Service, but graciously hosted an important segment of the symposium as well.

Dominique Danguy des Deserts, Director of the École Nationale du Génie Rural des Eaux et des Forêts (ENGREF), the French national forestry school at Nancy, enthusiastically hosted the other major segment of the symposium. He and the rest of the ENGREF faculty and staff warmly welcomed the delegation of Americans come to express their gratitude for ENGREF's role in educating Gifford Pinchot and several other early leaders in American forestry. Thanks to Dominique and to Librarian Jean-Marie

Lionnet, the bronze bust of Gifford Pinchot presented to ENGREF by the Forest Service and Pinchot Institute will have an honored place in the school's library, where it will remind future generations of the contributions that ENGREF made, through Pinchot, to the beginnings of American forestry.

As Pinchot himself noted in his autobiography, *Breaking New Ground* (Harcourt Brace, 1947), he gained important insights into the practice of forestry, following his studies at ENGREF, during visits to the Sihlwald, the city forest of Zurich, and to the Black Forest in Germany. Franz Schmithüsen, Professor at the Swiss Federal Institute of Technology; Konstantin von Teuffel, Director of the Forest Research Institute of Baden-Württemberg; and Heinrich Spiecker, Professor at the University of Freiburg helped symposium participants share in those insights, with valuable new perspectives gained in the intervening century. I am grateful to them for the time and resources they and their associates devoted to organizing and hosting these important segments of the symposium.

A number of our American colleagues were eager to demonstrate how the basic concepts of forestry have been adapted and applied in the United States. Pennsylvania State Forester Jim Grace and his staff at the Michaux State Forest, Cradle of Forestry Director Michael Milosch, Forest History Society President Steven Anderson, the Pisgah National Forest's Randy Burgess and Monica Schwalbach, Forest Service Southern Research Director Peter Roussopoulos, and Bill Alexander of the Biltmore Estate all helped ensure that our European colleagues came away from the U.S. segment of the symposium with a full appreciation of the richness and diversity of American forests and forestry.

I also thank His Excellency Jean-David Levitte, French Ambassador to the United States, for his vision and commitment to building on the past century of cooperation between our two countries on forestry matters, and to continuing this cooperation in the future as we address new challenges to forest conservation and sustainability. For reinforcing the need for this kind of cooperation and its essential role in improving global environmental stewardship, thanks are due also to John Turner, U.S. Assistant Secretary of State for Oceans and International Environmental and Scientific Affairs; Michael Johanns, U.S. Secretary of Agriculture; and Mark Rey, U.S. Undersecretary of Agriculture for Natural Resources and Environment. I am grateful to Peter O'Donohue of the U.S. State Department for his help in making this exchange possible.

The challenge of capturing all this in the book you now hold in your hands fell to an editorial team consisting of Franz Schmithüsen, Purdue University's Dennis LeMaster, Dominique Danguy des Deserts, ENGREF Visiting Professor Patrice Harou, Steven Anderson, and myself. Steven Anderson, President and CEO of the Forest History Society, very capably led the compilation and production of the book, with invaluable assistance from Sally Atwater and Jamie Lewis. To these my colleagues and friends, I want to offer a heartfelt thanks for all their investment of time and resources in selecting authors, translating documents, and editing manuscripts. Finally, I'd like to express my appreciation to the authors themselves, and the many others who contributed in ways both large and small to this effort to honor this legacy of encouragement that was given at the right place and time to make a difference, and to celebrate a renewed commitment to cooperation as we move forward to address the conservation challenges of a new century.

—V. Alaric Sample
President
Pinchot Institute for Conservation

FOREWORD

A Growing Transatlantic Convergence

Dale Bosworth
Chief, USDA Forest Service

Among the very hopeful developments we're witnessing on both sides of the Atlantic, I see a growing basis for international collaboration in the area of forest conservation. Conservation has deep roots here in Europe, where forestry has been practiced for centuries. But when Europeans settled America, many thought they'd discovered unlimited forest resources, a place where forestry wasn't needed.

For centuries, America's forests were routinely mined for timber, with no thought to the future. Gifford Pinchot summed it up this way: "To waste timber was a virtue and not a crime." Those who questioned the waste were thought of, according to Pinchot, as "more or less touched in the head."

But there were a few "crazy" visionaries in the 19th century, and Pinchot was among them. Yet there was no profession of forestry in the United States, and certainly no academy for producing foresters. Gifford Pinchot and other American forestry pioneers had to go to Europe for their training. Luckily, many of them found the training they needed here in France, at the National School of Forestry. Pinchot later wrote that without the European connection, "groping in the murk of American public indifference, we would have been lost indeed."

As part of his training, Pinchot traveled through French, German, and Swiss forests, where he was deeply impressed by the multiple-use management he saw. Wood and wildlife as well as hiking and hunting were all integrated into forest management. What Pinchot learned during this time laid the groundwork for his vision of sustainable forest management in America.

20th-century Divergence

So we owe Europe a debt of gratitude for helping to get forestry and conservation going in the United States. France in particular helped us pull ourselves up by our own bootstraps. But what happened next, during the 20th century, was a long period of divergence between forestry practices in Europe and in America. There are at least a couple of reasons for that:

- American foresters had to adapt European approaches to their own ecological, economic, and social circumstances. You have to remember that, in forestry terms, we were still a developing country. And conditions in America are just so different that we've sometimes had to develop completely different management approaches. Wilderness protection and the history of fire use in America are good examples.
- Social and political events also carried our countries in different directions. Following World War I, the heavy reparations demanded of Germany led to strains on German forests. And following World War II, the focus on federal forestland in the United States shifted to timber production while state and private stocks recovered from the war. By the 1990s, delivering enormous quantities of timber along with clean air and water, habitat for fish and wildlife, outdoor recreation, and all the other uses had strained our ability to meet public expectations.

Renewed Convergence

But now I think we see a growing transatlantic convergence again. For one thing, we all confront the same global challenges, including climate change, invasive species, biodiversity loss, and ecological decline. Through global markets and consumption choices, the challenge of sustainable forest management has become increasingly international.

For another thing, we now share a common vision of sustainable forest management. We also share a common language for sustainability on the ground, through things like certification and community-based forestry. Before closing, let me just say a few words about community-based forestry.

Eighty percent of the world's poor depend on forest resources, and more than a billion poor people live in the world's 19 biodiversity hotspots. If we want to protect biodiversity worldwide, then we have got to give local communities a stake in the land. More and more governments are engaging communities in managing their local forest because they see that the best caretakers are those who know and depend on the land the most. We're seeing a global trend toward community-based forestry.

In Europe, many forests are communally managed; you have long had strong traditions of community-based forestry. In several countries, for example, the Pan-European Forest Certification system has established special criteria for cooperatives formed from groups of smallholders. It also has special criteria for community-based forest enterprises in France.

Maybe we can learn something from that. There's much we've learned in the past from France and other countries, and I believe that there's much we can learn in the future.

Opportunity to Work Together

That brings me to my final point: Since we face common global challenges —and since we share a common vision of sustainable forest management— it only makes sense for us to look for common solutions. None of us want to have to reinvent the wheel. We ought do everything we can, based on the values we share, to cooperate in developing new forest science, new forestry practices, and new management policies that will promote sustainable forest management on both public and private lands.

Peter O'Donahue of the U.S. State Department mentioned several recent initiatives in which France and the United States have cooperated to assist developing nations in Africa, Asia, and elsewhere in the world to better conserve and manage their forests. It was not so long ago that the United States was a developing country and benefited from France's assistance. So we know how valuable the right kind of help at a critical moment can be. We at the U.S. Forest Service look forward to being a part of such efforts, together with our European colleagues in forestry.

That's why we're here, and I'm full of hope for the future. In this time of transatlantic convergence, there's much to be gained for all of us from thinking together globally—and maybe even acting together locally in some places—for the benefit of future generations around the world.

FOREWORD

Commemoration of the Centennial of the U.S. Forest Service

Jean-David Levitte
French Ambassador to the United States

I am very pleased to welcome your group and celebrate with you a century of forestry friendship across the Atlantic. I would like first to thank the American institutions that participated in the success of the various events the European delegation of foresters enjoyed on this side of the Atlantic: the USDA Forest Service, the Pennsylvania Bureau of Forestry, the Society of American Foresters, the Forest History Society, and the National Forest Foundation. And I'd also like to express special thanks to the Pinchot Institute for Conservation, which has generously contributed to this farewell gathering.

The theme of your talks this week was "Working toward Common Goals in Sustainable Forest Management." A very modern concept, it seems. Not so. The first mention of sustainable management dates back to 1346: It appears in one of the founding texts of "modern" French forestry, *l'Ordonnance de Brunoy*, in which the then-king of France, Philip VI of Valois, ordered the forest officers to manage their forests in a sustainable manner. And I quote: *pour que lesdites forêts et bois se puissent perpétuellement soustenir en bon état* ("so that these woods and forests may forever sustain themselves in good shape"). You will have noted with interest that the English language is thus indebted to French for two key words of the economic vocabulary: entrepreneur, of course, but also sustainable.

Over the course of the centuries, as a result of demographic pressure, the amount of forest land in France declined, often giving way to farming.

Natural disasters multiplied as a consequence of erosion. Over time, scientists came to realize the role of forests in maintaining the overall ecological balance. The state passed much stronger legislation protecting forests and reviews them on a regular basis.

All French forests have been managed in a sustainable way for a long time now according to simple guiding principles along a three-pronged approach—protection, production, and social function:

- Protection—that is, conserving the environment—is the basis of forestry management and dictates all the rest.
- Production is developed wherever it is compatible with forest protection.
- The social function allows people to keep in touch with nature.

All three functions are carried out simultaneously. This is perhaps one of the differences between France and the United States, due to the smaller size of the forests in France.

Finally, if I may, I'd like to touch very briefly on the international action taken by France to address the important issues involving forests worldwide, because, as we all know, while French and American forests are sustainably managed, this is not the case everywhere. France has taken a strong stand on illegal logging and will be actively involved in this issue, advocating much tougher controls and sanctions. This issue is a priority in the French facilitation of the Congo Basin Forest Partnership initiative.

France also participates in the European Union's program for forest law enforcement, governance and trade, prioritizing a partnership-based approach, and has recently adopted a wood products procurement policy as part of its national strategy on sustainable development. Enhanced forest law enforcement and better governance are key components of sustainable forest management. Producing and consuming countries must work on this jointly, making use of all means, including trade, dialogue between stakeholders, cooperation, and implementation of best practices.

This meeting between French and American foresters to honor the memory of Gifford Pinchot, who worked so hard to build an awareness of the necessity of sustainable management, proves that his efforts have been successful. As he famously said, "A nation deprived of its liberty may win it; a nation divided may reunite; but a nation whose natural resources are destroyed must inevitably pay the penalty of poverty, degradation and decay."

He was fortunate enough to be honored as a prophet in his own land. But much remains to be done. It will be easier if two great nations, France and the United States, both with a splendid forest heritage, pool their efforts, based on mutual comprehension and friendship. After this week, I have every confidence this will be the case. Thank you.

INTRODUCTION

V. Alaric Sample

On July 1, 2005, France and the United States pledged to expand their cooperation on forest conservation and management, with one another and in other parts of the world still striving to shift from unsustainable exploitation of their forest resources to sustainable forest management. Signing the agreement were Jean-Jacques Benezit, Director of International Affairs in the French Ministry of Agriculture, and Dale Bosworth, Chief of the U.S. Forest Service in the Department of Agriculture. Strong support for this renewed, high-level cooperation on forestry matters was voiced by Jean-David Levitte, French Ambassador to the United States; John Turner, U.S. Assistant Secretary of State for Oceans and International Environmental and Scientific Affairs; Michael Johanns, U.S. Secretary of Agriculture; and Mark Rey, U.S. Undersecretary of Agriculture for Natural Resources and Environment.

This pledge of renewed cooperation between France and the United States is symbolic in many ways. It was signed on July 1, the 100th anniversary of the signing of the Transfer Act of 1905, which established the U.S. Forest Service in the Department of Agriculture and transferred the responsibility for managing the federal forest reserves (now national forests) to the Forest Service from the Department of the Interior. Further, the agreement was signed at the very desk used by Gifford Pinchot when he served as the first Chief Forester of the United States and founding chief of the Forest Service.

There being no forestry school at any university in the United States during Pinchot's time, he had received his professional education at the French national forestry school, the École Nationale Forestière in Nancy, in 1889. Like Pinchot, many of the other early leaders in forestry in the United States received their training at European forestry schools. They brought back with them the sum of experience, and knowledge of forest science and forestry practice, developed in Europe over more than a thousand years.

Adapting this knowledge to the unique ecological, economic, and social circumstances in the United States at the time, Gifford Pinchot and his contemporaries launched not only the U.S. Forest Service but also the profession of forestry itself in the United States. Through Pinchot and others, Europe made a major contribution to accelerating the transition in the United States from our own unsustainable exploitation of forests to conservation and sustainable forest management as we know it today.

The signing of this agreement, on the centennial anniversary of the founding of the U.S. Forest Service, was in many ways a recognition of this important contribution and an acknowledgment of thanks to our European colleagues and forestry education institutions. It marked the culmination of an international colloquium organized by the Pinchot Institute, the U.S. Forest Service, and the École Nationale du Génie Rural des Eaux et des Forêts to examine the common roots of forest science and forestry practice, the divergent paths followed by European and American forestry during the 20th century, and the reconvergence that is taking place in the 21st century around common concerns, such as conserving biological diversity, protecting water quality, and promoting sustainable forest management in both developed and developing countries. Other sponsoring organizations included the Swiss Federal Institute of Technology (ETH), Forest Research Institute of Baden-Württemberg, University of Freiburg, Forest History Society, Pennsylvania Bureau of Forestry, National Forest Foundation, Society of American Foresters, Stihl, Blooming Grove Club, American Chestnut Foundation, and Biltmore Estate.

Divergence and Reconvergence

Forestry in Europe and the United States shares common roots, not only in terms of the practice of silviculture but also in the institutional, legal, and policy framework that forms the basis for sustainable forest management. Sustainable forest management, as the term is currently applied, explicitly incorporates ecological and social considerations as well as economic. European forestry institutions, especially educational institutions such as the École Nationale Forestière in Nancy, France, contributed in important ways to the introduction of basic principles of forestry to the United States in the late 19th century, and catalyzed the nation's transition from unsustainable exploitation of its forest resources to the conservation and sustainable management of those resources.

Early American forestry leaders who received their training in Europe, such as Gifford Pinchot, quickly recognized that the silviculture and forest science they had been taught there would have to be adapted to the very different circumstances prevailing in the United States, not only in terms of different forest types, but also to respond to important social, economic, cultural, and political differences. The institutional, legal, and policy framework for forestry in the United States developed along distinctly different lines than in Europe and continued to diverge throughout the 20th century. It also evolved at a far faster rate, such that during the second half of the 20th century, forestry in the United States was already struggling to address significant changes in social values and perspectives regarding forests and forestry—changes that are only now sweeping through forestry in Europe.

At the start of the 21st century, European and American forestry institutions are focusing on many of the same concerns—sustainable wood production, biodiversity conservation, protection of water quality, climate change mitigation, and sustainable economic development in rural communities as a few examples. This reconvergence is resulting in increased cooperation in the development of new forest science and technologies among individual scientists and forestry practitioners, and new strategic alliances among forestry institutions involved with research, technical assistance, and forest management. Increasingly, this cooperation is taking place not on European or American soil, but in developing countries in Africa, Asia, and Latin America that are at the point in their own history where they are struggling to make the transition from unsustainable resource exploitation to resource conservation and sustainable use.

Historical Framework

Sustainable forest management in Europe developed over a period of more than a thousand years, dating back to medieval edicts governing woodcutting and the taking of game animals in royal forests. As chronicled in the chapter by David Adams, the framework of legal principles underlying forest use and management goes back at least to the *Corpus juris civilis* compiled by the Roman emperor Justinian in the sixth century. The Romans introduced the concept of privately owned forests (*res in patrimonio*) to lands that previously had been treated as commons. Following the fall of Rome, the barbarians of central and northern Europe enacted the first Germanic forestry laws between the fifth and seventh centuries, promulgating fines and punishments for forest trespass and declaring all forests, except royal territories, commons subject to free public use.

Canute, ruler of England, Denmark, and Norway in the 11th century, established laws granting private ownership and use of forests and reserved royal forests for the protection of both wild game and the woods themselves. As populations in Europe increased, impacts on the forests also increased, prompting the enactment of forest protection laws in Europe and in Norman England. Tensions over the enforcement of these notoriously strict laws governing the use of forest lands helped give rise to the Magna Carta and its accompanying Magna Foresta in the 13th century. In the early 19th century, the Napoleonic Code swept away many of the remaining vestiges of feudalism and opened forests throughout the portion of Europe once conquered by Napoleon to private ownership and use.

Forest science and the practice of silviculture also came of age in Europe in the 18th and 19th centuries, as described in chapters by Marie-Jeanne Lionnet, Heinrich Spiecker, and Yves Birot and François Houllier. The concept of managing forests for a sustained yield of wood arose out of economic and social problems created by forest exploitation for shipbuilding, charcoal making, and other uses that made it difficult for local communities to meet their needs for fuelwood, fodder, and food. Selective harvesting systems based on coppicing (regeneration through stump sprouts), coppicing-with-standards (leaving occasional large trees to provide for forest regeneration from seeds as well as sprouting), and high forests (regenerated primarily through seeding and planting) helped sustain forests for a variety of uses, theoretically in perpetuity. Tree breeding and the introduction of new species brought about higher forest productivity, along with the use of even-age silvicultural systems involving the periodic clearing and regeneration of large areas of forests under the "regulated forest" concept.

Changing Science and Social Values

In much of Europe during the 20th century, preferred species of trees, such as Norway spruce and European beech, came to be planted over large areas, often with only a single species represented. In recent years, many problems with this approach have become apparent, including insect outbreaks, severe weather damage, and disease problems. These problems have had major economic impacts and have caused European forestry to shift back toward mixtures of commercial and native species. As noted in chapters by Franz Schmithüsen, Christian Barthod, and Konstantin von Teuffel, forestry in Europe is also changing in response to evolving social values and cultural perspectives regarding forests, and the need to provide greater protections to natural characteristics not usually found in large monocultures of nonnative tree species.

Ironically, forest scientists and forest managers have for most of the past two centuries focused on methods by which to maximize wood production; having been highly successful in accomplishing this, however, European foresters are finding that social goals relating to forests have changed in the meantime. An entirely new set of social and economic challenges has arisen in European forestry, and the traditional institutions of forestry research, forest management, and forestry education are struggling to meet these new challenges.

These kinds of challenges are not new to forestry in the United States, where, interestingly enough, environmental, economic, and social concerns arose decades earlier than in Europe. Chapters by Michael Williams and Char Miller describe some of the unique frontier values that shaped the early American view of forests and helped drive the wave of deforestation and forest exploitation that swept across America during the 19th century. It was the widespread environmental and economic damage from this exploitation that caused scientists and authors such as George Perkins Marsh, Charles Sprague Sargent, and John Aston Warder to sound the alarm and call for government action to halt the devastation of the nation's forest resources.

It was into this set of circumstances that young Gifford Pinchot was thrust, urged by his father to go to Europe to study forestry and bring back to America a more enlightened approach to utilizing its forests. The notion that a forest could be cut and at the same time preserved was indeed a foreign concept to 19th-century America, as it rushed to open its last frontiers, capitalize on its natural assets, and join the Industrial Revolution sweeping through Europe at the time. Pinchot's family itself had once made its fortune in the lumber business, clearing timber and abandoning the land in the style that was customary and accepted at the time, a fact that may have had some bearing on Pinchot's choice of profession.

Pinchot's conservationist tendencies are compared in a chapter by John Perlin with those of another icon of the American conservation movement, John Muir. Conventional wisdom holds that Muir regarded Pinchot's utilitarian approach to forests as anathema to his own preservationist approach, and that the hostility was mutual. In fact, Perlin points out, many of Pinchot's writings and public pronouncements at the time reveal a strong tendency toward forest protection. Use of the federal forest reserves and other public lands by local individuals was inevitable, Pinchot reasoned, so the most practical approach was to allow such uses but regulate them to prevent resource depletion or long-term damage to the land's productivity. Perlin likewise examines Muir's writings at the time and finds that he, too, understood this approach but also regarded some landscapes as almost sacred in their pristine form, to be held inviolate by any human exploitation. Unfortunately for history, and for the relationship between these two early conservation leaders, they differed over one particular landscape—Hetch Hetchy Valley in Yosemite. Particularly ironic is that, long after Muir's death in 1914, Pinchot increasingly favored strong governmental intervention to protect forests on private as well as public lands in the United States, eventually becoming highly critical of the close association between the lumber industry and his beloved U.S. Forest Service.

The evolution of the Forest Service into the nation's largest single timber producer by the mid-20th century had a major impact on the national forests, but also on the public perception of the Forest Service itself, as noted in the chapter by Paul Hirt. With Europe still reeling from the devastation of World War II, the U.S. economy was the fastest growing in the world at midcentury. The American spirit was one of unflagging optimism and confidence that, with a combination of economic resources and technological know-how, anything was possible. Forest science and the practice of forestry in the United States focused almost entirely on maximizing wood production and were very successful in achieving that goal. But as in Europe, social values and public preferences had shifted in the meantime. Forestry found itself out of step with the rest of society and subject to a storm of public criticism that foresters—most of whom considered themselves conservation-minded—struggled to understand. Now, after nearly four decades of controversy over timber harvesting and other forest practices, forestry in the United States seems to have come full circle, observes Hirt. Timber harvesting on the national forests has declined from previous unsustainable levels, and the focus has shifted more toward what it was a century ago—watershed protection, ecological restoration, forest health, maintaining forest extent, and wood harvesting mostly by regional and community-based firms for local processing and economic development.

Anticipating the Future

European and American forestry are facing similar challenge and opportunities in the 21st century. At no time in history has there been greater public interest in the conservation and sustainable management of forests—in Europe, in the United States, and throughout the world—than at present. There is widespread recognition that maintaining forests is an essential prerequisite to conserving biological diversity, including habitat for threatened or endangered plant and animal species as well as for game species. Protecting water quality from forested watersheds has become a critical need in many parts of the world as an increasing proportion of the population becomes concentrated in urban centers. Increasingly urbanized populations also mean that forests and other natural areas are becoming more important for outdoor recreation and relief from the pressures of urban life. More people are coming to understand the value of wood as a renewable resource, and that it can substitute for other kinds of building materials whose mining or manufacturing have a far greater impact on the natural environment. Most recently, there is growing recognition of the important role forests play in mitigating global climate change, either through sequestration of carbon dioxide or by substituting "biofuels" for fossil fuels in energy production, a major source of greenhouse gases.

The controversies and public debates over timber harvesting and other forest practices in Europe and the United States have stimulated many efforts to define sustainable forestry. The chapter by Al Sample describes how these many separate efforts have led to a remarkably consistent identification of "generally accepted principles of sustainable forest management." Increasingly, these principles are finding their way into international trade in forest products through new mechanisms like independent, third-party certification. "Green" certification gives confidence to the purchaser of a wood product that it is from a well-managed forest, whether the purchaser is an individual consumer or a company intent on demonstrating its commitment to environmental stewardship. Over time, certification will reward conscientious forest managers with greater market share, while gradually eliminating market access to wood from exploited or endangered forests.

These basic principles are becoming the core of forest management planning for the future, in both Europe and the United States, as described in chapters by Jean-Jacques Bénézit and Cyrille Van Effenterre, and Michel Vernois. Internationally, they are increasingly being manifested in the influences that European and American forestry professionals are having on key institutions such as the United Nations Food and Agriculture Organization and its forestry program, as described in the chapter by Jean-Paul Lanly. Gérard Buttoud expands further upon this in his discussion of the new concepts and policies emerging from the broader international dialogue on forestry, which increasingly involves private and nonprofit organizations as well as government entities. Author Jeffery Burley gives further emphasis, stressing that multilateral and multisectoral (e.g., private enterprises and nonprofit organizations, not just governments) cooperation and action will be needed if forest conservation and sustainable forest management are ever to be achieved at the global level. Sustainable forest management has a crucial role to play in poverty alleviation in developing countries, not only through maintaining community supplies of fuelwood and fodder, but also in protecting water supplies and water quality in rural areas often devastated by drought and water-borne diseases. As local economies are increasingly drawn into the global economy, developing countries are becoming the fastest-growing sources—and markets—for wood and wood-fiber products. Ensuring that these develop in ways that can be sustained over the long term will be a major challenge for the developing countries themselves, but also for multilateral development banks and sources of private capital that fund major forest development projects.

All of this has significant implications for forestry education in Europe, in the United States, and throughout the world. This colloquium was inspired by the important contributions that forestry education at European universities made to the United States at a critical stage of its development as a nation, by educating Gifford Pinchot and other early forestry leaders. Having recognized that the United States needed to develop forestry education programs of its own, Pinchot helped establish a new forestry school at Yale University. More than 50 additional forestry programs have since developed in the United States, mostly at state universities. But as Patrice Harou, Ed de Steiguer, and Terry Sharik observe, the enrollments in forestry programs at universities in both Europe and the United States have been steadily declining for several years. Forestry programs at many universities have been blended into broader programs in agriculture or environmental sciences. At other universities, the forestry programs have simply disappeared altogether.

What is particularly ironic is that this decline in university-based forestry programs is coming at a time of unprecedented worldwide public concern over forest conservation, when there has never been a greater need for competent, well-trained forestry professionals. These professionals are needed in the field, where they can develop a first-hand understanding of resource problems and their underlying causes and find effective ways to address them. But experienced, knowledgeable, and articulate forestry professionals are also needed at the highest levels of governments and private enterprises, to guide policy making so that it is practical and effective, and so that unintended negative consequences are avoided.

Forestry education in Europe and the United States has made important contributions to sustainable forestry over the past century. But in many cases, these institutions are not capable of preparing the next generation of forestry professionals for the new set of challenges they will be facing. How can forestry education adapt to these changing needs? Creative partnerships and strategic alliances that allow university-based forestry programs to combine their strengths and share resources internationally—such as distance learning programs that allow students around the world to take on-line classes with top professors at many different universities in a single degree program—will be essential to meeting the world's changing needs for forestry education.

Conclusion

This colloquium marked the centennial of the U.S. Forest Service and acknowledged the important role of European forestry educators in fostering the development of forest science and the practice of forestry in the United States. As François Le Tacon points out in his chapter, scientific and technical exchange in forestry between the United States and Europe has been going on for more than a century. But today's challenges in forest conservation and sustainable forest management will require far more than developed countries' assisting one another and learning from one another's experiences. There are many countries in the world that are today striving to make that same transition that was so important to the United States at the time of

Gifford Pinchot—from unsustainable exploitation of their forests to conservation and sustainable management.

The continuing differences in perspective between European and American forest scientists and managers are essential to being able to learn from one another, and to joining together to provide a diversity of expertise and experience to other countries. LeMaster and Schmithüsen point out that, although there has indeed been a convergence among European and American forestry professionals, important differences remain—as well they should. The use of different approaches to similar challenges allows us to continue learning from one another's successes, and one another's failures. We are thus less prone to the errors that sometimes arise from monolithic thinking. We remain open and receptive to the idea that one can always learn new and better ways of accomplishing the enduring goals of sustainable forest management, especially in a scientific and social environment that is ever changing.

In his 1911 book, *The Fight for Conservation*, Pinchot wrote, "A nation deprived of its liberty may win it; a nation divided may reunite; but a nation whose natural resources are destroyed must inevitably pay the penalty of poverty, degradation and decay." In our interconnected global society, no individual nation can suffer such a fate without affecting other nations halfway around the world. On the other hand, a nation that achieves success in sustaining its resources and its people becomes a positive force in the global economy and a contributing citizen in the global community.

Addressing the new and growing cadre of forestry professionals in the United States that he helped to inspire, Gifford Pinchot also wrote, "Our responsibility to the Nation is to be more than just good stewards of the land. We must be constant catalysts for positive change." Today, our responsibility is to the global community, and it is in part through expanded international cooperation that we will fulfill that responsibility to be constant catalysts for positive change and continue to advance conservation and sustainable forest management.

ABSTRACT

The story of forests in Europe and the British Isles begins with abundant forest resources. Overutilization, prompted by population growth and industrial development, resulted in degradation and deforestation. Realization of the situation brought such civil action as the Magna Carta and the related Carta de Foresta in England, and various laws in France, Germany, and other countries on the Continent. Necessity—the continued demand for wood, for which few substitutes existed—eventually gave rise to silvicultural practices. Thus many of the forest conditions and management practices of today have ancient roots.

CHAPTER 1

European Forest Conditions Prior to 1805

David Adams

Sometimes we look at ourselves and our activities as the be-all and end-all of forestry. We developed this. We invented that. We did it. Yet we and what we do are simply spots on an evolutionary path. If your definition of management is broad enough, forests were being managed since the first primitive hominoid broke limbs off trees to use as weapons or fuel. And many of the forest conditions and management practices of today have ancient roots. Deforestation of our eastern forests must have resembled the "wastage" of English forests; aesthetics and recreation are becoming more important in developing management objectives; and government regulations and restrictions are playing a greater role. So sometimes history *does* repeat itself, particularly if one disregards the lessons of the past.

We don't know much about the early uses of the forests. Consumptive use of forest products probably had little impact upon natural systems, but the wide-ranging fires of the aborigines surely did. Just when forest management—the institutional control of forest-related activities—started is unknown, but there seems to have been some during the days of King David (c. 1000 b.c.), for he, speaking as God, states, "For all the animals of field and forest are mine" (Psalm 50, v. 10). Phillip of Macedonia (400 b.c.) maintained forests for his hunting and other recreational pursuits. But the greatest forest notoriety probably came from development of the English royal forests during the Middle Ages.

"Forest," in the medieval sense, is a difficult concept for modern westerners to comprehend. As far back as Roman times, there was confusion between *sylva* or *saltus*, meaning woods, and *foresta*, meaning "a safe abiding place for wild beasts" (Manwood [1615] 1976: 32). The word "forest" is derived from the Latin *ferae*, meaning "wild," and *statio*, meaning "to remain," and thus referred to uncultivated, unoccupied, and undeveloped areas that might or might not be wooded. Lands devoted to the production of wood were, of course, "woods," and their custodians were "woodwards." The medieval concept of landownership beneath the level of sovereign monarchs was far different from what we are accustomed to today and consisted more of a *usufruct*—a right to use without title having been conveyed—granted by the sovereign; such rights were readily revocable. We tend to view landownership as virtually complete and resent any intrusion by government into private land management activities. Americans in particular tend to consider the rights of private ownership to be all-encompassing and exclusive and have the feeling that "Nobody is going to tell me what I can do on my land." These postulates are of course false, but they are deeply rooted in our population and form the reference against which other forms of ownership are compared.

Many forestry terms in common usage during the Middle Ages—justiciar, verderer, agistor, venery, pannage—have long since departed from our vocabulary. Add to these complexities the fact that the most comprehensive treatment of English lands during that time, Manwood's *Lawes of the Forest* [1615], is printed in Old English characters and partially written in Latin (I have difficulty reading Old English and don't read Latin). Much of the literature on continental conditions is French or German, and I don't read those either. And finally, for every provision, requirement, or restriction concerning forests and their use, means of circumvention exist—usually through payment of money. Thus there are many opportunities for error and ambiguity.

One of the first codifications of existing law was that compiled by the Roman emperor Justinian between a.d. 529 and 546 (Buckland 1963: 39–47). The code was written more as case law than as statutory law, citing cases tried before Roman judges and their decisions. The code did not address forestry concerns directly, but it did include legal concepts that affected forest ownership and use and were to become the cornerstones of much of present western law. Land was either outside private ownership (*res extra patrimonium*) or privately owned (*res in patrimonio*) (J. Inst. 2.2.1). If outside private ownership, it could be held in common, to be shared by all citizens (*res communis*), or owned by the state as public property (*res publicae*). Land in private ownership could be protected against trespass, but the animals living on it were owned by no one (*res nullius*) until or unless lawfully possessed (Dig. 41.1.5.1; Buckland 1963: 182, 204–206). Although the Justinian Code contributed little to forest management, it forms the basis for much of our concept of public trust and littoral, riparian, and maritime law—and for American wildlife law.

The next section of this paper, describing conditions in England, is based largely on the first chapter of *Renewable Resource Policy: The Legal-Institutional Foundations*, by David A. Adams (Washington: Island Press, 1993).

England

Following the dissolution of the Roman Empire, western Europe and the British Isles became the domain of what we today would call warlords, split into numerous fiefdoms, each under the control of its own king (Swindler 1965: 9). The controlling monarch owned all the land and the wildlife upon it, granting the privilege of use to loyal followers. Pursuit of game animals was a valued right of royalty, and fiercely protected. About a.d. 1018 the Danish king Canute established special laws regarding the forests of England, which he then ruled, and declared,

> *I will and grant that each one shall be worthy of such venery [hunting places] as he by hunting can take either in the plains or in the woods within his fee or dominion, but each man shall abstain from my venerie in every place, where I will that my beasts shall have firm peace and quietness, upon pain to forfeit as much as a man can forfeit.* (Manwood [1615] 1976: 30)

During the 49 years between Canute's reign and the arrival of William the Conqueror, the English forest laws were repeatedly expanded and amended (Manwood [1615] 1976). In 1615, Manwood colorfully described the evolution of the countryside thusly:

> *As I doe take it, a great part of our most ancient forests in England had there first being a wilderness full of great huge woods, because it was not inhabited with people, the same was also full of wild beasts almost of all sorts that are commonly known in England, and after*

> the same beganne to be inhabited with people, they did more and more destroy the woods and great thickets, that were neere unto the places which they did inhabit, so that will as the land increased and flourished with people, whose nature could not endure the aboundance of savage beasts, so cruelly annoy them as they then did, they sought by all means possible how to destroy such great woods and coverts, as were in any way neere unto their places of habitation, thereby to drive the wild beasts further from them. And so by that means the wild beasts were all driven to resort to those places, where the woods were left remaining … (Manwood [1615] 1976: 29, 30)

After the Battle of Hastings in 1066, Norman rulers voided all English landholdings, consolidating governance under William and strengthening the existing feudal system. They then returned much of the land to Normans, the clergy, and loyal Saxons contingent upon the landholders' vowing allegiance to the king and agreeing to provide him with men and provisions; in return, the tenant received protection (Swindler 1965: 10, 11). About this time, the concept of *res nullius* subtly also changed. Under Roman law, wild animals were truly owned by no one until lawfully possessed, but in medieval England they increasingly became the property of the landowner.

> And although men may kill such wild beasts in their wilderness, when they are found wandering, being out of any Forest, Chase, or Warren: Yet no man hath any property in them, untill they have killed them, for, during the time in their wilderness, they are Nullius in rebus, and then they must needes be said to be In manu domini regis, *in the king's possession:* And then ye king may priviledge them in any place where he will appoint, and so prohibit any man to kill or destroy them. (Manwood [1615] 1976: 26, 27)

Chases, warrens, and parks were "hunting preserves" of lesser status than forests. Royal game were protected on all, but only forests were subject to the Forest Laws. Thus wildlife became more the property of the king than of no one, and this property was conveyed to freemen along with the land.

Nobles and some citizens of lesser status did hold land, but these "freemen" constituted less than 10 percent of the English population. The concept and vocabulary of royal forests, brought to England by the Norman conquerors (Young 1979: 4), combined the feudal system with that of wild lands. The objective was to protect game for the king's use and enjoyment, not to grow trees. Initially, royal forests included only the king's own land, consisting largely of scattered parcels of unoccupied wasteland, but they grew to encompass at least one-fourth of England, including privately owned woods, pastures, agricultural fields, and even whole towns (Young 1979: 5). By the early 1600s, some 128 royal forests existed, at least one in almost every county, not counting royal chases, warrens, and parks (Cross 1928: 37). At one time or another, the entire counties of Devon and Essex were afforested (Cross 1928: 63; Ecott 2004).

> *[A] Forest is a Certain Territory of wooded grounds and fruitful pastures, privileged for wild beasts and fowls of the Forest... to rest and abide in, in the safe protection of the king, for his princely delight and pleasure...And therefore a Forest doth chiefly consist of four things... vert, venison, particular laws and privileges, and of certain meet officers appointed for that purpose. (Manwood [1615] 1976: 18).*

Vert was the habitat (all green vegetation); venison was the wildlife (not just the deer). Particular laws were a collection of royal proclamations that operated outside English common law, and the officers were a complicated hierarchy of enforcement and judicial personnel who managed the royal forests under the Forest Law.

The royal forest was a legal, not a geographical, entity. The king could create or extend a royal forest in or across farmland, other privately held land, or even villages and towns. Once a royal forest was established, freemen's rights to clearing new land, protecting and cultivating existing fields, pasturing livestock, and cutting timber or firewood within the forest were greatly restricted. In most cases, persons living within the forests retained a right of commons and were permitted to graze cattle—but not geese, goats, sheep, and swine (Manwood [1615] 1976: 97). However, swine were permitted in the forests to feed on acorns (upon payment of a fee, of course).

Royal forests might be considered somewhat analogous to present-day land-use concepts of conservation overlays or zoning classes but were far more restrictive. Furthermore, the king could create forests by proclamation, and affected landholders received no compensation for their loss. Just how onerous were the Forest Laws?

- [By] the Laws of the Forest, no man may cut down his woods, nor destroy any coverts, within the Forest, without the view of a Forester and license of [the chief judicial officer] of the Forest. (Manwood [1615] 1976: 60)
- In one instance, hedges were ordered burned and ditches leveled, so that animals [boar] could feed on crops. (McKechnie 1958: 416)
- Freeholders within a Forest could not clear their lands, nor plough wastelands, nor root out trees, nor build a pond or mill, nor enclose space within a hedge (for Forests were "unenclosed") nor cut a tree or lop off branches without permission (and payment). (McKetchnie 1958: 425)
- During "fence month" (31 days beginning 15 days before midsummer), all livestock in the forest were required to be confined; swine, sheep, or goats found in the forest were confiscated and became the king's property. Persons could not even go "wandering up and down in the Forest...for fear of troubling or disquieting of the wild beasts in the time of their fawning." (Manwood [1615] 1976: 92)

The royal forests were managed by a special government-within-a-government. Forest officers included magistrates who tried lesser cases and reported forest conditions to the king, knights who inspected the forests every three years, enforcers of grazing restrictions, and special courts (*eyres*) in which transgressors were tried. When such courts were in session, all persons living within two leagues (about 9.7 kilometers) of the forest were required to attend (Young 1979: 22, 51). Every three years, a panel of 12 knights was to inspect all forests and report violations. Transgressors were then tried before eyres.

> *[If] woods are so severely cut that a man, standing on the half-buried stump of an oak or other tree, can see five other trees cut down round about him, that is regarded as 'waste'.... Such an offence, even in a man's own woods, is considered so serious, that he in no way be quit of it by his session at the Exchequer, but must all the more suffer a money penalty proportionate to his means. [Note: not proportionate to the offense.]* (Dialogus de Scaccarrio, 61, cited in Young 1979: 35)

Penalties for violating the Forest Laws could be extreme: fining or confiscation of entire towns (McKechnie 1958: 428), mutilation, or even death (Young 1979: 30), but the Forest Laws may not have been so severe in practice. Eyres were held only infrequently (sometimes at intervals exceeding 25 years, and in the case of Pickering Forest, more than 50 years (Cross 1928: 6), and penalties frequently were not imposed on defendants "because some are dead, some are fugitives, and some have nothing" (Young 1979: 39, 89). Furthermore, the enforcement system was a setup for graft and corruption—each officer depended upon his superior for employment, some paid for the right to hold office (Young 1979: 14, 51), and many were unpaid and "lived upon the country" (McKechnie 1958: 418).

There also were ways around the harsh Forest Laws. Rights could be purchased, and, with such fines as may have been collected, this commerce provided an important source of funds to a government struggling from the cost of funding the crusades, a high rate of inflation, and a depressed economy (Young 1979: 39; Swindler 1965: 71, 72). One could purchase a right to hunt, pasture domestic animals, clear and cultivate areas, salvage windfalls of timber, cut hay and firewood, collect nuts, in fact, acquire almost anything the forest produced except royal game—red deer, fallow deer, roe deer, wolves (extirpated during the time of Henry VII, 1485–1509), and wild boar (which became extinct in England during the reign of Charles II, 1660–1685) (Cross 1928: 8; Young 1979: 46, 47, 55–58, 115; Manwood [1615] 1976: 19). During the reigns of Richard and John (1189–1216), the total area of the royal forests was reduced by sale of forest privileges and exemptions; even disafforestation was possible if the payment were high enough (Young 1979: 20).

Compounding the loss of land rights taken by the royal forests was the practice of placing land *in defensum*. Under this concept, access to lowlands bordering rivers was restricted or prohibited in the name of national security, and local residents were required to build and maintain bridges. Lands *in defensum* grew until they included most of the English bottomlands, further adding to the king's hunting domain (Swindler 1965: 296).

After 150 years of this treatment—extreme restrictions unfairly and inequitably enforced, bribes and payoffs, and particularly the creeping acquisition of freeholds—the freeholders had had enough. The catalyst for action was an August 25, 1214, statement by Archbishop Stephen Langton that "a charter of Henry the first [successor to William the Conqueror] has just now been found by which you may, if you wish it, recall your former rights and condition." Just what that charter said I don't know, but it was sufficient to set off an armed rebellion (Swindler 1965: 71–79).

After almost a year of fighting, apparently both sides got tired of the struggle. On June 15, 1215, the year before the end of his reign, King John notified the opposing barons that he was ready to come to terms, and the disputants met on the south bank of the Thames River, in a meadow called Runnymede (Swindler 1965: 71–79). Within four days the parties had agreed on the articles to be addressed and sealed an agreement, but it took ten years of wrangling and nitpicking before the "Great Charter of 1225" was issued. The Magna Carta is one of the most important documents in western history and was precipitated, at least in part, because of forest conditions. It guaranteed the "liberties of the subject," limited punishment for small offenses, and protected the right of due process—provisions that were later incorporated into the first ten amendments of the U.S. Constitution. It and its companion document, the Carta de Foresta of 1217, issued during the reign of Henry III, also had a profound effect upon the royal forests and lands *in defensum*.

Several provisions in the Magna Carta concerned the forests:

- All lands afforested or placed *in defensum* in "our time" (Henry I, 1100–1135) were returned (Articles 16, 47).
- People no longer were required to attend the eyres unless they were party to an action (Article 44), nor did they have to build and maintain bridges (Article 16).
- Groups of knights (inquisitors) were to investigate all forest officers "and all the bad customs concerning the forests" and take remedial action (Articles 45, 48).

The Carta de Foresta also addressed use of the forests:

- Prohibitions against killing deer and using falcons within the forests were relaxed (Cap. 11, 13).
- Death and mutilation as penalties for Forest Law violations were eliminated, but stiff penalties for unlawfully taking deer were retained (Cap. 10).

The new forest laws of the early 13th century did much to correct civil injustice (at least on paper), but they did little to enhance English forest land. Unprotected by the forest laws, the newly returned lands frequently became targets for wildlife and timber poaching and for overharvesting (Cross 1928: 11, 12).

There seems to be a dearth of forest literature from the 13th to the 18th century (a pretty long time). Conditions during the 1700s are fairly well described in the Shelburne documents, collected by Sir William Petty, second Earl of Shelburne and first Marquess of Lansdowne (1737–1805), and published in part by the University of Michigan (Cross 1928: vii, viii). Although these papers shed little light on the individual happenings during these 700 years, they do give us a picture of conditions in the forests at the end of this period.

The Shelburne Manuscripts contain numerous accounts of trespass and misuse in the forests; I shall cite only a few. Under certain conditions, freeholders living within Dartmore Forest were permitted "the Liberty of Inclosing Eight Acres of Land on any part of the Forest upon paying one Shilling Yearly for the same." A 1766 inspection found that "instead of Eight Acres or thereabouts.... The late Newtakes or Inclosures contain some of them near Three Hundred Acres each, and not one of them Less than One Hundred Acres" (Cross 1928: 71).

By 1725, in an effort to retain timber in the New Forest, a navy board recommended that all trees to be felled be marked by a hammer "with two faces, the one to mark such Trees as are judged proper to be felled for use of the Navy, and the other for defective Timber, or such as is not fit for use of the Navy." This board also recommended a complicated recordkeeping system that, if implemented, would have accounted for all forest products, from logs to tops and branches (Cross 1928: 54, 55).

About 1763, the following comments were made concerning the Chase of Enfield:

> [S]ome years ago a Scheme was formed for the taking and inclosing at certain Periods particular Quantities of the said Chase not only with a View the better to preserve the Growth of the Young Trees and Saplings growing spontaneously upon the said Chase but to plant such Inclosures as Occasion might be and when the Wood should arrive at a proper Growth and be no longer liable to the Damage and Spoil of Cattle such inclosures to be laid open again.... But the same still remains to be carried into execution. (Cross 1928: 99)

In some cases, the needs of the Royal Navy competed directly with those of private interests. After citing the dire shortage of timber for the navy, a 1769 report on conditions in the Dean Forest cited timber poaching by colliers, who wanted timbers to reinforce mine shafts, and the East India Company, which was building private trading vessels. An act of Parliament passed in 1680 provided that 11,000 acres of wastelands within the New Forest be "inclosed for the growth and preservation of Timber" and "because of the good Effects of this Law" this policy be extended the Dean Forest (Cross 1928: 105, 106). But a report submitted a year earlier cited the proposal to regenerate oaks through enclosures and planting in the New Forest and observed,

> *this has been so shamefully executed by those who were employed in the doing it.... That tho' the Inclosures have been perfectly well and very expensively fenced off for about 13 years past, there is not the least appearance of timber coming up, not could I see, even one single Oak in the compass of several hundred Acres. (Cross 1928: 116)*

Although by far the majority of writers decried the destruction of the royal forests and the shortage of wood for the navy, at least some evidence to the contrary exists. A 1768 report to the Treasury stated that the New Forest contained "a sufficiency of Timber" to provide "Three Hundred and fifty Loads of Oak Timber and fifty Beeches, in addition to the Timber already provided from that Forest... For eighty, or an hundred years to come," but this letter apparently never reached the navy board (Cross 1928: 111). Some recent authors (Michael Flynn and George Hammersley, cited in Williams, 2003, 170) also have disputed the extreme timber shortage in England, believing that it existed only in areas easily accessible from ports.

A 1770 report again recommended seeding oak, beech, hazel, ash, and other hardwoods within enclosures in the forest to prevent wildlife and livestock damage. It also advocated practicing coppice management, thinning, and selective cutting to favor oaks (Cross 1928: 128). These recommendations, though probably beneficial if implemented, did little to bring about forest recovery, and by 1793 "only twenty forests which could supply timber to the navy existed" (Cross 1928: 11). By 1773, suggestions were made to begin or increase oak imports from Canada, Germany, and the American colonies (Cross 1928: 126).

Deforestation became so severe that the Royal Navy had insufficient supplies to maintain the fleet and increasingly turned to the American colonies for this resource. England even sent foresters to the colonies to mark superior trees—those greater than 24 inches (61 cm) diameter 12 inches (30 cm) above the ground and declare them "royal property," reserved for the Royal Navy (Dana and Fairfax 1980: 5, 6).

The results of disafforestation in Hainault Forest in the County of Essex are typical. In 1851, the 3,000-acre Hainault Forest was clearcut and converted to agriculture. The responsible person summed up his attitude thus: he hoped to see the adjoining forest cut in the same way soon because "it would be a great service to me, as well as to the whole neighborhood" (Scott 2004).

But there were some successes. A letter written in 1770 gives a long and detailed account of 45 years' experience on some private estates. On those lands, oak was successfully regenerated and grown through a sequence of cattle exclusion, coppice encouragement, selective and sequential coppice thinning, and selective harvesting (Cross 1928: 130–34). Another success was achieved by the iron industry, which, after experiencing the effects of overharvesting, used crop rotation and coppicing to achieve sustained, long-term yields on more than 200,000 acres (Williams 2003: 203).

If one could summarize the history of the royal forests of England, it would begin with the sovereigns' desire to protect wildlife for their own hunting pleasure, through enactment of severe forest laws, subsequent graft and corruption, followed by rebellion. It continued with disafforestation and the concomitant development of manors and estates. Failure to protect both the vert and the venison led to a change of management emphasis from wildlife to sale of natural resources. A countrywide wood shortage and fears that the Royal Navy could not be supported prompted increasing reliance on timber imports from the American colonies and other countries, and ultimately to efforts to manage forests. Today, royal forests are limited to Crown lands, and disafforested lands are now in chases, private ownership, public parks, and reserves.

On the Continent

I was unable to find much documentation of continental forest conditions during medieval time (at least in English). Most of the following was gleaned from Williams's *Deforesting the Earth* (2003) and the web pages of the European Forest Institute (FINE), without which I would have been hopelessly lost.

The sequence of events on the Continent generally parallels that in England. Most authors describe the continental forests during early Roman time as being vast and largely unbroken. Some others, however, point out that large areas even then must have been cleared for agriculture to support the human population (Williams 2003: 84). At the time of the Roman invasion, most of Bavaria probably was covered by dense, dark forests. Forests were treated as commons (*res communis*), free to all, and no thought was given to depletion or conversion (FINE n.d.(e)).

After the fall of the Roman Empire, successive population increases and decreases corresponded to deforestations and reforestations. Between a.d. 600 and 1000, the European population doubled, Scandinavians migrated into western Europe, and the forests suffered. During the next 200 years, the population doubled again (Williams 2003: 107). During this period, the clerics also were busy—acquiring land, clearing it for monasteries, and awarding land grants to followers. This process created nuclei of deforested settlements within the forest, and the nuclei spread and multiplied (Williams 2003: 113).

The first fragmentation of Belgian forests occurred during Roman times. The Romans also introduced the concept of privately-owned forests (*res in patrimonio*) to lands which previously had been treated as commons and began arboriculture for cherries, pears, plums, quince, walnuts, and medlar. Following ejection of the Romans, the barbarians enacted the first Germanic forestry laws between the fifth and seventh centuries, imposing fines and punishments for forest trespass and declaring all forests, except royal territories, commons subject to free public use (FINE n.d.(c)).

Prior to the 12th and 13th centuries, Spanish forests were treated as unrestricted commons, unfettered by any kind of law. Much of the wood harvested was used by foundries for fuel, but by the 14th century increasing population brought the towns into direct competition with metallurgical industries and greatly increased the drain upon the forests. By then, primary wood users were foundries, house construction, and shipbuilding, in that order. In 1548, each town in Gipuzkoa, Spain, was required to plant 500 oaks or chestnut trees per year, and in 1593, in Bizkaia, two trees had to be planted for every one cut (FINE n.d.(a)).

The history of the Pyrenean forests is one of continual encroachment and diminution, first from pasturage, then agriculture (which took the most fertile soils), and overcutting (particularly of oaks). By 1467, regulations had been enacted that prohibited conversion of forests to agriculture, and in 1560 and 1784 addition restrictions were added, but by the early 19th century the only woodlands that survived in relatively good condition were those administered by the navy for production of shipbuilding materials (FINE n.d.(b))

The centuries from 900 to 1400 also saw large areas of German forest deforested, for various reasons: impoverished nobles sought to generate income through forest exploitation, monastic lands increased in area, the general population was growing, free farmers received monastic land in return for services, and forest owners were granting concessions to others for cutting wood, grazing livestock, and collecting tanbark and litter collection; the last product was used to improve cropland tilth, but its collection deprived forest soils of nutrients (FINE n.d.(c)).

Then came the Black Death, and it took another 200 years to regain the population of the early 13th century (Williams 2003: 107). The plague reduced the European population by about one-third and in some places by as much as one-half, causing village and cropland abandonment and permitting natural forest regeneration over much of the land (Williams 2003: 136).

By 1350, however, the needs of an increased population had converted about two-thirds of Bavaria to agriculture—a ratio that generally exists today. Somewhere during this time, all "unowned" property became the property of the king, and numerous wars "devastated the land and the forest" (FINE n.d.(e)).

By this time, also, the combined effects of population growth and need for agricultural land may have reduced the French forests by about 17 million ha (though perhaps a quarter of the country remained forested). In central Europe, about half of the forest cover may have been removed prior to the late Middle Ages (Williams 2003: 123).

Just as in England, continental royalty and nobility coveted their hunting rights and territories and attempted to protect them against encroachment. These efforts conserved some of the remaining original forest land, but in the long run the forest was doomed to fall before the unrelenting pressures of an expanding population (Williams 2003: 130, 132). Continental timber shortages were caused by four activities: "clearing for agriculture and for fuel wood supplies, which were local and domestic; shipbuilding, which was basically foreign and strategic; and charcoal supplies for iron making and industry in general, which were a bit of both" (Williams 2003: 171–72). Williams (2003: 171,) questions whether a general timber scarcity existed throughout Europe, believing instead that shortages were largely logistical, a function of demand proximity (similar to the debate over English forests). Whether perceived or real, timber shortages prompted enactment of forest protection legislation.

By the 16th century, overharvesting in Galicia, Spain, had led to creation of strips two (and later three) leagues wide along the coastlines and banks of other navigable waters within which no wood could be cut or slash-and-burn agriculture practiced without a license, and no wood could be exported. At this time, forest owners were largely absentee nobles and ecclesiastical institutions, but residents were given a usufruct in common for firewood, wood, pasturage, and other forest products (FINE n.d.(d)).

In European Russia, just as in England, concern focused on wood for naval uses. During Peter the Great's time (1682–1725) owners of private forests were prevented from harvesting wood suitable for shipbuilding. A 1703 edict required such persons to "describe all forests within 50 km from large rivers and 20 km from small ones." From such areas, no oak, maple, elm, larch, or pine could be removed (FINE n.d.(f)).

During the 17th century in Germany, the glass, potash, resin, and salt industries expanded, greatly increasing the pressure upon forest resources, and the first Germanic forest ordinances were enacted (FINE n.d.(e)).

The Duchy of Luxemburg's Edikt of 1617 probably was the first comprehensive forest legislation. It showed a concern for forest conservation and the public interest in forests and covered the customary laws, sustainable forest management, private forest ownership, forest personnel, and police and justice. It was followed in 1754 by enactments of the Austrian Empress Maria-Theresia, which contained even more restrictive provisions (FINE n.d.(c)).

The French Forest Ordinance of 1669 was landmark continental forest legislation. It

> sought to rationalize and codify the mass of ancient French forest law ...and to make forestry an autonomous branch of the state economy while ensuring a supply of timber for the navy. It was designed for the royal forests but was soon being applied to all forestland—ecclesiastical, community, and private.... It stipulated that kilns, furnaces, and charcoal making were to be restricted in forests, and a range of other woodcutting occupations—such as coopers, tanners, and woodworkers—had to be located at least 1.5 mi (2.4 km) away from the forest edge. Cattle, sheep, and goat grazing was absolutely prohibited in forests,

pannage was regulated annually by common agreement, and large trees and seed-bearing trees… Were marked and reserved. (Williams 2003: 204–205)

Great for the forests, but a bit rough on the citizenry!

During the 17th and early part of the 18th century, several other legislative measures were taken to improve forest conditions, but in the later half of that century a series of wars in Spain led to "massive and indiscriminate use of the forests." Communal forest lands were sold, tree nurseries ceased to operate, and cutover forests did not recover (FINE n.d.(a)).

By the 18th century, biennial forest inspections were mandated (though not necessarily conducted) in Galicia, and mature oak stands were being converted to pine plantations (species that provided greater production). By this time also, the coastal forests were being administered by the navy, and naval officers began procuring wood, primarily oak, for ship construction. During (roughly) the 18th century, between 400,000 and 700,000 m^3 of oak and 250,000 m^3 of pine were consumed, the products of approximately 666,000 ha of oak and 224,000 ha of pine (FINE n.d.(d)). At this time, Flanders contained about 155,000 ha of forests, about the same as today. But a dramatic shift from the more fertile soils of the south and west to the more sandy soils of the north and east has occurred (FINE n.d.(c)).

The French Revolution and the rule of Napoleon Bonaparte brought drastic changes in the way French citizens and French forests were treated. His Code of 1804 "swept away feudal privileges, establishing equality before the law, guaranteed individual liberty and protected private property" (Holmberg n.d.). It also changed the Roman concept of wild animals as *res nullius* to that of private property (*res in patrimonio*). The Napoleonic Code contained the following articles:

544. Property is the right of enjoying and disposing of things in the most absolute manner, provided they are not used in a way prohibited by the laws or statutes.

545. No one can be compelled to give up his property, except for the public good, and for a just and previous indemnity.

546. Property in a thing, whether real or personal, confers a right over all which it produces, and over all connected with it by accession, whether naturally or artificially.

547. The natural or artificial fruits of the earth, and the increase of animals, belong to the proprietor by right of accession.

578. Usufruct is the right of enjoying things of which the property is in another, in the same manner as the proprietor himself, but on condition of preserving them substantially.

582. The usufructuary has a right to the enjoyment of every species of benefit, whether natural, or artificial or civil, which the object of usufruct is capable of producing.

590. If the usufruct comprehend underwood, the usufructuary is bound to observe the order and proportionate quantity of his cuttings, conformably to the established usage and custom of proprietors, without indemnity however to be made to the usufructuary or his heirs, for ordinary cutting, whether of underwood, poles, or timber, which may have been committed during his enjoyment.

591. The usufructuary receives the benefit likewise, always conforming to the seasons and custom of the ancient proprietors, of those parts of a wood of tall trees as have been placed in regular cuts, whether such cuts are made periodically over a certain extent of ground, or whether they are made of a certain number of trees taken indiscriminately over the whole surface of the domain.

592. In no other cases can the usufructuary touch full grown timber-trees: he can only employ trees blown down or broken by accident, for the reparations to which he is bound: he may however for that object cause some to be felled in case of necessity, but on condition of making such necessity appear to the proprietor.

Conclusion

So is there any central theme to all this mess? I think so, but it's a bit shaky.

The story begins with abundant forest resources, the value of which is unrecognized and unrealized (European forests in Roman times, eastern North American forests during early colonization). Upon realization of forest product or alternative values came utilization of the forest (shipbuilding in Europe, export from the colonies, agriculture in both places). Overutilization resulted in degradation, decimation, and destruction of the resource (wastage in Europe, destruction of the eastern forests in the United States). Realization of the situation brought civil, then legislative action (Magna Carta, Carta de Foresta, and other laws in Europe, and creation of the National Forest

System under Pinchot in the United States). We saw overshooting, perhaps, in the environmental legislation of the 1970s. And finally, we have retrenchment to reasonable and effective measures for forest ecosystem management.

We aren't there yet, but at least we're going in the right direction.

References

Britannia. n.d. "Monarchs of Britain." http://www.britania.com/history/h6f.html.

Buckland, W.W. 1963. *Textbook of Roman law from Augustus to Justinian*. Cambridge: Cambridge University Press.

Cross, A.L. 1928. *Eighteenth century documents relating to the royal forests, the sheriffs and smuggling*. Selected from the Shelburne Manuscripts in the William L. Clements Library. New York: Macmillan Company.

Dana, S.T., and S.K. Fairfax. 1980. *Forest and range policy: Its development in the United States*. New York: McGraw-Hill Book Company.

European Forest Institute: Forest Information Services Network for Europe (FINE). n.d.(a). Forest history in Basque Country. http://www.efi.fi/fine/Spain/Euskadi/history/history_e.htm.

———. n.d.(b). Forest history of Aragon. http://www.efi.fi/fine/Spain/Aragon/history/history_e.htm.

———. n.d.(c). The forest is a connection between the past and the future. http://www.efi.fi/ fine/Belgium/ Policy/E_BelHist.htm.

———. n.d.(d). Galician forest history. http://www.efi.fi/fine/Spain/Galicia/history/history_e.htm.

———. n.d.(e). The history of Bavaria's forest. http ://www.efi.fi/fine/Germany/Bavaria/English/history.htm.

———. n.d.(f). General information about Mari El. http://www.efi.fi/fine/Russia/mari_el/ history_e.html.

Holmberg, T. n.d. Reader's articles: Politics and government. The civil code (the Code Napoleon). http://www.napoleonseries.org/articles/government/civil code.cfm.

Manwood, J. 1976 (1615). *Treatise of the lawes of the forest*. Published in facsimile. Norwood, NJ: W.J. Johnson.

McKechnie, W.S. 1958. *Magna Carta: A commentary on the Great Charter of King John*. New York: Burt Franklin & Company.

Scott, B. 2004. Hainault Forest web site: The Forest of Essex. http://hainaultforest.co.uk.htm.

Swindler, W. F. 1965. *Magna Carta: Legend and legacy*. Indianapolis: Bobbs-Merrill Company.

Williams, M. 2003. *Deforesting the earth*. Chicago: University of Chicago Press.

Young, C.R. 1979. *The royal forests of medieval England*. Philadelphia: University of Pennsylvania Press.

ABSTRACT

The forests of Europe have long been shaped by the presence of man and serve as both a land reserve and a multipurpose natural resource. For many centuries, human use of the forest focused on harvesting. Firewood, timber, and game were extracted from the forest, and husbandry of domestic animals depended on forest products as well. Beginning in the ninth century, public authorities intervened to regulate harvesting and conserve resources. In the 18th century, a scientific approach to forest management was taken, in both France and German regions, but only on paper in France. Not until after the French Revolution did the need for professional training programs in forest management become apparent. L'École Forestière de Nancy opened its doors in 1825. Early on, the curriculum was based on German advances in forestry, but it quickly came into its own. Natural regeneration and replacing coppice-with-standards with high forest became the organizing principles. When Gifford Pinchot arrived in Nancy at the end of the 1800s, he benefited from the cross-fertilization of two forestry cultures—the German and French schools—as well as the analytical achievements of Lucien Boppe and, above all, training in the field.

CHAPTER 2

European Silviculture and the Education of Gifford Pinchot in Nancy

Marie-Jeanne Lionnet and Jean-Luc Peyron

*I*n this chapter, we retrace the European and French tradition of forestry through historical snapshots: European forestry before 1800, the development of French forest science and education in the 19th century, major 19th-century advances in silviculture, and forestry education during Pinchot's time in Nancy.

European Forestry Before 1800

Although it expanded considerably during the 19th century, European forestry came into being much earlier, and its heritage should be examined in its proper perspective. The last ice age came to an end 18,000 years ago. With the warming that followed, forests began expanding again throughout Europe over the next few thousand years. The migration of plant species occurred less readily in Europe than in America because of east-west barriers formed by mountain ranges and seas. Nonetheless, given that this reforestation was relatively fast, it is clear that long-distance migration was carried out by animals, waterways, and most certainly, man. Without a doubt, human beings instituted a kind of silviculture through their activities.

When humans began to settle 8,000 years ago, they started clearing forested areas to create fields and pastures. In France and elsewhere, population growth led to an unrelenting reduction in forest cover; this occurred earlier and more rapidly around the Mediterranean, at an accelerated pace in the British Isles, to a lesser extent in central Europe, and relatively little in Scandinavia.

But the forest is more than just a land reserve. It has always supplied countless goods and services. Population growth, with the ensuing reduction in forest areas and demand for firewood and timber, resulted in widespread shortages. Later, when Americans visited Europe, they were astonished by the urgency with which the tiniest twig was harvested from forests.

The most widely used method of forest management in premodern times was a system called *tire-et-aire*. This system of gradual cutting dated back to ancient times and brought order to the practice of coppice harvesting. It is characterized by two features that played, and continue to play, a major role in forests: operational control well suited to the method and the unquestionable initiation of relevant forest planning.

Forest supervision became increasingly necessary as wood shortages grew. At first, judges arbitrated disputes, but in a *capitulaire*, or ordinance, issued by Charlemagne in 813, the term *forestari* ("forester") appeared. And in 1291, a royal ordinance of Philippe IV le Bel referred to the *Grand-Maître des Eaux et Forêts* ("grand master of waters and forests").

Very quickly, the ordinances extended beyond legal matters and began to create true forest management, beginning with the royal domain. The 1346 ordinance signed in Brunoy by Philippe VI de Valois is notable from this point of view. It states that "forest masters will survey and visit all the existing forests and woodlands and realize any appropriate cuttings, with reference to what the said forests and woodlands can hold sustainably in good condition in the long term."

As shortages grew and the forest became a national concern, the scope of application of the ordinances extended more and more to community and even private forests. Perhaps it was for that reason that Gifford Pinchot returned to the United States with the conviction that the private forest must be integrated into forest policy. After a period of lax regulation, the ordinance of 1669 demonstrated the political will to reintroduce order into the forests of the French kingdom and at the same time develop important ideas. These included the desire to extend the area managed as coppice-with-standards and even to prevent from cutting one-fourth of any forest, which would later facilitate the transition to the high forest.

In the 18th century, silviculture expanded throughout Europe, especially in France, through the work of Réaumur, Buffon and above all, Duhamel du Monceau, often considered the father of silviculture. His celebrated tome, *De l'Exploitation des Bois* ("Forest Management"), published in 1764, was immediately translated into German, then published in that language in 1766. This new science would gain momentum in Germany, where it complemented the German experience in the field, and then return to France in the early part of the 19th century, after the turmoil of the French Revolution and Napoleonic Wars had abated.

Forest Science and Education in the 19th Century

The unmistakable growth of forest science in the 18th century significantly expanded in the 19th and, above all, became accepted practice through educational efforts. Higher education in forestry was first offered in Germany in 1763. Numerous other educational programs would follow, either in specialized institutions or as part of existing institutions of higher learning.

During the 1820s, Jacques-Joseph Baudrillart, revisiting a project developed during the French Revolution and keeping in line with the movement in Germany, strongly supported the creation of a forestry school, arguing that there existed a gap between theory and practice. He won his case. In August 1824, it was decided to create l'École Forestière. Its establishment in Nancy became official on December 1, 1824, with its inauguration scheduled for the beginning of 1825.

The school was created in Nancy for several reasons: the Lorraine origins of the French general manager of forests, the proximity to Germany—the focal point of forest science at the time, and the simple fact that Nancy had the largest number of forests in northeastern France. Finally, other than the large amount of conifers in the Vosges Mountains, Nancy was well forested, making day-long field trips possible.

Six principals led the school during the 19th century. The school's first principal, Bernard Lorentz, was named in early December 1824. His bust, housed at the school, was later produced by Frédéric Bartholdi, better known for creating the Statue of Liberty.

Alfred Puton was the principal who welcomed Gifford Pinchot to Nancy in 1889. Mrs. Puton also hosted the young American.

Lucien Boppe, the vice principal who would later succeed Puton, taught silviculture. Pinchot greatly valued his teachings, choosing to keep and make use of Boppe's *Traité de Sylviculture*[1], published in 1889. A few days after his arrival in Nancy, he had already bought Bagnéris' book on silviculture. Bagnéris held the chair before Boppe but was no longer teaching. Pinchot was less enthusiastic about Boppe's assistant, Gustave

Huffel, who had just arrived and, without doubt, had relatively little experience. However, Huffel then became an important teacher in Nancy and his work was broadly analyzed in Theodore Woolsey's book, *Studies in French Forestry*, written in 1912.

Pinchot did not travel to Nancy intending to study natural history. Nevertheless, he participated in outings with Fliche, who succeeded Matthieu; both educators stressed the importance of instruction in this field. Fliche's assistant, Henry, joined an excursion in the Alps that Pinchot appeared to have appreciated despite his frustration with the inadequate commentary on the sites visited. Also on this trip were Petitcollot, who taught forest management—one of Pinchot's required courses—and Guyot, a professor of forest law. Guyot would later become the seventh principal of l'École Forestière and also the author of a book on a much-valued history of the school.

Understanding of forestry was evolving, and to maintain pace with advancing knowledge, the school's library became one of the best stocked in the world in this field. Over the course of the 19th century, its collections grew through gifts, exchanges, acquisitions, and compensation after the universal expositions of 1867, 1878, and 1889, all held in Paris. Pinchot visited the exposition of 1889, the first time he learned of French forestry. He returned to France for the exposition of 1900 and participated in an international congress on silviculture. The school's collections were originally stored in a pavilion at the rear of its campus; this building was rebuilt in 1896, shortly after Pinchot's departure. Today, it is named Daubrée, after a general manager of forests whom Pinchot met on at least two occasions.

Arboretums had also been constructed at Nancy, and starting in 1866, experiments, most notably in the field of forest meteorology at the initiative of Matthieu, were carried out at the arboretum in Bellefontaine; Pinchot spent some time here.

Research centers gradually appeared throughout Europe, starting in Baden and Saxony in Germany in 1870. In 1882, a center was set up in Nancy and grew at the impetus of Bartet, with whom Pinchot participated in several outings, particularly in Bellefontaine and in the Haye Forest. As soon as the research center was created, ten experimental plots were installed in the Haye Forest, at the edge of Nancy, to analyze the production and silviculture of

Pinchot's Disciple in Europe: Theodore Salisbury Woolsey, Jr.

Some decades after Pinchot, Theodore Salisbury Woolsey, Jr., continued the study of European forestry that had begun with Pinchot. In 1905, after graduating from Yale, he traveled to Europe to complete his forest education—probably following Pinchot's advice. First he went to the Black Forest, which he described as not only commercially valuable but also a natural recreation ground. He observed German people's close relationship with their village forests, which were full of paths and trails. He was particularly interested in the records that German foresters kept on the history of the stands. He then stayed three months in France before leaving for India.

Woolsey returned to France for six months in 1911–1912. By then, he had some professional forestry experience, and his intent was to study French soil conservation on sand dunes, in mountain areas, and around the Mediterranean (including the French colonies in Africa). He went also to Nancy, where he was impressed with the practical nature of the forest research and the holdings of the library. In Franche-Comté he gave a talk on American forests and what the American forester could borrow from French management.

He went to France a third time during the First World War with the American expeditionary forces. In his position as a member of the Allied committee for wood procurement, he was frustrated with the constraints imposed on wood sales, intended to limit the overexploitation due to war efforts and ensure long-term forest planning.

Woolsey wrote two books on his French experience. The first, *French Forests and Forestry*, was published in 1917, just before his wartime travel to France. In its foreword, Pinchot calls France a destination worthy of study for young American foresters because with its colonies, it offered biogeographic diversity (including the borders of the Sahara desert), exemplified many of the same forest problems as in the United States but on a smaller scale, and demonstrated a range of technical solutions.

The second book, *Studies in French Forestry*, was published in 1920 and is based on all three of his experiences in France; it runs some 550 pages. Pinchot undoubtedly supported this thorough assessment of the French forestry. French participation in the international Congress of Silviculture of Paris, in 1900, had revealed the progress achieved since the time of his own studies at Nancy ten years earlier, and in Woolsey, Pinchot found the intelligent observer to convey French lessons in forestry for the profession in the New World.

the beech tree. The trees in these plots provided invaluable data until they were felled by storms at the end of 1999.

Such scientific developments led to the decision in 1891, one year following Pinchot's trip to Europe, to create the International Union of Forest Research Organizations (IUFRO).

French Advances in Silviculture

A combination of advances in science and education led to a number of innovations in the field of silviculture. Undoubtedly, its overall philosophy was inspired by a text written by Dralet, but that philosophy was largely spread through the courses taught by Bernard Lorentz and Adolphe Parade and their successors: *Imiter la nature, hâter son oeuvre* ("Imitate nature, accelerate its work").

Lorentz, the first principal of the École Forestière, declared himself the sworn enemy of the coppice in all its forms and advocated converting coppice-with-standards into high forest. He and his students put his ideas into action starting in 1825 in the Amance Forest, close to Nancy.

The *tire-et-aire* system was a straightforward scheduled cutting system, based solely on forest area. Increasingly sophisticated forest planning methods developed around 1800 in Germany. Hartig recommended planning by volume, and Cotta preferred a mixed system organized by volume in the short term but by forest area in the long term. This latter method was imported to France and taught in Nancy by Parade, a former student of Cotta's. He called it the natural reseeding and thinning method, which fits well with the two adaptations he applied to the original method. Parcels of land to be cut were not necessarily contiguous, as with the tire-et-aire system. In the end, this method was called the German method in France and the French method in Germany, at least until the war of 1870.

The importance given to thinning was in keeping with the French tradition of Duhamel du Monceau and well expressed by Philibert Varenne de Fenille in 1791: "To maintain very high densities, we produce more trees and less wood."

In mountain forests, a system based on permanent forest cover and dispersed tree cutting had been in place for a long time. At the end of the 19th century, Adolphe Gurnaud most famously theorized about silviculture involving the selection system; he had developed a well-regarded *méthode du contrôle*, exhibited at the universal expositions of 1878 and 1889, the latter of which was attended by Pinchot.

But in mountainous areas, forests were often stressed by competitions with crops, overgrazing, and overexploitation. A series of catastrophes in the 1840s and 1850s led to a program of measures called *restauration des terrains en montagne* ("restoration of mountain lands"). Starting in 1880, work was undertaken that essentially combined two techniques. The first consisted of repairing the streambeds. Reforestation of the hillsides came next—and today the French mountain landscape is quite different from its condition one century ago. Pinchot had the opportunity to tour the Alps, which impressed him, and where he was undoubtedly able to observe such work up close.

Gifford Pinchot in Nancy

By the time Gifford Pinchot arrived in Nancy in 1889, l'École Forestière enjoyed an excellent reputation. Two sentiments are apparent in Pinchot's writings: a very critical view of his fellow students in general, as well as certain professors and the teaching methods; and in retrospect, the impression of having nonetheless acquired a general philosophy of forestry.

Despite criticisms of l'École Forestière de Nancy, several non-Frenchmen encouraged Pinchot to go to Nancy. Among them was Dietrich Brandis, who favored training English forest managers in Nancy for their service in India. Despite tensions between France and Germany at that time, there was a sense of community between the foresters of the two countries.

In Nancy, Pinchot rubbed elbows with two graduating classes and numerous students and fellows from outside France. Moreover, he arrived at a time of great change in recruitment practices and, therefore, in the profile of the student body. Regular students were housed on campus; Pinchot rented a room in a nearby house. Pinchot forged a friendship with one student, Joseph Hulot, who was pursuing independent studies. The two took many walks together, and Hulot also invited Pinchot to his family home near Nancy.

But in general, Pinchot found the French students relatively undisciplined and unmotivated by their studies. Besides what seemed to be marginally interesting courses, Pinchot took advantage of a number of forest excursions. Ultimately, Pinchot always acknowledged his time spent at l'École Forestière de Nancy and even declared himself an alumnus.

Notes

1. This exemplar with numerous annotations is in the Grey Towers library.

References

Bagnéris, G. 1878. *Manuel de sylviculture*, 2nd edition. Nancy: Berger-Levrault.

Bartet, E., et E. Reuss. 1884. *Etude sur l'expérimentation forestière. Organisation et fonctionnement en Allemagne et en Autriche*. Nancy: Berger-Levrault. (First published in *Annales de la science agronomique*, vol.1, 189–396).

Baudrillart, J.J. 1823–1825. *Dictionnaire raisonné et historique des eaux et forêts*. 2 vols. Paris: Arthus Bertrand. (See the article "Écoles forestières.")

Boppe, L. 1889. *Traité de sylviculture*. Paris-Nancy: Berger-Levrault. (The introduction takes a very new point of view about silviculture.)

Dralet, F. 1807. *Traité de l'aménagement des bois et des forêts appartenant à l'Empire, aux communes, aux établissements publics et aux particuliers…* Toulouse: Douladoure.

Duhamel du Monceau, H.L. 1764. *De l'exploitation des bois ou moyen de tirer un parti avantageux des taillis, demi-futaies et hautes futaies*. 2 vols. Paris: Guérin et Delatour.

Guyot, C. 1898. *L'Enseignement forestier en France: l'École de Nancy*. Nancy: Crépin Leblond.

Lorentz, J.B., et A. Parade, 1837. *Cours élémentaire de la culture des bois*. Paris: Huzard, et Nancy: Grimblot.

Peyron, J.L. 1999. *L'École forestière au XIXe dans son contexte historique*. Conférence du 4 mars 1999. Nancy: École nationale du génie rural, des eaux et des forêts.

Varenne de Fenille, Ph. 1792. *Mémoires sur l'administration forestière et sur les qualités individuelles des bois indigènes*. 2 vols. Bourg: Philipon.

Woolsey, Th.S. 1920. *Studies in French forestry, with two chapters by William B. Greeley*. New York: Wiley, and London: Chapman and Hall.

ABSTRACT

Cooperation between French and American botanists, conservationists, and foresters predates the founding of the United States. Frenchmen were among the first to explore and catalogue the flora of the North American continent. In the 19th century, Americans led the way in protecting sites of extraordinary value; the U.S. national park system became the model for French parks. American tree species, notably Douglas-fir and poplars, were introduced to the continent to assist in French reforestation efforts, and European species of poplar have been used in American hybrids. In recent decades, French and American researchers have collaborated on reducing the risks from accidental introductions of insects, fungi, bacteria, and viruses that can damage trees; oak wilt is a focus of these efforts. And in recent years, scientists on both sides of the Atlantic have been sequencing the genomes of forest tree species and biotrophic fungi to enhance understanding of ecosystem functioning.

CHAPTER 3

A Historical Perspective on French-U.S. Forestry Cooperation

François Le Tacon, Jean Pinon, and Francis Martin

The present divergences between the United States of America and France cannot erase more than two centuries of friendship and cooperation. French-American ties are extremely deep, based on a solidarity that has never failed, from the battlefield of Yorktown to Utah and Omaha beaches. France will never forget the sacrifice of the boys who came twice to our country and gave their lives for our freedom. Our friendship is rooted in shared values. The American dream is close to the ideals of the French Revolution. The Universal Declaration of Human Rights was written more than 50 years ago by an American lady, Eleanor Roosevelt, together with a Frenchman, René Cassin. This community of values is today our most precious good. France, Europe, and America are now facing the same threats, and only cooperation will allow us to address these important common challenges.

The ties between France and the United States are manifold and include the economy, culture, science, and technology. France's number-one scientific and technical partner is the United States, the world leader in science. Scientific cooperation between our two nations takes many forms, from direct collaboration between scientists and laboratories to institutional agreements.

France is the fourth-largest scientific collaborator of the United States, after Canada, Japan, and Germany. Each year, more than 5,000 joint publications are produced between our two countries. As you know, in forestry the cooperation between the United States and France has always been very strong, and it is becoming stronger. In this article, we shall summarize the main areas of cooperation in forestry between our two countries, past to present.

Conservation and Preservation of Nature

Biological diversity is the living wealth of Earth. During the past 3.5 billion years, evolution has led to the rise and fall of many species. Over time, conditions on Earth slowly change. Only the species best able to adapt to these changes survive. Human beings have dominated life on Earth through advancement in technologies, growth in population, development of land, and exploitation of natural resources. The impact of man upon biodiversity has been greater than that of any other species. Global biodiversity is now being threatened, and it is imperative that we change our ways of thinking and living if we want to preserve it.

Before we can preserve nature and biodiversity in a country, it is necessary to explore the land and to know the plant and animal diversity by determining what species are to be found, and then categorizing and naming them, if they have never been described before.

In 1620, in a book published in Paris, Vespasien Robin (1579–1662) described plants discovered in Virginia and introduced in the King's Garden (Le Jardin du Roi) in Paris by himself or his father, Jean Robin (1550–1629), botanist of Kings Henri III, Henri IV, and Louis XIII. Among the plants described is the black locust, *Robinia pseudoacacia* L., introduced in Paris about 1600. In the same period, another French botanist, Philippe Cornut (1606–1651), described numerous plants from North America in his book, *Canadensium plantarum aliarumque nondum editarum historia* (Paris, 1635).

More than one century later, two French botanists and horticulturists, André Michaux (1746–1803) and his son, François André Michaux (1770–1855), made a major contribution to the knowledge of American flora. André Michaux, born in Satory, near Versailles, was a student of Bernard de Jussieu (1869–1777), who was a professor of botany at the King's Garden in Paris and later director of a botanical school in the Trianon Garden in Versailles. He developed the first natural system of plant classification based on morphological similarities. André Michaux was employed in the King's Garden in Paris under the supervision of the Count of Angiviller.

From 1782 to 1785, in very difficult conditions, André Michaux collected seeds and plants in the Middle East, from the Caspian Sea to the Indian Ocean, and introduced numerous eastern plants into French botanical gardens. On his return, he was chosen to lead a scientific mission to North America. His main goal was to explore American forests for new species of trees that could be used to restore French forests. Appointed King's Botanist, André Michaux arrived in New York on November 13, 1785, with his 15-year-old son, François André; a gardener, Pierre-Paul Saunier; and a servant, Jacques Renaud. Michaux first established a nursery near Hackensack, New Jersey. He met George Washington at Mount Vernon and became a friend of William Bartram's, a botanist who had explored the American South. André Michaux established a nursery in Charleston and shipped American plants to France. He also introduced exotic species to America, such as tea and the silk tree. In 1787, accompanied by the Scottish botanist John Fraser, he followed Bartram's route up the Savannah River and explored Cherokee territory. Later, he explored unmapped areas east of Mississippi, Spanish Florida, and Quebec. He tried unsuccessfully to reach Hudson Bay.

In 1792, he proposed to the Philosophical Society of Philadelphia to explore the west coast of America. This project was approved by George Washington and funded by about 30 members of the Philosophical Society of Philadelphia, among them Thomas Jefferson, John Quincy Adams, and James Madison, all later presidents of the United States. But Michaux cancelled his trip. Having lived for nine years in the young American republic, he embraced the French Revolution, although he did not return to France in 1789. In 1792, he was charged by the new French Republic with assisting a revolt in Louisiana against the Spanish.

Later, he returned to botany and explored the Carolinas and Georgia. On August 30, 1794, he ascended Grandfather Mountain and wrote in his diary, "I sang the Marseillaise and shouted, Long live America and the Republic of France! Long live liberty!" On December 23, 1797, after almost 12 years of exploration in North America and using his own funds to pay for his trips, André Michaux came back to Paris. He published two books in France, *Histoire des chênes de l'Amérique septentrionale* (*History of oaks from North America*) (1801) and *Flora boreali-Americana* (1803). This second publication, probably written in collaboration with Louis-Claude Richard, described 1,700 American species. In 1803, on the way to Australia, he stopped in Madagascar and died while exploring the island.

Michaux's son, François-André, sojourned twice more in America, from 1801 to 1802 and from 1806 to 1808. In 1805, he wrote an account of his exploration of America with his father, *Voyage à l'ouest des Monts Alleghanys* (*Travel west to the Allegheny Mountains*). From 1810 to 1813, François-André published in Paris in three volumes under the title *Histoire des arbres forestiers de l'Amérique septentrionale* (*History of the forest trees from North America*). They were translated into English in 1817–1819 as *The North American sylva*.

André and François-André Michaux described and named many North American species and made a very valuable contribution to the knowledge of forest biodiversity in North America. At his death, François-André Michaux bequeathed all his property to America. The André Michaux International Society was created in 2002.

The work of André and François André Michaux was continued in the United States by another French botanist, Michel-Victor Leroy (1754–1842). Originally from Lisieux in Normandy, Leroy settled in Hispanolia, known as Saint-Domingue in French. After the revolt of the slaves in 1791, France lost this island, which became independent and is now shared between Haiti and the Dominican Republic. Michel-Victor Leroy was obliged to flee to Boston. He became professor of botany at the University of Boston and then retired to Baltimore. He explored Tennessee, the area around Lakes Erie and Ontario, and the Allegheny Mountains. He corresponded with François-André Michaux, who was then back in France, and continued to ship American forest species to Europe. In 1820, he sent to France 24 varieties of American oak. The acorns were sown in the Bois de Boulogne in Paris, near the Auteuil pond, by François-André Michaux. In 1831, Michaux wrote to Leroy that the American forest was growing very well. Unfortunately, almost all the oaks were destroyed during the war of 1870 and the insurrection of the Communes in 1871. Other American oaks were introduced by Leroy and Michaux in Normandy in the Harcourt Forest, which is now managed by the French Academy of Agriculture.

After those inventories of the North American flora, France and the United States did not much cooperate on forest preservation or conservation. But the two countries had deeply influenced each other.

The idea that it was neccessary to preserve biodiversity on Earth first emerged in the United States in the 19th century, and the first American national park, Yellowstone, was created in 1872. When Theodore Roosevelt became president of the United States in 1901, the movement promoting natural resources management divided into two concepts, which were actually not so far apart. John Muir favored preservation and advocated zoned public land use, in which a large proportion of public lands would be preserved in a pristine state for low-intensity recreation. Gifford Pinchot, educated in forestry in France, Germany, and Switzerland, favored multiple resource use on public lands. He believed all resources should be developed to their full potential. Pinchot's philosophy gained the greatest following within the Progressive Movement. However, Muir's philosophy was very influential in establishing the National Park System. Since 1872 the National Park System has grown to include more than 370 areas covering more than 80 million acres.

The U.S. National Park System claims some of the most complete ecosystems and biodiversity remaining in the United States. More than 60 percent of the endangered species in the United States are found within national parks. The habitat preserved within park boundaries affords many species an oasis for survival and some of the last bastions of intact ecosystems found today. But these islands of wildlife may not be enough to ensure the preservation of natural diversity in the United States. Now, 154 million acres of U.S. land are protected, including 77 million acres of forests, the equivalent of about 10 percent of the total forest area. Timber harvesting has been considerably reduced in the national forests.

In December 2000, Undersecretary of Agriculture Jim Lyons ended his term with a request to the timber industry to stop logging virgin forests:

The values of old growth, as the public is coming to understand them and as we understand them scientifically, far exceed their value solely for timber. I think it would be a feather in the cap of industry to engage in a dialogue to end old-growth harvest and, at the same time, work with mills that are dependent on old growth to develop the technologies to harvest second growth and smaller-diameter material and, frankly, capitalize on the technology and the markets that are out there.

The debate is still very vigorous, but it seems that the cause of preserving ancient forests is gaining ground.

In France, the appearance of the first protected areas is linked to the feudal system. The objective of the lords, however, had been to protect game in large forests for their own use, and not to protect trees. Colbert's Forest Ordinance of 1669 established a new method of management in the French forests, allowing the protection of trees for special purposes. It was only at the end of the 19th century that a need to protect nature was felt in France. Curiously, the first militants were artists—the famous Barbizon painters. They started a battle against the French administration to preserve the landscapes that they were painting. In 1853, almost 20 years before the creation of Yellowstone National Park, they won complete protection for 624 hectares in the National Forest of Fontainebleau, near Paris. The Barbizon School is now famous in France as much for its success in protecting nature as for its art. But progress was slow, with phases of regression. Decades passed before the law on protection of nature was promulgated, on May 2, 1930. The first French national park was not created until July 6, 1963.

The creation of the French national parks was greatly influenced by the U.S. national parks. Indeed, France has adopted the American model. There are now seven French national parks that protect a total of 3,710 km^2 in central zones and 9,162 km^2 in secondary zones. This puts more than two percent of the total area of France under some level of protection. The creation of new national parks in the Iroise Sea, in French Guyana, and on Réunion Island is still under discussion.

As in the United States, more and more people in France and in Europe began to think that the creation of national parks or natural reserves was insufficient. In 1992, the European Union approved the Habitats Directive (92/43/EEC), promoting the protection of European Community natural heritage and completing the Birds Directive (79/409/EEC), which has promoted wild bird conservation since 1979. The Natura 2000 European network is an outgrowth of the Habitats Directive. The idea is that, in defined areas, agriculture and forestry can be compatible with nature conservation while remaining economically viable. The choice of Natura 2000 sites is based exclusively on scientific criteria. In these defined areas, the objective is to promote the conservation of habitats for wild fauna and flora, while taking into account the economic needs of the local population.

The Natura 2000 network promotes a policy of contracts drawn with all the local partners and provides legal protection in a yet-to-be-defined percentage of European land; the proportion could reach 15 percent of European Union territory. There are two types of Natura 2000 sites: areas of conservation and areas of special protection for birds.

In forestry, the concept of sustainable management emerged 20 years ago throughout the world—in Europe and in France after innumerable debates. These debates are leading to new methods of forest management of native forests. Nevertheless, artificial plantations still have a place. We need diversity in methods of forest management. Extensive forest management does not exclude intensive silviculture—the two methods are complementary.

Cooperation in Reforestation and Tree Breeding

At the beginning of the 19th century, French forest managers faced the immense task of rehabilitating forests that had been overexploited for centuries. The idea of introducing exotic forest species from the New World and Asia arose at the end of the 18th century with André Michaux and several other botanists and foresters. In the middle of the 19th century, several arboretums were created to study new species that might be useful for reforestation. Nevertheless, at that time, support for the introduction of exotic species in French forests was far from unanimous. Auguste Mathieu, a professor in natural history at the Forestry School of Nancy and author of a book on forest flora that was reprinted three times during the second half of the 19th century, strongly disapproved of the introduction of exotic species. In the preface to the third edition of his book, he wrote,

> *I am more and more convinced that only native plants can constitute true forests capable of perpetuating and regenerating themselves naturally, and that only native plants are suitable for the creation of new varieties with a sustainable future. Thus, in my opinion, foresters should study these species in preference to all others.*

Later in the book, referring to native species, he added,
> *Perfectly well adapted to the environment in which they originated, they are more suitable than any other plant of foreign origin that one might be tempted to introduce or to use as a replacement. Considering the range of different wood they offer, native species are fully able to satisfy a wide range of different needs ....*

Another professor at the Forestry School of Nancy, Paul Henri Fliche, was also skeptical about the advantage of using foreign species. He wrote,
> *As you can see, my conclusions show that both naturalization and acclimatization are extremely rare in France; that the latter is*

impossible, and that, as far as the former is concerned, it is unlikely that future experiments will be any more successful than those undertaken in the past ….

That pessimistic view has altered considerably over the years, given the successful introduction into France of several species native to North America, including red oak, several conifers, and poplars. We shall take only two examples, Douglas-fir and poplars.

Douglas-fir. Douglas-fir (*Pseudotsuga menziesii* Franco S.L.). is named after Archibald Menzies, a Scottish physician and naturalist who saw the tree on Vancouver Island in 1791, and David Douglas, a Scottish botanist who identified this species in the Pacific Northwest in 1826. Douglas-fir is North America's most plentiful coniferous species, accounting for one-fifth of the continent's total coniferous reserves.

It was first introduced to France in the middle of the 19th century and has become the most popular species for reforestation in France since 1980. Now there are in France almost 400,000 hectares of pure Douglas-fir stands with a standing volume of 62 million m^3. The annual crop was 100,000 m^3 in 1980, and 20 years later, it reached 1.5 million m^3. The annual crop expected in 2035 is 5.6 million m^3. Today, by comparison, the production of coniferous sawtimber is about 7 million m^3. This success is partly due to the strong cooperation established between France, Europe, and the United States in the past 30 years.

From 1980 to 1995 a collection of 540 progenies, mainly from Washington State, were planted in Germany, Belgium, France, and Spain. In 1991, the U.S. Forest Service provided more than 600 additional plus-tree open progenies. This second collection was shared among United Kingdom, Belgium, France, and Italy. One objective is to create in western Europe a broad-based population for a long-term Douglas-fir breeding program. Another objective is to evaluate the evolution of the genetic diversity after transfer to Europe and after natural regeneration.

More than 50 trials of progeny comparison were established in France by the Institut National de la Recherche Agronomique (INRA), Cemagref, and Office National des Forêts. Two Douglas-fir seed orchards have started to produce elite seeds. Four others will be available soon. Bulk propagation of improved families is also in progress.

Poplar hybrids. Poplar hybrids (*Populus* sp.) can result from natural crosses among poplar species. Most poplar hybrids, however, result from artificial hybridization and subsequent planting. At the end of the 18th century, André Michaux and later Victor Leroy brought to France the North American eastern cottonwood (*Populus deltoides*), which crossed naturally with poplars in Europe (*P. nigra*). Poplars were used as windbreaks around fields, and fast-growing hybrids were selected by farmers. Hand-pollinated poplar hybrids were first produced in Britain in 1912, and many European countries established plantations of hybrid poplars after World War II in response to shortages of timber. Some of the European hybrids were introduced to North America during the early 20th century. The first large-scale hybridization project with poplars in the United States began in 1925. Similar projects began in Europe mainly in Italy, the Netherlands, and France. Poplar hybrids are now the fastest-growing trees in Europe and North America. They are a short-rotation woody crop that can typically be planted and harvested in fewer than 20 years. Uses include pulpwood, plywood, other composite construction materials, pallets, and shipping lumber. Better-quality wood can be used for furniture. In France, hybrid poplars produce up to 2.5 million m^3 of wood per year. This represents 30 percent of France's annual crop of hardwood timber. It is produced on 1.4 percent of the total French forest area, or 2.7 percent of the hardwood forest area.

Cooperation for Reducing Phytosanitary Risks

Accidental introductions of insects, fungi, bacteria, or viruses from one continent to another could lead to catastrophic damage to trees. Numerous cases of terrible damage have already occurred. The Dutch elm disease in Europe and the chestnut blight in North America are among the most striking examples; both led to the near-extinction of two forest species. Only strong cooperation between Europe and North America may prevent the risks.

A very good example of cooperation is evaluating the risk to European oaks affected by oak wilt from the United States. This work was conducted by Jean Pinon from France, William MacDonald and Mark Double from West Virginia University in Morgantown, and Frank Tainter from Clemson University in South Carolina.

Oak wilt is due to a fungus, *Ceratocystis fagacearum* (Bretz) Hunt. Adult red oaks can be killed within a growing season, but white oaks are tolerant.

The disease is transmitted through root grafts and beetles. Because American oak logs are exported to Europe, this disease may threaten European oak species. To reduce the risk, a European regulation was established for better control of wood importation from North America, and a joint research program began in Great Britain, France, and Germany to evaluate the risk. British and German scientists demonstrated that *C. fagacearum* could potentially be disseminated in Europe via a native beetle, *Scolytus intricatus* (Ratz).

In 1981, two experimental designs were set up, in Morgantown and Clemson, to estimate the susceptibility of European oaks. Acorns from European species were collected in various European countries and sown in the two American locations; American white and red oaks were the controls. In 1996, 15-year-old trees were inoculated with *Ceratocystis fagacearum* on a branch or the stem. An assessment of symptoms was conducted in 1996 and 1997. American white oaks (*Quercus alba* and *Q. prinus*) confirmed their good level of tolerance, but no red oaks (*Q. rubra*) were still alive. In Morgantown nearly all inoculated European oaks (*Q. robur* and *Q. petroea*) were also killed. Around 30 percent of the uninoculated controls died after natural root-graft transmission. In Clemson, European oaks (including *Q. pubescens*) appeared also very susceptible.

Because of the sensitivity of European oaks to *Ceratocystis fagacearum* and the existence of *Scolytus intricatus* as a potential vector, the introduction of the fungus to Europe must be prevented, and *S. intricatus* must be prevented from entering the United States. European regulations that require U.S. fumigation of potentially dangerous wood before transportation to Europe have to be strictly enforced.

The recent outbreak of sudden oak death on the Pacific Coast of the United States is due to an oomycete, *Phytophthora ramorum*. In Europe, *P. ramorum* was recently detected on ornamental rhododendrons but has not yet been detected in European oak forests. However, *P. ramorum* is a serious potential danger on both sides of the Atlantic. Active research involving both the United States and Europe is needed to produce more information about *P. ramorum*.

In Europe and North America, premature defoliation of hybrid poplars is caused by leaf rust fungi belonging to the genus *Malampsora*. Epidemics of leaf rust are favored by growing susceptible hybrid clones in monocultural stands. Constant cooperation is necessary between European and American scientists in poplar breeding and leaf rust biology programs.

Cooperation in Tree Genome Sequencing

Sequencing of a tree genome was a dream five years ago. In 2002, the decision to sequence a poplar (*Populus trichocarpa*) was made by the U.S. Department of Energy, Genome Canada, and Umeå Plant Science Center in Sweden. The leader of the project was Jerry Tuskan, from Oak Ridge National Laboratory. The sequencing was performed by the Department of Energy's Joint Genome Institute (JGI), associated with the University of British Columbia, Umeå Plant Science Centre, and four French INRA laboratories, three in France and one in Belgium. Poplar was chosen because of its rapid growth, small genome size, and widespread use in areas of interest to the forest industry. The *Populus* genome was released by JGI in 2004. More than 200 scientists throughout the world are continuing the work of gene annotation. This effort will furnish scientists in the United States and abroad with a catalog of genes and knowledge about their functions, and it offers an exciting opportunity to better understand how poplar grows and interacts with its environment and with its associated microorganisms.

The sequencing of two mycorrhizal fungi associated with poplar, *Laccaria bicolor* and *Glomus intraradices*, is under way at JGI. In March 2005, JGI released the first assembly draft of *L. bicolor* S238N-H82, only one year after commencing the work. Fifty scientists from the United States and Europe are now working on the annotation of the *Laccaria* genome. The *G. intraradices* genome will be available in the following months. The main participants to this project are Francis Martin (INRA, France), Gopi K. Podila (University of Alabama), R. Hamelin (Natural Resources Canada), Pierre Rouzé (University of Ghent, the Netherlands), and Steve DiFazio and Jerry Tuskan (Oak Ridge National Laboratory, Tennessee).

JGI will soon start sequencing of *Melampsora larici-populina*, which causes widespread economic losses in poplar plantations worldwide. The comparison of the genomes and transcriptomes of symbiotic (*Laccaria bicolor, Glomus intraradices*) and pathogenic (*M. larici-populina*) basidiomycetes interacting with poplar will provide insights into mechanisms developed by the different types of biotrophic fungi. Moreover, the sequencing will allow an unprecedented view of the complex mutualistic and pathogenic communities associated with poplar. The study of these interactions at the genome level will add a new dimension to research on ecosystem functioning.

Conclusions

Cooperation between the United States, France, and Europe in the field of forestry has a long history and is rooted into two centuries of friendship and cooperation. The exchange of ideas, concepts, and living material has been extremely beneficial for the two continents. Today, cooperation in forestry research has never been so strong. With sequencing of the poplar genome and its symbiotic or pathogenic associates, we are entering in a new era that will provide unprecedented insights into the molecular bases of symbiosis, pathogenicity, and adaptation of trees to their environment. It will be a major step toward moving genomic research into ecology and developing a new discipline, ecogenomics. But we have to address other common challenges, one of the most important being the control of greenhouse gas emissions, with its consequences of global climate change for forests and water availibility. Increases in energy costs and exhaustion of fossil fuel energy sources also pose a new challenge to foresters.

And more generally, the duty of North America and Europe is to improve our knowledge of sustainable forest management and to share it with the rest of the world. The future of our planet is in the hands of each of its inhabitants.

References

Brasier, C.M., J. Rose, S.A. Kirk, and F. Webber. 2002. Pathogenicity of *Phytophthora ramorum* isolates from North America and Europe to bark of European Fagaceae, American *Quercus rubra*, and other forest trees. Sudden Oak Death, a Science Symposium: The State of Our Knowledge, 30–31. December 17–18, Monterey, CA.

Davy de Virville, A. 1954. Botanique des temps anciens. In *Histoire de la botanique en France*. Paris: SEDES.

De Gruyter, H., R. Baayen, J. Meffert, P. Bonants, and F. van Kuik. 2002. Comparison of pathogenicity of *Phytophthora ramorum* isolates from Europe and California. Sudden Oak Death, a Science Symposium: The State of Our Knowledge, 28–29. December 17–18, Monterey, California.

Delatour, C., M.-L. Desprez-Loustau, C. Robin, and C. Husson. Les Phytophthoras en Europe: Risques liés à l'existence des Phytophthoras en chênaie et à l'apparition de la "Sudden Oak Death" aux USA? Académie d'Agriculture de France, séance du 10 11 03.

Gibbs, J.N., W. Liese, and J. Pinon. 1984. Oak wilt for Europe? *Outlook agric.* 13: 203–207.

Jung, T., E.M. Hansen, L. Winton, W. Osswald, and C. Delatour. 2002. Three new species of *Phytophthora* from European oak forests. *Mycol.* Res. 106: 397–411.

Pluchet, R. 2005. André Michaux, le laboureur explorateur, Hommes et plantes 5: 4–12.

Michaux, A. 1801. *Histoire des chênes de l'Amérique septentrionale*. Paris.

———. 1803, *Flora boreali-Americana*. Paris.

Pinon, J., W. MacDonald, M. Double, and F. Tainter. Les risques pour la chênaie européenne d'introduction de *Ceratocystis fagacearum* en provenance des États-Unis. Académie d'Agriculture de France, séance du 10 11 03.

Robin, J., and P. Vollet. 1608. *Le jardin du roy Henry IV*. Paris.

———. 1623, *Le jardin de Louis XIII*. Paris.

Tissot, A. 1877. *Note sur la biographie et les travaux de Victor Leroy, Botaniste et horticulteur, né à Lisieux, introducteur en Europe d'un grand nombre d'Arbres, Arbustes et Végétaux d'origine américaine*. Lecture faite à la séance de la Société Linnéenne de Normandie, tenue à Lisieux le 24 juin 1877.

Thivolle-Cazat, A. 2004. Le Douglas en France, une ressource en pleine expansion, AFOCEL, laboratoire économie et compétitivité.

International Scientific Contacts

André Michaux International Society, Box 220283, Charlotte, NC 28222.

Department of Plant Systems Biology and INRA-associated laboratory at Gent University: http://www.psb.ugent.be/; Yves Van de Peer, yves.vandepeer@psb.ugent.be; Pierre Rouzé Pierre.Rouze@psb.ugent.be.

DOE Joint Genome Institute (JGI): http://www.jgi.doe.gov; David Gilbert, gilbert21@llnl.gov.

Genome British Columbia: http://www.genomebc.ca/ Linda Bartz, lbartz@genomebc.ca.

Genome Canada: http://www.genomecanada.ca/; Anie Perrault, aperrault@genomecanada.ca.

INRA-Bordeaux Christophe Plomion, plomion@pierroton.inra.fr.

INRA-Nancy: Francis Martin, fmartin@nancy.inra.fr tél: 03 83 39 40 80.

INRA-Orléans: Gilles Pilate, pilate@orleans.inra.fr; Catherine Bastien catherine.bastien@orleans.inra.fr; Philippe Label label@orleans.inra.fr.

International Poplar Consortium: http://www.ornl.gov/sci/ipgc/; Gerald Tuskan, gtk@ornl.gov. Oak Ridge National Laboratory: http://www.ornl.gov; Ron Walli, 865-576-0226; wallira@ornl.gov.

Stanford Human Genome Center: http://www-shgc.stanford.edu; Ruthann Richter, richter1@stanford.edu.

Umeå Plant Science Centre: http://www.upsc.nu/; Stefan Jansson, stefan.jansson@plantphys.umu.se.

ABSTRACT

Using selected examples from Europe and the United States, this chapter illustrates three main stages in the development of forest science during the past half-century. Forest science first focused on the characterization and description of the spatial and temporal variations of forest ecosystems for numerous traits, in particular their tree and stand components. In a second phase, researchers investigated functions and processes related to physical, biological, ecological, and social factors and their interactions. Most recently, emphasis has been on predicting the dynamics of forest ecosystems in relation to environmental, social, and economical changes, and identification of sustainable management methods for adapting to new contexts. The examples cited reveal progress in various scientific disciplines, as well as their applications to practical forestry, and also point the way to some major scientific challenges for the future.

CHAPTER 4

Science and the Forest: Achievements, Evolution, and Challenges

Yves Birot and François Houllier

On both sides of the Atlantic, forest science has evolved under internal and external driving forces, comparable but also different in nature and intensity. The following external trends, inter alia, have been observed: 1) a widening of the range of functions, from production to societal and environmental functions, 2) a transition from tree and stand management to integrated ecosystem management, 3) appraisal of forest issues at the scale of the landscape and larger areas, 4) emergence of forest issues on the political international agenda, and 5) implementation of sustainable forest management principles. Forest-related research agendas in Europe and the United States show some similarities and differences because of the local context, including history, culture, know-how, and forest ecosystem and ownership types. The mandate of forest research has traditionally been—and still is—to provide knowledge for innovation, but today, knowledge is also needed to generate expertise for the political decision-making process, particularly in a context of acknowledged uncertainty about climate changes and risks.

Advancing European and North American forest research has in general followed similar approaches as science and technology have evolved:
- the capture and handling of georeferenced data (satellite imagery, global positioning systems, and geographic information systems);
- methods based on the use of stable isotopes (N, O, C) allowing the understanding of many ecophysiological and biogeochemical processes through in situ and ex situ experimental studies, and the quantification of fluxes;

- the development of molecular biology and bioinformatics, opening new avenues for genetics and physiology (genomics and postgenomics) but also forest ecology with all its associated living organisms; and
- advances in statistics, biometrics, mathematics, and informatics, including data analysis, simulation and modeling, and virtual trees, stands, landscapes and experiments.

Looking back over the past 50 years, we see that forest science has been characterized by a three-step process, with some overlap on the time scale. The first step was based on a descriptive approach aimed at observing the phenomena and quantifying them through data analysis, starting with graphical techniques and continuing into more advanced statistical methods. Such an approach is still used to a large extent. The second step has focused on functions and processes in trees and forest ecosystems. It can be referred to as a deterministic approach that leads to an understanding of the functioning of complex systems. The third step aims at organizing available knowledge into methods and instruments allowing researchers to predict the dynamics of forest ecosystems in relation to evolving conditions and adapt to this new and changing context.

The Prerequisite: Description and Characterization

Historically, forest science has tried to quantify the variations observed in the characteristics of forest ecosystems—their composition, environmental

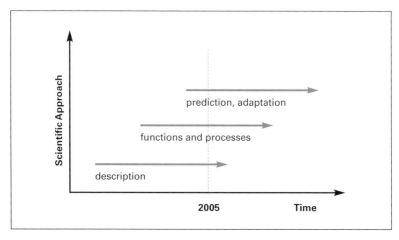

Figure 1. Steps in advancing forest science: a conceptual framework.

conditions (soil, climate, elevation, aspect), dynamics, production of ligneous matter, and wood traits. Through assumptions, observations, and experiments, the goal was to identify patterns and rules on which forest planning and silvicultural prescriptions could be based.

Forest growth-and-yield and spatial variations. A first goal of forest research was to describe and quantify the variations in forest ecosystem features in relation to their physical environment. Disciplines such as biogeography, plant ecology and phytosociology, pedology (soil science), site typology, and mapping have led to coherent bodies of knowledge for characterizing forest ecosystems and their (bio)diversity, with numerous applications in forest planning and management at various geographical scales. Studies by Cajander (1926) in Scandinavia played a pioneer role in this area.

Practical applications included determining the fitness of tree species to site and quantifying site-productivity relationships. As an example, Eichhorn's (1904) rule expressed the invariance of the relationship linking stand height and stand volume or basal area of well-stocked stands, regardless of site quality. Forest scientists further tried to quantify the yield and dynamics of forest stands in relation to silvicultural treatments and stand density and structure. The first attempts in Europe, especially in Germany but also in France and other countries, were based on the monitoring of long-term experimental plots, where somewhat different silvicultural treatments were applied. Such approaches led in North America to the establishment of data cooperatives, where various silvicultural treatments (such as initial stand density, intensity, nature and timing of thinnings, and sometimes of fertilization or vegetation management techniques) were applied across a wide range of site types.

These empirical approaches first resulted in the elaboration of yield tables related to a fixed silvicultural schedule. The tables were based on a few empirical relationships, such as Eichhorn's rule or height-growth equations indexed by site quality. Further developments led to the notions of stand density and tree-to-tree competition (as defined by comparisons of extreme situations, such as open-grown trees compared with very dense stands with self-thinning). The introduction of such concepts in yield studies opened the way to the development of true dynamic models operating at the level of the stand, and then of the tree. Such models were first developed in North America (Stage 1973), using experimental data but also national forest inventory data; they were then extended to Europe (and other parts of the world), such that simple growth-and-yield models are now available for the most important species.

Another major development in this field was related to the conception and utilization of rigorous forest inventory methods. The first national forest inventories based on such techniques were carried out in the 1920s in Scandinavia. The three major drivers were the sophistication of statistical and sampling techniques (e.g., sampling with partial replacement, Cunia and Chevrou 1969; kriging techniques, Matern 1960, Houllier and Pierrat 1992), the development of new techniques of capturing georeferenced data (aerial photographs, satellite imagery, global positioning systems), and the development of information systems (databases, geographic information systems).

Genetic architecture of forest tree species. The intuition that interpopulation variations existed in tree species dates back before the French Revolution, to Duhamel du Monceau in France. In the middle of the 19th century, Philippe-André de Vilmorin was the first to test this idea by introducing into the Arboretum des Barres trees from several populations of a single species, planted in separate plots. And in the first years of the 20th century, the International Union of Forest Research Organizations organized concerted networks of provenance trials in Europe for various species. By matching genotypes with various environments, researchers hoped to discover geographic variation within a tree species and apply the results to the selection of "best seed sources" for plantations. These investigations have for decades occupied a large community of forest geneticists of both sides of the Atlantic. The development of the concepts of quantitative genetics has allowed the emergence of efficient individual tree selection programs. Despite the "black box" aspect of such an approach, breeding work proceeded, and large seed orchard projects delivering significant genetic gains were established. In the United States, Gene Namkoong has been a pioneer in the optimization of selection methods based on the multiple populations concept (Namkoong et al. 1988).

The possibility of appraising "neutral genes" through isoenzymes and even more so through DNA studies has opened the door to powerful analyses of population genetics and spatiotemporal genetic diversity evolution, from the tree level to the entire range of a species. Kremer and his European partners (2002) investigated 2,600 oak populations in the largest genetic survey ever undertaken of living organisms and obtained a good understanding of the past and present evolution of oaks. They have also shown the possibility of tracing oak products on these bases. Such results pave the way to an integrated management of forest genetic resources.

Trees and the forest in interactions with their physical environment. The importance of the link between forest cover and water resources was recognized long ago in France—in 1291, when an office of "master of waters and forests" was established. Much later, the issue of water was a main reason for the establishment of the national forests in the United States. In France, Surell (1841) conducted a pioneering study relating the hydrological regime of torrents to forest cover in mountainous watersheds. Extensive hydrological research has been carried out in United States on experimental forest watersheds (USDA 2000).

Although forest management involves low inputs compared with agriculture, the mineral nutrition of trees as it relates to the vitality of forest ecosystems is quite important. Nevertheless, site studies and studies of nutrient cycling have been carried out for a long time without proper consideration to tree nutrition, leading to underestimates of the role of nutrients as regulators of processes in ecosystems (Tamm 1964, 1995).

In the 1980s, growing concern about forest decline associated with atmospheric pollution and acid rain was a catalyst for vast, coordinated research programs on ecosystem function (and dysfunction), such as DEFORPA in France (Landmann and Bonneau 1995). In addition to elucidating the main causes of the disturbances, these programs have contributed substantially to our knowledge of ecosystem processes.

Trees and the forest in interactions with their biotic environment. Forest health has been (and still is) a main concern of foresters, and many investigations have dealt with forest pests and pathogens. For decades, research programs have focused on the description and understanding of biological cycles of harmful insects and microorganisms, as a prerequisite for developing control and suppression methods. In the United States and Canada, ambitious cooperative research programs have been undertaken on the spruce budworm and bark beetle (Alfaro et al. 2000; Berryman 1974). In France, Pinon and Frey (1997) have carried out comprehensive work on the interactions between poplar cultivars and the rust foliar disease with its different pathotypes.

Studies on symbiotic microorganisms of the rhizosphere, in particular the endo- and ectomycorrhizae and their ecology and interactions with tree host species, have also been carried out, leading to applications in plantation forestry, as demonstrated by Don Marx (1977) in the United States.

Characterization of wood properties in relation to processing. X-ray technology was a key to developing the systematic investigation of the variations in wood density associated with tree growth patterns and the links between density and other wood mechanical properties (Polge 1978); it paved the way to nondestructive testing. Coupling tree growth models, wood quality models, and computer simulation techniques provided a way to predict the consequences of silviculture on the quality of end-products (Leban and Duchanois 1990; Maguire et al. 1991; Houllier et al. 1995; Dreyfus and Bonnet 1996) from simple tree measurements in the forest without using destructive methods or even increment cores.

A limit to the descriptive and statistical approach: nature is not immutable. Most technologies employed in current forest management are based on the fruits of many years of descriptive research. However, many concepts (e.g., climax forest, site index, genotype x site interaction) assumed an equilibrium in natural processes that would lead to a stable state of forest ecosystem. Rapid environmental changes due to anthropogenic effects (increased carbon dioxide in the atmosphere, rising global temperatures, nitrogen deposition) over the past decades have resulted in numerous demonstrations of evolution, including increased forest productivity in Europe (Spiecker et al. 1996), altitudinal migration of plant species in the Alps, and changes in insect species distribution. The fact that nature is not immutable shows the limits of the descriptive approach and highlights the need for investigating functional aspects and processes.

Functions and Processes: The Key to Multifunctionality

Forest ecosystems are extremely complex in their structures and functions because so many environmental, biological, and human factors are involved. The expanding roles assigned to forests have necessitated management based on integrated knowledge of their ecosystemic dimension. The concept of ecosystem management, developed on both sides of the Atlantic, requires increased understanding of the functions and processes in forest ecosystems.

Basic biological functions and processes in trees. The development of molecular biology tools over the past 20 years has led to many applications, particularly in the field of genomes. The genome sequencing of poplar—the first tree species to be thus studied (Tuskan et al. 2004)—was achieved in 2004 by an American-Canadian-European consortium (including French teams from the Institut National de la Recherche Agronomique, INRA), illustrating the efficiency of transatlantic scientific cooperation. This breakthrough is the prerequisite for understanding gene expression and function. Ongoing research on the genomics of poplar-associated pathogens (*Melampsora*, poplar rust) and symbiotic ectomycorhiza (*Laccaria*) will contribute to elucidating their interaction with their host (http://genome.jgi-psf.org/Lacbi1/Lacbi1.home.html). Genetic modification is also a powerful tool for investigating gene expression and function.

The integration of molecular, genetic, physiological, and environmental mechanisms regulating the functioning of the cambium and the formation of wood cells, as well as the elaboration of heartwood, is being investigated by INRA teams in France, opening interesting perspectives for the future. Systems biology is still in its infancy but will certainly be a major area of scientific inquiry in the future.

In the field of tree physiology, ecophysiology research has been a priority. The understanding of tree behavior and its response to environmental constraints has become absolutely essential. Melvin Tyree (1991, 2003) in the United States has conducted pioneering research in the area of water in trees. Working with a French collaborator, he investigated the phenomenon of cavitation or embolism in wood vessels after a dry period and the mechanisms of restoration (Cochard and Tyree 1990). Moreover, the development of study methods based on stable isotopes of H, C, O, and N has permitted a better understanding of basic physiological processes in trees and of ecosystem functioning. As in other types of ecosystems, stable isotopes have turned out to be useful tracers of the functioning of trees and of forest ecosystems in relation to their physical environment. Isotopes have been employed for describing where root mineral uptake takes place in the soil (e.g., Dambrine et al. 1997, using Sr isotopes), assessing the influence of past land uses on present forests (e.g., Koerner et al. 1997, using N isotopes), tracing the temporal variations of carbon dioxide (Buchmann et al. 1997), and analyzing the functional diversity of tree species in complex tropical canopies (Guehl et al. 1998).

Forest-environment interactions. Forest-atmosphere exchanges are related to stocks and fluxes of energy—matters whose quantification has become a major goal. Biogeochemical cycles for major elements and nutriments are ever-better documented. Many research programs have focused, inter alia, on the cycle of water, nitrogen, and carbon (in relation to the political issue of the Kyoto Protocol). The need for a continuum of long-term assessment and observations, ranging from simple observation networks to coordinated sophisticated and instrumented research facilities, has been clearly identified, and the response includes the LTER (Long-Term Ecological Research) network in United States and the ORE (Observatoire de Recherche en Environnement) program in France.

Ecosytem functioning. New knowledge about individual mechanisms of tree physiology (photosynthesis, water relation, nutrient uptake, growth) and the dramatic increase in computing capacity have led to the development of integrated simulation models that mimic natural processes taking place in trees and forests. Today's generation of such models, called process-based growth models, may reduce the need for experimentation and testing yet enhance our ability to predict forest ecosystem evolution under environmental evolution. The models also allow the identification of "black boxes" still to be deciphered.

Population biology and dynamics. These disciplines have undergone impressive growth in the last decades. The possibility of assessing gene flows and simulating population dynamics has opened new perspectives in ecology at various spatial scales, including the landscape. In particular, this allows the investigation of the functional aspects of biodiversity—that is, the mechanisms that affect biodiversity in its structures and evolution. An excellent example of the functional approach of biodiversity is given by Jactel et al. (2002) and the ISLANDES project. This research has shown that planting small hardwood stands within vast pine monocultures reduces the impact of some pest insects because their natural enemies (birds, parasitoids, predatory arthropods) are accommodated in the clumps of hardwoods.

Social sciences. The diverse functions assigned to forests, as well as man-induced environmental changes, are an indication of the interaction between forests and society. Human attitudes, behavior, traditions, economic choices, and institutions are the major driving forces behind these evolutions. It is essential to understand the human forces, the values underlying them, and their interactive effects on natural resources. Forest managers face trade-offs that can be summarized in one question: How do we maintain the ecological health of forest ecosystems while satisfying human needs and desires for the goods and services they provide? By studying human needs and impacts on forests, the social sciences can contribute to answering this question and inform natural resources management plans. In addition, the social sciences provide useful information on society's perception, beliefs, and values, allowing managers to understand what citizens want and how to make rational, effective decisions for long-term, sustainable forest resources. The growing importance of the social sciences as they relate to forests is seen in Europe as in the United States, as illustrated by the social science research agenda of the USDA Forest Service (2004).

Prediction and Adaptation: The Key to Sustainability

The evolution of the biophysical as well as the human environment of forest ecosystems and the recognition of the need for their sustainability call for a two-pronged approach, with two basic questions: How will forest ecosystems respond to changes, including silvicultural factors? And how can we help forest ecosystems adapt to the new context?

1. Predicting the evolution of forests. This is a huge challenge, since experimentation is obviously not possible. Among the various research approaches, two examples are illustrated below.

First, the development of simulation and modeling tools, from the level of the individual tree to complex landscapes, including process-based modules, opens the way to virtual experiments. In France, the software platform developed by AMAP in Montpellier (Reffye et al. 1995; Reffye and Houllier 1997) can predict the individual growth of trees competing within a stand in relation to silvicultural prescriptions and determine their internal wood characteristics.

Second, advances in evolutionary biology and ecology, whose integrated form can be referred to as ecogenetics, offer interesting perspectives in developing "demogenetic" approaches. The objective is to combine genetic and demographic information into simulation models that spatially describe the vegetation dynamics and composition over succession stages. Gene flow (pollen dispersal), parent tree stocking, mating regime, and genetic diversity influence the genetic characteristics of the seedlings and saplings, and consequently their ability to survive, compete, and grow.

2. Adapting forest ecosystems to new environmental and economic conditions. A main characteristic of recent environmental changes is their relatively fast speed, which gives natural systems only a limited time to adapt through evolutionary processes. It may well be that in some cases, actions must be taken by man to enhance the adaptation of forest ecosystems to the new contexts. "Tailoring" forest resources so that they better meet future industry needs also requires a strong scientific commitment.

Designing trees for the future, based on the advances of genomics (with or without using genetic modification), is certainly a challenging objective for plantation forestry. Research teams are already working on breeding for drought tolerance and better water use efficiency, among other topics. Candidate genes related to wood and fiber properties are being investigated before their integration into breeding programs.

The conservation of forest tree genetic resources and their diversity is clearly important, since diversity will provide the capacity and the flexibility to adapt to the new context. Two situations can be distinguished: the populations within a given species that are endangered (e.g., because they grow at the margins of the species' range) and require specific measures for conservation; and "ordinary" populations, which should be silviculturally managed in such a way that their diversity is not decreased. A rational conservation strategy implies extensive knowledge of the genetic architecture of the species over its entire range, as exemplified for oaks by the research European teams coordinated by Kremer et al. (2002).

Biodiversity can also be seen as exceptional (e.g., "hot spots" of endangered habitats and species) or ordinary, and again, conservation requires specific methods. However, it will be important to emphasize the mechanisms and processes that are responsible for the biodiversity, rather than protecting or putting into sanctuaries the elements to be preserved, although this can be justified in certain cases. Working with the processes requires a good understanding of evolutionary biology and ecology.

Specific vegetation management can be required for specific objectives, such as water production, fire risk prevention, and creating or restoring habitat for certain plant or animal species.

Scientific Challenges for the Future

Forest research must address the issue of sustainable forest management in its three dimensions: economic, environmental, and social. One major problem is setting up a coherent, long-term research strategy that prioritizes the research themes and allocates research forces and funding. A conceptual framework provides support for developing and balancing various research

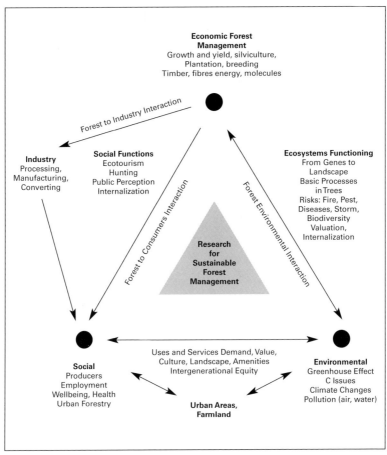

Figure 2. A conceptual framework for future forest research.

programs (Houllier et al. 2005). Foresight studies based on scenario analysis, for example, can usefully complement this approach (Sebillotte et al. 1998).

1. Ecosystem functioning. This will remain a vitally important area: it is the basis for understanding forest evolution and the rationale of sustainable forest management.

Some projections show that the potential range of tree species could be profoundly modified (see http://www.nancy.inra.fr/extranet/com/carbofor/carbofor-D1-resume.htm). It is therefore particularly crucial to investigate the processes that allow trees and forests to adapt to environmental changes. However, since these changes are operating at an unprecedented speed, it may well be that natural processes of adaptation are insufficient. Some proactive measures, such as designing corridors and using spatially explicit planning or even plantations, could be needed.

Because of the complexity of forest ecosystems, scientific understanding of their functioning has been based mainly on investigations that consider biological and physical processes separately. An important scientific challenge is integrating both types of processes.

Global change and the transcontinental movement of people, goods, and living organisms have made the biological invasion of plants, insects, and microorganisms a real concern in many parts of the world. In California, sudden oak death (Rizzo et al. 2001) has become a critical issue. In Europe, emerging pathologies are also cropping up; a striking case is the threat of pine nematode (Mota et al. 1999). More generally, the increased frequency of catastrophic events—windstorms, fires, floods, pest outbreaks—calls for a better integration of risk into forest management. This implies new research efforts on risk assessment and risk management methods (Birot and Gollier 2000; Arbez et al. 2002).

The scientific community involved in forest ecosystem research confronts a major challenge: how to scale up our understanding of processes—for example, gaseous exchange—from the level of an organ to the whole tree and beyond, to the forest, the landscape, the region, and even the globe. A first step is increased investigation of forest ecosystems at the landscape level (Andersson et al. 2004).

2. Forest multifunctionality. Managing economically important forest resources for multiple goods and services will be more and more important. Multifunctionality has become an overarching principle in forestry. Yet the economic profitability of forest management, in particular for small forest holders, is going down in Europe, raising the question of a "green desert" versus a living forest (Leonard 2004). The Ministerial Conference on Forest Protection in Europe, in Vienna, has addressed this issue in one of its resolutions (MCPFE 2003). Research can contribute to better profitability of forest management in several ways:

- developing decision support tools with improved capacities of prediction regarding forest ecosystems dynamics under management regimes and evolving environmental conditions, by coupling empirical and process-based models;
- studying the utilization of wood and developing innovative wood-based products and industrial processes, such as the pioneering work in France

on wood welding (Leban et al. 2005) and tannin-based, environment friendly adhesives (Pizzi et al. 1995);
- advancing forest-based bioenergy with different approaches, such as biorefineries and wood gasification; and
- ensuring high-quality water resources from proper forest management.

The multifunctionality of the forest is often invoked in a way that implies a strict balance between, and integration of, the three functions at the stand level: this view has been dominating in Central Europe. It is more and more obvious, however, that a better multifunctionality can be achieved at a broader scale, such as the landscape level. This in turn will require some specialization of forest areas (i.e., spatial segregation of functions versus integration) and new efforts in land-use planning. Nevertheless, a proper valuation of forest externalities linked to goods and services not marketed today and the development of mechanisms for internalizing these externalities represent important scientific challenges (Mogas et al. 2005). This is even a prerequisite for true, sustainable multifunctionality.

3. Forest science and society. An increasingly urbanized society will be more and more demanding of the uses and services provided by forests; research in this field must therefore be strengthened. The demands of the various stakeholders may sometimes be in conflict. Nevertheless, it is of paramount importance to appraise social perceptions of forests (including their cultural and spiritual values), their products and services, and their temporal evolution. Improving the science-society dialogue on forest-related issues should also be a priority. Other emerging topics, such as the relationship between the forest environment and human health, likewise deserve further investigation.

References

Alfaro, R.I., S.P. Taylor, R.G. Brown, and J.S. Clowater. 2000. Susceptibility of Northern British Columbia forests to spruce budworm defoliation. *Forest Ecology and Management* 145: 181–90.

Andersson, F., Y. Birot, and R. Päivinen R. (eds.). 2004. Towards the sustainable use of Europe's forests—forest ecosystem and landscape research: Scientific challenges and opportunities. *European Forest Institute Proceedings* 49.

Arbez, M., Y. Birot, and J.M. Carnus (eds.). 2002. Risk management and sustainable forestry. *European Forest Institute Proceedings* 45.

Birot, Y., and C. Gollier. 2000. Risk assessment, management and sharing in forestry with special emphasis on wind storms. 2001 Proceedings of the 14th CAETS Convocation, Espoo, Finland, 233–65.

Berryman, A.A. 1974. Dynamics of bark beetle populations: Towards a general productivity model. *Environmental Entomology* 3: 579–85.

Buchmann, N., J.-M. Guehl, T.S.Barigah, and J.R. Ehleringer. 1997. Interseasonal comparison of CO_2 concentrations, isotopic composition, and carbon dynamics in an Amazonian rainforest (French Guiana). *Oecologia* 110(1): 120–31.

Cajander, A.K. 1926. The theory of forest types. *Acta Forest. Fenn.* 29(3): 1–108.

Cochard, H., and M.T. Tyree. 1990. Xylem dysfunction in *Quercus*: Vessel sizes, tyloses, cavitation and seasonal changes in embolism. *Tree Physiology* 6(4): 393–407.

Cunia, T., and R.B. Chevrou. 1969. Sampling with partial replacement of three or more occasions. *Forest Science* 15: 204–24.

Dambrine, E., M. Loubet, J.A. Vega, and A. Lissarague. 1997. Localisation of mineral uptake by roots using Sr isotopes. *Plant and Soil* 192(1): 129–32.

Dreyfus, Ph., and F.R. Bonnet. 1996. CAPSIS (Computer-Aided Projection of Strategies in Silviculture): An interactive simulation and comparison tool for tree and stand growth, silvicultural treatments and timber assortment. Connection between silviculture and wood quality through modelling approaches and simulation software. Actes du colloque IUFRO: WP S5.01-04 second workshop, Berg-en-Dal, Kruger National Park, South Africa, August 26–31, 57–58.

Dreyfus, Ph., C. Pichot, F. de Coligny, S. Gourlet Fleury, G. Cornu, S. Jésel, H. Dessard, S. Oddou Muratorio, S. Gerber, H. Caron, C. Latouche-Hallé, F. Lefèvre, F. Courbet, and I. Seynave. 2005. Couplage de modèles de flux de gènes et de modèles de dynamique forestière. Un dialogue pour la diversité génétique: Actes du 5ème colloque national BRG, Lyon, 3–5 novembre 2004. *Les Actes du BRG* 5 (sous presse).

Eichhorn, F. 1904. Beziehungen zwischen Bestandshöhe und Bestandsmasse. Allg. Forst- u. Jagd-Zeit., 45–49.

Fourcaud, T., F. Blaise, P. Lac, P. Castera, and Ph. de Reffye. 2003. Numerical modelling of shape regulation and growth stresses in trees II. Implementation in the AMAPpara software and simulation of tree growth. *Trees Structure and Function* 17: 31–39.

Guehl, J.M., A.M. Domenach, M. Bereau, T.S. Barigah, F.A. Casabianca, and J. Garbaye. 1998. Functional diversity in an Amazonian rainforest of French Guyana: A dual isotope approach (d15N and d13C). *Oecologia* 116(p): 316–30.

Houllier, F., and J.-C. Pierrat. 1992. Application des modèles statistiques spatio-temporels aux échantillonnages forestiers successifs. *Canadian Journal of Forest Research* 22: 1988–95.

Houllier, F., J.M. Leban, and F. Colin. 1995. Linking growth modelling to timber quality assessment for Norway spruce. *Forest Ecology and Management* 74(1): 91–102.

Houllier, F., J. Novotny, R. Päivinen, K. Rosén, G. Scarascia-Mugnozza, and K. von Teuffel. 2005. Future forest research strategy for a knowledge based forest cluster: An asset for a sustainable Europe. Discussion paper. European Forest Institute.

Jactel, H., M. Goulard, P. Menassieu, and G. Goujon. 2002. Habitat diversity in forest plantations reduces infestations of the pine stem borer *Dioryctria sylvestrella*. *Journal of Applied Ecology* 39: 618–28.

Koerner, W., J.L. Dupouey, E. Dambrine, and M. Benoit. 1997. Influence of past land use on the vegetation and soils of present day forest in the Vosges Mountains. *Journal of Ecology* 85(3): 351–58.

Kremer, A., J. Kleinschmit, J. Cottrell, E.P. Cundall, J.D. Deans, A. Ducousso, A.O. Konig, A.J. Lowe, R.C. Munro, R.J. Petit, and B.R. Stephan. 2002. Is there a correlation between chloroplastic and nuclear divergence, or what are the roles of history and selection on genetic diversity in European oaks? *Forest Ecology and Management* 156(1–3): 75–87.

Landmann, G., and M. Bonneau. 1995. *Forest decline and atmospheric deposition effects in the French Mountains*. Springer Verlag.

Leban, J.M., and G. Duchanois. 1990. SIMQUA Un logiciel de simulation de la qualité du bois. *Annales des Sciences Forestières* 47(5): 483–93.

Leban, J.M., A. Pizzi, M. Properzi, F. Pichelin, and R.C. Gelhaye. 2005. Wood welding: A challenging alternative to conventional wood gluing. *Scandinavian Journal of Forest Research* 20(6): 534–38.

Leonard, J.P. 2004. *Forêt vivante ou désert boisé; la forêt française à la croisée des chemins*. L'Harmattan.

Maguire, D.A., J.A. Kershaw, and D.W. Hann. 1991. Predicting the effects of silvicultural regime on branch size and crown wood core in Douglas-fir. *Forest Science* 37: 1409–28.

Marx, D.H. 1977. Tree host range and world distribution of the ectomycorrhizal fungus *Pisolithus tinctorius*. *Canadian Journal of Microbiology* 23(3): 217–23.

Marx, D.H., J.L. Ruehle, D.S. Kenney, et al. 1982. Commercial vegetative inoculum of *Pisolithus tinctorius* and inoculation techniques for development of ectomycorrhizae on container-grown tree seedlings. *Forest Science* 28: 373–40.

Matern, B. 1960. Spatial variation. *Lecture notes in statistics*, vol. 36. Berlin: Springer Verlag.

MCPFE. 2003. Fourth Ministerial Conference, Vienna. http://www.mcpfe.org/resolutions/Vienna.

Mogas, J., P. Riera, and J. Bennett. 2005. Accounting for afforestation externalities: A comparison of contingent valuation and choice modelling. *European Environment* 15(1): 44–58.

Mota, M.M., H. Braasch, M.A. Bravo, A.C. Penas, W. Burgermeister, K. Metge, and E. Sousa. 1999. First report of *Bursaphelenchus xylophilus* in Portugal and in Europe. *Nematology* 1(7–8): 727–34.

Namkoong, G., H.G. Kang, and J.S. Brouard. 1988. *Tree breeding: Principles and strategies*. New York: Springer-Verlag.

Pinon, J., and P. Frey. 1997. Structure of *Melampsora larici-populina* populations on wild and cultivated poplar. *European Journal of Plant Pathology* 103(2): 159–73.

Pizzi, A., N. Meikleham, B. Dombo, and W. Roll. 1995. Autocondensation-based, zero-emission, tannin adhesives for particleboard. *Holz als Roh Werkstoff* 53(3): 201.

Polge, H. 1978. Fifteen years of wood radiation densitometry. *Wood Science and Technology* 12(3): 187–96.

Reffye (de), Ph., F. Houllier, F. Blaise, D. Barthélémy, J. Dauzat, and D. Auclair. 1995. A model simulating above- and below-ground tree architecture with agroforestry applications. *Agroforestry Systems* 30: 175–97.

Reffye (de), Ph., and F. Houllier. 1997. Modelling plant growth and architecture: Some recent advances and applications to agronomy and forestry. *Current Science India* 73(11): 984–92.

Rizzo, D.M., M. Garbelotto, J.M. Davidson, G.W. Slaughter, and S.T. Koike. 2001. A new Phytophthora canker disease as the probable cause of Sudden Oak Death in California. *Phytopathology* 91(6 Supplement): S76.

Sebillotte, M., B. Cristofini, J.-F. Lacaze, A. Messéan, and D. Normandin. 1998. Prospective: La forêt, sa filière et leurs liens au territoire. INRA, 2 vols. (collection Bilan et Prospective).

Spiecker, H., K. Mielikäinen, M. Köhl, and J.P. Skovsgaard (eds.). 1996. Growth trends in European forests: Studies from 12 countries. EFI research report 5. Springer-Verlag.

Stage, A.R. 1973. Prognosis model for stand development. USDA Forest Service research paper INT-137. Ogden, Utah: Intermountain Northwest Forest and Range Experiment Station.

Surell, A. 1841. *Etude sur les torrents des Hautes-Alpes*. Paris: Cariline-Goeury et Victor Dalmont.

Tamm, C.O. 1964. Detemination of nutrient requirement of forest stands. *Int. Rev. For. Res.* 1: 115–70.

———. 1995. Towards an understanding of the relations between tree nutrition, nutrient cycling and environment. *Plant and Soil* 168–69(1): 21–27.

Tuskan, G.A., et al. 2004. Poplar genomics is getting popular: The impact of the poplar genome project on tree research. *Plant Biology* 6: 2–4.

Tyree, M.T. 2003. The ascent of water. *Nature* 423: 923.

Tyree, M.T., and F.W. Ewers. 1991. Tansley Review No. 34: The hydraulic architecture of trees and other woody plants. *New Phytologist* 119(3): 345–60.

USDA. 2000. Water and the Forest Service. Washington, DC: U.S. Department of Agriculture.

USDA Forest Service. 2004. Social science research agenda. Washington, DC: U.S. Department of Agriculture.

ABSTRACT

In colonial America, clearing the forest—an eerie place full of savages, wild animals, and evil spirits—was first an almost sacred duty, then a secular responsibility tied to progress. Jefferson's agrarian ideal meant turning America's vast forests into productive agricultural land through honest labor. Soon, however, this rural, agricultural economy was transformed into an urban, industrialized one that demanded tremendous amounts of timber for buildings, ships, fences, railroads, and mines—not to mention domestic and industrial fuel. As lumber prices rose, energetic capitalists engaged in a ruthless exploitation of land and labor, aided by new machines, new methods of transport, and new systems of business organization. Simultaneously, appreciation of the aesthetic qualities of the forest became stronger: it was picturesque to some, sublime to others, and even a source of moral improvement. Poets, novelists, naturalists, philosophers, and artists reveled in the wildness of American scenery. But the most influential critics of exploitation were concerned about the environmental effects of clearing on rainfall and soil—and thus on agriculture. Growing unease about the country's resources at last prompted Congress, in the 1870s, to consider resource management a duty of government, setting the stage for federal forestry in the United States.

CHAPTER 5

Forest Policy in America as a Developing Nation: Jeffersonian Democracy, the Taming of the American Wilderness, and the Rise of the Conservation Movement

Michael Williams

When we think of the humanly induced global environmental problems of today, deforestation is high on the list, together with climate change, atmospheric and oceanic pollution, genetic modification, and biodiversity loss. Deforestation is commonly thought to be a relatively recent phenomenon and largely a problem of the tropical world. But nothing could be further from the truth. Ever since humans have been on earth they have been chopping, hacking, grubbing up, and setting fire to trees in order to make new land, provide raw materials for buildings and implements, and obtain fuel for energy for keeping warm, cooking, smelting metals, and baking clay for ceramics. Deforestation is a basic and ubiquitous process of all developing societies, and many others also (Williams 2003).

In the annals of past deforestation, the medieval clearing of western and central Europe, the 17th- and 18th-century clearing of Ming China, and the mid-19th- to mid-20th-century clearing of India must rank as high points. But an equally spectacular episode of clearing occurred in the American continent from the mid-17th to mid-19th centuries. And it is that American deforestation that forms the indispensable background to our deliberations at this and the Grey Towers meeting. We have to ask what were the social, economic, and cultural values that shaped American attitudes toward the vast expanse of forest that confronted them, and how did these societal values in turn shape the public policy regarding the use and acquisition of the public domain. These were events and attitudes that led to concern about clearing and ultimately and protractedly to the foundation of the USDA Forest Service in 1905, the centenary of which we are celebrating here today.

The Forest Setting

The forest that confronted the first European settlers in the mid-17th century was immense. Estimates vary, but it is commonly thought that there were 820 million to 850 million acres (332 million to 344 million hectares) of dense, "commercial" forest, with another 100 million to 150 million acres (40.4 million to 60.7 million hectares) of more open woodland of little commercial value. Thus, about half of the area of the coterminous United States was in thick forest and another 10 percent was in sparse forest.

Of course, the forest was not uninhabited. The preconquest Indian population is conservatively calculated as being 3.8 million. More important than numbers alone is the fact that these many millions could not have practiced intensive hoe cultivation and widespread hunting without radically altering their environment. Periodic and frequently extensive clearing for agriculture and extensive forest firing were features of Indian life. Consequently, the forest was anything but pristine, and it is a reasonable speculation that had the Europeans not landed on the eastern seaboard in the early 17th century, the increase in Indian numbers might, within another couple of centuries, have reduced the forest to a fragment of what it was. But of course, that was not to be. Warfare and particularly disease reduced Indian numbers to about a tenth, and they retreated before the advance of superior numbers and technology.

Settlers adopted not only Indian crops (potatoes, maize, tomatoes, pumpkins, squash, beans) but also Indian methods of clearing, including girdling (ring barking) and burning, followed by hoe cultivation between the stumps. In addition to learning new methods of cultivation and adopting new crops, the settlers also benefited from the many extensive clearings they found. These clearings, abandoned by the depopulated tribes, provided ready ground for crops. Consequently, in many places, the forest was anything but the impenetrable mass of trees beloved by early writers. It was more a mosaic of forest in different stages of succession. It is entirely possible that due to the decrease in Indian numbers and subsequent forest regrowth, there was actually more forest in 1750 than there had been in 1650 (Denevan 1992: 375).

Pioneer societies everywhere regarded the forest in both negative and positive terms, and it is impossible to understand early clearing without appreciating some of the cultural and folk values of those who did it. What follows is a broad-brush, chronologically structured view of major shifts in perceptions and responses, which can be represented diagrammatically. In noting the onset of these shifts in emphasis, one should be aware that in the real world of 17th- to 19th-century America, these did not fall into such neat categories or time slots. Many were overlapping and interdigitating, so that previous emphases do not disappear but simply become less dominant than succeeding ones. (Figure 1)

Before 1800: The Forest as Obstacle

In looking at the forest as obstacle, we can detect three subthemes—redemption, progress, followed by the early exploration of the resource.

Redemption. For the early Americans the forest was repugnant, forbidding, and even repulsive. Some of these feelings had roots that went back a long way into their European past. Forests were dark and dangerous places, the abode of wolves and bears; they were alien and were called *wilderness*, a word that etymologically speaking was literally "the place of wild beasts." And in the eeriness and horror of the forest one could become "bewildered." From Greek mythology to postmedieval popular folk culture, forests were the abode of satyrs, sprites, witches, ogres, trolls, werewolves, and child-eating monsters. Early Christianity extolled the culture of clearing the forest, which was equated with the devil and the abode of natural sin. Yet if the forest was to be cleared, its products were to be valued. Piety and economic progress went hand in hand, and in the clearing of the forest lay the creation of a new land that was akin to Paradise.

To the American pioneer the forest was all or some of these things. It was the "enemy" that needed to be "conquered" to make it safe. However, the dangers were real enough, since the forest was the haunt of wild animals and, of course, the Indians. In the settlers' minds, if the Indian was "savage" (a word derived from the Latin for "woodland"), it was because he inhabited the forest, and both had to be eliminated. It was something the Puritan New Englanders took seriously, for they passionately believed that social order

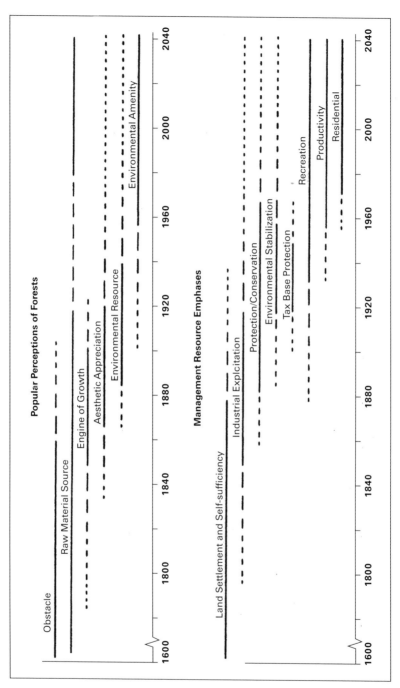

Figure 1. Popular perceptions of forests and management resource emphases through time.

and the Christian concept of morality stopped on the edge of the clearing. The only value of a forest was when it was felled, and the settler had little time or value for the beauty and novelty of the untouched forest.

The attitude was summed up well by the perceptive traveler Alexis de Tocqueville when he described pioneer life in Michigan. The pioneer, he said, "living in the wilds... only prizes the works of man. He will gladly send you off to see a road, a bridge or a fine village. But that one should appreciate great trees and the beauties of solitude, that possibility completely passes him by." Generally, settlers were "Insensible to the wonders of inanimate nature and they may be said not to perceive the mighty forests that surround them, till they fall beneath the hatchet." Their eyes are fixed upon another sight, "people solitudes and subduing nature" (Tocqueville 1831: 335, and 1838: II: 74). The Promised Land or Paradise that the early settlers had hoped to find was seen as covered by an obstacle that needed backbreaking work to convert it to the merest essentials of living. The cleared patch, the made ground neatly fenced in, was a symbol of order and civilization.

Progress. Although the clearing of the forest had an almost sacred, redemptive motive, it increasingly came to have a secular one. The concepts of progress, development, and ultimately civilization itself had never been far beneath the surface from the beginning, since they were the logical outcome of the fight for survival. From the 18th century onward, however, the concept of controlling nature and making it useful gained strength. The ideal was the rural, domesticated, agrarian landscape. It was a "made" landscape, said Crèvecouer, that had been converted from forest to "fair cities, substantial villages, extensive fields, an immense country filled with decent houses, good roads, orchards, meadows, and bridges where a hundred years ago it was a wild woody and uncultivated" (Crèvecouer 1782: 2: 39). It was a sentiment shared by practical farmer settlers. When William Cooper reminisced about his life in the northwestern portion of New York State in the late 18th century, he was sure that his "great Primary aim" had been "to cause the wilderness to bloom and fructify" (Cooper 1810: 6).

And this "middle landscape," as Leo Marx (1964) has called it, was neither wilderness nor city but the "garden" and even "paradise," and it was visible evidence of progress, expansion, moral effort, and industry. It had a long lineage in American thought and culture, and successive prominent Americans declared its virtues in one form or another. When William Penn said that "the country life is to be preferred for there we see the works of God, but in the Cities little else but the works of man," he was extolling the virtues of the made landscape and the rural pioneer life—its simplicity, sobriety, frugality, freshness: he was not extolling the untouched forest (Penn 1663, I: 220). The agrarian ideal reached its apogee in the pronouncements of Jefferson, who saw no need to turn the fledging country toward manufacturing when "we have an immensity of land courting the industry of the husbandman," especially since those "who labor in the earth are the chosen people of God, if ever He had a chosen people, whose breasts he made his peculiar deposit for substantial and genuine virtue" (Jefferson 1955: 164–65). For Benjamin Franklin, agriculture was the only "honest" way of acquiring wealth, and moreover, it kept people virtuous and morally righteous (Franklin 1905: 1: 245). The link with virtue was pointed out explicitly in Andrew Jackson's second annual address, in which he asked rhetorically,

What good man would prefer a country covered with forests, and ranged by a few thousand savages to our extensive Republic, studded with cities, towns, and prosperous farms, embellished with all the improvements which art can devise or industry execute, occupied by more than 12,000,000 happy people, and filled with all the blessings of liberty, civilization, and religion? (Richardson 1897: 3: 1084)

To promote the ideal of the pioneer backwoodsman, Jefferson was instrumental in devising the first public system of land disposal in the United States. In 1784 he chaired a committee charged with initiating a rectangular survey system across the new public lands west of the eastern seaboard states, whatever their nature. The Land Ordinance of 1785 established its main outlines, and the systematic survey of almost all the remainder of America followed the rectangular grid of square-mile blocks divided into four quarter-sections, the object being to facilitate simplicity of ownership and title and the orderly disposal of public lands. It was a "striking example of geometry over physical geography" (Pattison 1957: 1), but it was also a triumph of Jefferson's democratic ideal of promoting a national body of farm owners, who were hard working, frugal, sober, independent, resourceful, God fearing, and democratic. "Adam in Paradise," said Thoreau, "was not so favourably situated on the whole as is the backwoodsman in America" (Nash 1967: 90).

Exploring the forest resource. As positive feelings of progress began to override the negative perceptions of a hostile forest, the settlers became aware increasingly that the forest was their most valuable resource. It provided wood, the raw material that was central to their life and livelihood. Unlike western Europe, from where most of them had come, North America had enormous forests. Wood was abundant, it was ubiquitous, and consequently it was used prodigally. It entered nearly every aspect of life, quite literally from cradle to coffin. For about 300 years, between 1600 and 1900, it formed an essential element in human needs that it is difficult to understand from statistics and almost impossible to express in figures, though we will try later on in this paper. When James Hall (1836: 100–101) said, "Well may ours be called a *wooden country*; not merely from the extent of its forests but because in common use wood has been substituted for a number of necessary and common articles—such as stone, iron, and even leather," he was thinking of the multifarious human needs in the life and livelihood of the people that were supplied by wood. Quite simply, pioneer life in America revolved around wood—its removal, consumption, and fabrication.

Admittedly, there was a lifetime's labor and unremitting toil involved in clearing the forest, but the end result justified the effort so that gradually, utility overcame redemption as a dominant theme. There were two preferred methods of clearing: clearcutting and leaving the stumps to rot or grubbing them out later, and girdling the tree, thereby cutting off the flow of sap until ultimately the leaves no longer shaded the ground and the branches fell. Broadly speaking, the former method was favored in the North—and was even called the Yankee method of clearing—and the latter in the South, but there was much variation within these broad divisions. Girdling was more economical on initial labor—the scarce commodity on the forest frontier—enabling a crop to be planted quickly. But ultimately, when stump removal was taken into account, there was little to choose between the two methods in terms of cost of time. At most ten to 15 acres (four to six hectares) could be cleared in a year, and few farmers wanted more than 25 or 40 acres (ten to 16 hectares) of cleared land. More affluent pioneers employed "setup" men, itinerant laborers who could be hired if cash was available, but more commonly groups of neighbors helped each other in a "logging bee," chopping trees, clearing brush, and even building a log cabin in a cooperative effort to prepare several acres for immediate cultivation.

Because the forest was so common, and because it was regarded as an obstacle to progress, it was not valued highly. Standing timber was waste material that "lumbered the landscape"; it was useless and cumbersome and had little value. Indeed, bare land was worth more than land with trees. And yet in the process of exploring the potential and using the raw material, settlers were indelibly altering the forest and creating a new resource of agricultural land, and a conservative calculation is that nearly 114 million acres (46 million hectares) of land was cleared before 1850.

The best way to understand this varied scene is still the four sketches of pioneer life in upstate New York by Orsamus Turner (1851). They are like four "stills" clipped out of a continually moving picture of the forest landscape. They portray the essential truth of an experience that was repeated a few million times in the northern and eastern portions of the United States until the closing years of the 19th century.

Few native-born Americans ever bothered to comment about the changes wrought by clearing, but they fascinated overseas visitors. Not only had they adjusted aesthetically to the "made" and well-tended landscape of Europe, but also they were used to a situation where wood was scarce and needed to be conserved. The prospect of a landscape's being made anew was exciting and novel, often in its sheer ugliness, and the prodigal waste astounded them. "Man gets used to everything," wrote de Tocqueville in 1831 (329),

> *[He] fells the forest and drains the marshes…. The wilds become villages, and the villages towns. The American, the daily witness of such wonders, does not see anything astonishing in this. This incredible destruction, this even more surprising growth, seem to him the usual progress of things in the world. He gets accustomed to it as the unalterable law of nature.*

The ultimate judge of American improvement was not the foreigners but the home folk. Touring North Carolina's Appalachians, the philosopher William James "passed by a large number of 'coves' [clearings] newly cleared and planted." He thought the impression was one of "unmitigated squalor": zigzag rail fences around the scene of havoc, irregular clumps of corn among the charred stumps of trees. He said, "What had 'improved' the forest out of existence was hideous, a sort of ulcer without a single element of artificial grace." But the mountain pioneer tells him, "We ain't happy here, unless we're getting one of those coves under cultivation." James realized that he had missed the inner significance of the scene: "To me the clearings were a mere ugly picture on the retina. But when *they* looked on the hideous stumps, they thought of personal victory. The chips, the girdled trees, and the vile split rail spoke of honest sweat, persistent toil and final reward" (James 1899: 150–52). The spectator's judgment was sure to miss the nub of the matter and, in a way, completely missed the truth of the scene.

After about 1830 two events occurred that altered the perception of the abundance of timber. Farmers had reached the open prairie, where there was no timber for fencing, construction, or fuel. In addition, the large and growing cities on the eastern seaboard began to experience fuelwood shortages, especially in particularly cold winters. In both locales prices rose as a response to scarcity, and the taken-for-granted resource of wood began to be valued more highly.

1800–1880: The Forest as a Positive Resource

Raw material resource. During the opening decades of the 19th century, the perception of the forest as a source of useful products went from strength to strength as the predominantly self-sufficient, rural, agricultural—even peasant—economy was transformed into a highly commercial, urban, industrialized one. With lumbering, clearing, and other domestic and industrial impacts, the original forest cover of the coterminus United States was reduced to 470 million acres, from roughly 820 million to 850 million (332 million to 344 million hectares) by 1920. Of this, about 303 million acres (129 million hectares), or 65 percent, had gone in the creation of new agricultural land.

But timber was more than new land. As a measure of its importance in American life, we know that from 1850 to 1920 the value of products of the lumber industry outstripped all other forms of manufacturing except meat packing, iron and steel, and flour and grist mills. It was the second-ranking industry in 1850 and 1870, third in 1860 and 1910, fourth in 1880 and 1900, and fifth in 1890. It sank to ninth only in 1920, when modern manufacturing and value-added products, like automobiles and petroleum refining, shifted the ranking permanently.

But these rankings do not convey the true value or importance of wood in the economy. A whole technology based on wood existed from earliest times, as did a whole "society pervasively conditioned by wood" (Hindle 1975: 3), such that wood and wood products, as Hall observed correctly, entered every walk of life. To the lumber industry rankings of value should be added the products of the wood planers, the packing box manufacturers, the coopers, the tanners, the carriage makers, the house builders, the shipbuilders, and furniture makers, whose products rose to be at least as valuable as, if not more so than, the products of the lumber mills. The vast majority of buildings were made of logs or planks, and wood was an essential element even in those made of brick or stone. Wood was the principal material of the transportation system, being essential in the majority of ships, riverboats and barges, carriages and railcars, bridges, and railroad ties, in plank and corduroy roads, and even in road surface blocks and canal locks. In rural areas it was the principal fencing material until the latter 1890s.

Production of lumber was enormous. The slow increase during the opening decades of the 19th century suddenly went into a new upwardly sloping curve after 1859. Consumption of all sorts of wood (fuel, lumber, pulp, veneer, poles, etc.) stood at 3.76 billion cubic feet per annum in 1859. It nearly doubled to 6.84 billion cubic feet in 1879, and very nearly doubled again during the next 27 years to reach 13.38 billion cubic feet in 1907, a figure it has never reached again, hovering as it did during the 20th century at 10 billion to 11 billion cubic feet. American dependence on wood was overwhelming and ubiquitous. Wood entered every walk of life.

Most importantly, wood was the fuel for more than two-thirds of households until as late as 1880 and for a quarter of households as late as 1920. It made the steam that drove the engines of the railroads, riverboats, and factories well into the 1880s, and wood was still used to make the charcoal for a surprisingly large quantity of pig iron smelted because of the high quality of the steel produced. The last charcoal iron automobile engine block was made by Ford in Dearborn as late as 1940.

Wood was also the source of many important chemicals, such as potash and early industrial alkali, and naval stores, which included pitch, tar, and turpentine, all essential ingredients in maritime transport. Bark was a source of tannin for the leather industry, and the sugar maple produced a sweetener.

Although much of the forest destruction and exploitation was still being managed by individual farmers as they cleared their blocks, the latter part of the 19th century saw a new form of management response that was highly industrialized and is best described as industrial capitalism. It was characterized by a thorough and ruthless exploitation of land and labor, and it went hand-in-hand with an innovative and dynamic use of new machines, new methods of transport, and new systems of business organization. Everything in the lumber industry grew larger, faster, more complex, and yet more specialized in the quest for greater production and great profits. An energetic minority of lumbermen existed who had a clear view of the short-run profit that could be made from rapid and exploitative cutting, a view that could be transformed into action without restriction by any regulatory agency. In addition, there was very little popular condemnation of the destruction because the public at large demanded abundant and cheap timber.

Although most North Americans thought that the forest and its timber were limitless, there were indications by the later decades of the 19th century that timber was scarce and getting dearer. The prospect of an impending "timber famine," promoted by Gifford Pinchot and his associates at the turn of the century, was instrumental in galvanizing public and popular awareness about the potentially finite nature of forest resources. The response from the lumber industry was to combat deficiencies by better communications to bring timber from untouched forests farther afield—first the South, and then the Pacific Northwest and British Columbia, and the installation of mass-production methods to keep down costs. But by the latter years of the century it was clear that some form of silviculture was desirable—that is to say, the management of the existing forest to maximize yield. The public response was to demand and to bring into being a large and skilled body of foresters trained to manage the new state, provincial, and federal forest lands.

An engine of growth. Although it was rarely articulated or perceived at the time, in retrospect we can see that wood played an essential role in the rapid industrialization of the country. There is a convincing case to be made that the abundance of wood and the land created by clearing were the starting points for many economic, social, and technological changes. Brooke Hindle hints at this when he says that it was during America's "wonderful Wooden Age" that the country began its "assent to industrial primacy and to the highest standard of living in the world" (1975: 12). American development has been greatly influenced by resources, and of those resources wood constituted a visible, abundant, easily worked, and easily exploited raw material and fuel, quite unlike minerals, which were underground and needed much more complex machinery and organization for their exploitation. Wood was an inexhaustible raw resource of unprecedented magnitude.

The example of how countries today in the less developed tropical world regard the forest resource as a means of achieving economic growth and breakthrough, even in a fossil fuel age, should be borne in mind because it underlines the importance of this versatile resource.

Aesthetic appreciation. Although the utilitarian appreciation of the forest as a positive resource was dominant, a parallel appreciation of the aesthetic qualities of the forest became stronger as the 19th century wore on. Rather than finding the forest repugnant, some people began to find that nature's roughness and rawness, untouched by the improving hand of humans, had a picturesque quality about it. Others went further and began to think that wild nature was sublime, and that wild scenes, be they forests, mountains, or rivers, could bring pleasure and exalt the human spirit, just as much as could comfortable, well-ordered landscapes of cleared and made land (Williams 1989).

Far from producing degeneracy and immorality, as the early settlers thought, life in the forest was beneficial, some now believed, and actually produced civilized thoughts and a morally improved person. The evolution of the cult of the simple and the solitary and the preference for the rural over the urban became stronger. The idea that man was more perfect when in touch with Nature (and hence God) had deep roots in western intellectual thought and was an important part of the cult of Romanticism. From Daniel Defoe's *Robinson Crusoe* (1719) to Jean-Jacques Rousseau's *Emile* (1762), there had been the suggestion that the primitive life, despite its many disadvantages, produced a happiness and wholesomeness that was not to be found in the cities; Nature was teacher. An early expression of Romanticism in American was by Chateaubriand, a French traveler who was seized with "a sort of delirium" when he found so little evidence of civilization in his travels during 1791–1792. In contrast to the situation in Europe, the imagination "could roam…in this deserted region, the soul delights to busy and lose itself amidst the boundless forests…to mix and confound…with the wild sublimities of Nature" (Chateaubriand 1816: 138–39). In later years indigenous evocations of the forest were common. Successive diarists and natural scientists like Bartram, Michaux, Flint, Audubon, Kalm, Philip Freneau, and the historian Parkman discovered that the forest could be regarded with something other than hostility. Patches of undisturbed nature were what Washington Irving called "the little nooks of still water" and were preferable to the "rapid stream of migration, improvement and change" that swirled around the country (Irving 1820: 63–64).

The shift in attitudes was reinforced by ideas of patriotism. After the War of Independence the question was continually asked, "What was it in this new country that was distinctively America?" The continent with its short history and ill-formed traditions could not produce anything like the rich cultural heritage and antiquities of Europe. But one thing America did have was large areas of untouched land—forest, prairie, and mountain—that were unique and something to be proud of. Again, we can turn to Chateaubriand (1828: 1: 98): "There is nothing old in America except the woods... they are certainly the equivalent for monuments and ancestors." Naturalists and diarists had led the way in "spreading a proper feeling of nationality," but the "boundless," "trackless," "incomparable," and "fresh" forests also became the object of literary and artistic attention. The writings of Washington Irving, the poems of William Cullen Bryant ("A Forest Hymn," 1825) and James Kirk Paulding ("The Backwoodsman," 1818), and above all the Leatherstocking novels of James Fenimore Cooper, written between 1823 and 1841, produced a new appreciation of the country's forested landscapes, and pride and confidence in the qualities of the American scene. The literary appreciation was paralleled by the visual appreciation, and Thomas Cole in particular, the leader of the Hudson River School of painting that flourished during the 1830s and 1840s, reveled in the wildness of the American scenery and deemed it a subject worthy of study and reproduction.

Transcendentalism. Finally, by the mid-19th century, the significance of the uniquely American character of the scenery took another positive turn with the writings of Emerson and, particularly, Thoreau. They both expounded the transcendentalist philosophy that the experience of nature in general, but of forests in particular, produced a higher awareness and sense of reality than (and thus transcended) one's immediate physical surroundings, which were materialist, being dominated by expansion, exploitation, and urbanization. Unlike the Puritan pioneers, who thought that morality stopped at the edge of the clearing, the transcendentalists thought it began there, for man was inherently good and not evil, and perfection could be approximated by entering the forest. By halting stages the argument went further. If the forest and other wilderness areas were uniquely American, and if God's purpose was made more manifest in such places, then the very spirit of America and its creativity could be found in the forests, from whence came, said Thoreau, "the tonics and barks that brace mankind" (Thoreau 1937: 672). Here perhaps are the intellectual seeds of Frederick Jackson Turner's celebrated frontier hypothesis, that the frontier was the crucible of the distinctive American traits and institutions of self-reliance, industry, and democracy. Quite simply, for Turner, "American democracy was no theorist's dream...it came directly out of the American Forest" (Turner 1920: 293). Perhaps more importantly in the story of this conference, the transcendentalists' ideas were articulated with an intensity and enthusiasm by John Muir, who thought that "The clearest way into the Universe is through a forest wilderness" (quoted in Wolfe 1938: 313). Muir had an enormous influence on public opinion and became the arch opponent of Pinchot's "wise use" policy in the West.

Thus, over the century there was a dramatic reversal of values, and the human meaning of the forest changed from having negative to positive values. Moreover, by the end of the period, appreciation of the forest and other wild places had passed from being the concern of a small and articulate group of writers, artists, poets, and politicians to becoming national cults. For many Americans who were now urban dwellers, the primitive conditions of nature no longer impinged on their lives and were no longer to be feared. Wild landscapes were sought out actively, and they could be seen in comparative comfort by vacationers. In the popular imagination, the forest and other such places were imbued with attributes, all of which were good.

1850–1880: The Forest as an Environmental Resource

By 1850, about 114 million acres (46 million hectares) of once-forested land had been cleared, to which a further 190 million acres (77 million hectares) was to be added during the next 60 years. As the pace of clearing increased, concern over the growing scarcity of timber and the need to preserve the "wild" areas was supplemented by a new concern. What was the clearing doing to the land itself? As one vegetation cover was replaced by another—usually cropland but often by weeds or other trees—the visual and protective content of the land changed, and not always for the good. Soils, rainfall, runoff, hydrology, wildlife, and a multitude of other ecological characteristics were upset or altered. The forest seemed to be crucial for the stability of other parts of the land; it was a sort of environmental glue that was integral to farming, which was still the most important occupation in the nation. Humans had always intervened, but the intrusion had always been thought to be beneficial or "improving," and rarely if ever detrimental. A new facet was opening up in the relationship of humans with the forest.

The "environmental" critics were not consumer-oriented and geared to short-term profits like the lumbermen, neither were they the elite of Romantics, philosopher-mystics, or literati who seemed divorced from the realities of everyday working life. They cared about practical, down-to-earth matters of the farming environment and ultimately were probably the most crucial critics of excessive clearing. Most of the earliest commentators, like Benjamin Rush (1786), Samuel Wells (1794), Count Volney (1804), and John Lorain (1825), were concerned about the effect of clearing on rainfall and the seasons, but most outstanding among their number was George Perkins Marsh, who went further than all of these. He had told a gathering of farmers in his native Vermont in 1847 that "the signs of artificial improvement are mingled with the tokens of improvident waste." The denuded hilltops, the dry streambeds, and the ravines furrowed out by the torrents had changed the landscape. If a middle-aged farmer returned to his birthplace, he would look "upon another landscape than that which had formed the theatre of his youthful toils and pleasures" (Marsh 1848: 17–19). In subsequent years he continued to read (he was conversant with 20 languages), observe, and marshal information on changes to the earth, and drawing on his Vermont experience and stimulated by the devastated landscapes of the Mediterranean, where he was American ambassador to the newly created state of Italy, he finally published his seminal *Man and Nature: or, Physical Geography as Modified by Human Action* in 1864. It was a highly readable account of man's impact on the environment through time, and fully one-third was devoted to the forests. The theme of human injury to the earth dominates the book, and whereas earlier writers had emphasized the beneficial effects of clearing, Marsh emphasized the deleterious effects, pointing out "the dangers of imprudence and the necessity of caution in all operations which, on a large scale, interfere with the spontaneous arrangements of the organic and inorganic world" (Marsh 1864: viii).

Many people in America were influenced by Marsh. Outstanding was Frederick Starr of St. Louis, who wrote an article that appeared in the 1865 *Annual Report of the Commissioner of Agriculture*, titled "American Forests: Their Destruction and Preservation," in which he calculated that 10,000 acres (4,047 hectares) of forest was being destroyed daily, which was about right. He advocated a threefold solution to the impending crisis: to educate landholders about the properties of different types of wood for different purposes, to promote a specifically American literature on tree culture rather than relying on French and German manuals, and to influence Congress to take an interest in forest preservation. There was a need for immediate action because "growth is slow and restoration tedious, while destruction is rapid and injury instantaneous" (Starr 1865: 210).

Whereas Marsh had emphasized the environmental consequences, Starr had laid much more emphasis on the economic. It was left to Increase Allen Lapham and his coauthors to combine both in their *Report on the Disastrous Effects of the Destruction of Forest Trees Now Going On So Rapidly in the State of Wisconsin*; it predicted that clearing would not only lead to a timber famine but also alter climate so radically that southern Wisconsin's prosperous farms would be reduced to cattle ranges, which in time would be subject to "winds and droughts [which] shall reduce the plains of Wisconsin to the conditions of Asia Minor. Trees alone can save us from such a fate" (Lapham et al. 1867: 33).

Through the later 1860s and early 1870s, writers in every timber-growing state in the North began to voice their concern at the inroads of agriculture and the massive depredations made by the timber companies, which were becoming particularly evident in the Lakes States. They were now a joined by new voices of concern: the subject of aridity was beginning to force itself upon the national consciousness as settlers moved from the forested East into the subhumid plains west of the 96th meridian. The perennial question of whether tree planting would help increase rainfall was on many people's lips.

A particular concern was the idea of the correct balance between forested and nonforested land. This troubled Americans because if the environmental properties of forest were proven (and they seemed to be), and if this was coupled with the concept that the forest was a part of a distinctive national heritage, then the forests were of supreme importance in American life and livelihood. It seemed as though about a quarter of the land was still in forest, about the same as Germany, the country with which most comparisons were made. Americans heaved a sigh of relief. In short, perceptions and priorities were changing, and the ecological and moral-aesthetic arguments were ultimately to become major factors in the acceptance by the public of forestry in the United States.

"The Duty of Governments," 1870 and After

Early campaigners. It was perhaps fortuitous that the growing mood of unease about the country's resources, settlement future, and national ethos should have prevailed when Franklin B. Hough gave his address to the American Association for the Advancement of Science in 1873, "The Duty of Governments in the Preservation of Forests" (Hough 1874). Hough was a physician and amateur botanist, geologist, and historian of his native New York. His passion for statistics and gathering data found him engaged in the compilation of the New York census, and then later, in 1870, he was made superintendent of the U.S. census. There was nothing particularly new in Hough's address; it drew heavily on Marsh and all the other writers who had gone before, but he seemed to be responding to the sense of urgency that was prevalent in the country. The association set up a committee to inquire into the timber-cutting situation, and among other things it recommended that a federal commission be created to inquire into the amount and distribution of woodland, the rate of consumption, the influence of forests on climate, and the methods of forestry practiced in Europe.

Nothing was done until 1876, when Congress authorized the establishment of the Division of Forestry with Hough as its head. Hough then compiled three *Reports upon Forestry* (Hough 1878, 1880, 1882). These contained endless statistics on lumber cutting, clearing, and fuel consumption, bewildering in their millions and billions and lacking summary analysis or long time-series, but they were alarming enough in their magnitude to suggest widespread cutting. Carl Schurz, the secretary of the Interior, wanted to get some statistics on cutting into the 1880 census to provide a firmer basis for legislation, and through the good offices of the Smithsonian Institution got Charles Sargent to prepare his *Report on the Forests of North America*, which appeared as a massive three-volume work in 1883, the like of which had never been seen before (Sargent 1884). Neither Hough's reports nor that of Sargent paid much attention to the main culprit of clearing, agriculture; it was "natural." Instead they concentrated on industrial lumbering (which obligingly supplied what the public demanded), which in reality was making a much smaller impact than the chopping of millions of individual farmers. Time had made the past enchanting, but America was fast moving into an urbanized era. Sentimental glances backward to a golden age of rural living when all seemed simple and good were becoming increasingly common.

"The day is coming." One year after the Civil War, when people began to take stock of many aspects of the nation's situation, Andrew Fuller wrote a book on forest culture. In it he made a prophetic statement: "Every civilized nation feels more or less the need for an abundant supply of trees [and]

America has felt this need the least." But "the day is coming, if not already here, when her people will look back with regret to the time when the forests were wantonly destroyed" (Fuller 1866: 5).

By the early 1870s the day had come. Concern was no longer academic—if it had ever been—and a new urgency had entered into the discussion. By 1878 the prediction of a coming "timber famine," a phrase later intimately associated with Pinchot, had already been uttered by the commissioner for Agriculture, though the reality was still a long way off (USDA 1878: 245).

The whole debate was becoming vastly complicated because of the many and often conflicting voices of the different users and admirers of the forest. The lumbermen wanted to run a profitable business, the farmers wanted the land the forest covered, domestic and industrial users wanted the fuelwood, the settlers on the Plains wanted to plant trees to make rain, and the incipient foresters wanted to preserve forests and protect stream flow. In addition, there was now the federal government, which reluctantly realized that it owned and even had a responsibility for the land used by all of these. Above all, there was the new set of users and admirers who found peace and solace in the vast solitude of the forest, and there were those who thought the forest was a unique part of the American heritage and should be protected for future generations. Both the aesthetic and the patriotic users inclined to the notion that humans needed the forest in order to "re-create" themselves, and this concept of recreation grew stronger as America became more urbanized. Ultimately it was, perhaps, the major motivator in national forest policies. These voices and their opinions proceeded along parallel and nonconverging tracks with little cooperation and different vocabularies and assumptions.

Nevertheless, I think one can safely say that this babble of opinion, fact, and passion resolved itself into six basic questions:

1. Are the forests worth preserving?
2. How should the forest be managed, and who should do it?
3. Who owns the forest?
4. How much forest is left?
5. How should the forest be protected?
6. How should the forest be used?

Perhaps with the exception of the fourth question—How much forest is left?—these questions were rarely formulated in such clear terms because of the aforesaid conflicting voices and because in reality the five issues could not be separated so neatly. Nevertheless, as I point in the 12th chapter of *Americans and Their Forests*, prominent in providing answers to these questions were the opinions and actions of Carl Schurz, Franklin Hough, and Nathaniel Egleston in the fledging Division of Forestry; Bernhard Fernow, who took over from Egleston in 1886; J. Bird Grinnell the recreationist; and John Muir, whose missionary zeal over forest preservation in the Sierras has become a byword (Williams 1989: 394–411). By 1891, 17.5 million acres of forest land, mainly in the western states, had passed into federal reserves, though the Congress was not sure what to do with them. It was at this point that perhaps the most prominent and perhaps controversial voice of all, that of Gifford Pinchot, began to be heard. He had just returned to America, having finished his training in France and Germany. And that, as they say, is another and new story.

References

Note: Only the sources of direct quotations and citations to publications are listed below, since the body of this chapter is based on Michael Williams, 1989, *Americans and Their Forests* (Cambridge and New York: Cambridge University Press), in which full references are given.

Chateaubriand, François Auguste René, vicomte de. 1816. *Recollections of Italy, England, and America on various subjects*. Philadelphia: M. Carey.
———. 1828. *Travels in America and Italy*. 2 vols. London: Henry Colburn.
Cooper, William. 1810. *A guide to the wilderness: or The history of the first settlements in the western counties of New York, with useful instructions to future settlers*. Dublin: Gilbert and Hodges. Reprint, Rochester, NY: G.P. Humphrey, 1897.
Crèvecouer, J. Hector St. John de. 1782. *Letters from an American Farmer*. 2 vols. London: Thomas Davies.
Denevan, William. 1992. The pristine myth: The landscape of the Americas in 1492. *Annals of Association of American Geographers*. 82: 369–85.
Franklin, Benjamin. 1905. *The writings of Benjamin Franklin*. 10 vols. Collected and edited with a life and introduction by Albert Henry Smyth. New York: Macmillan, 1905–1907.
Fuller, Andrew S. 1866. *The forest tree culturist: A treatise on the cultivation of American forest trees*. New York: Geo. and F.W. Woodward.
Hall, James. 1836. *Statistics of the West*. Cincinnati: J.A. James.
Hindle, Brooke (ed.). 1975. *America's wooden age: Aspects of its early technology*. Tarrytown, NY: Sleepy Hollow Restorations.
Hough, Franklin B. 1874. On the duty of governments in the preservation of forests. *Proceedings of the American Association for the Advancement of Science*, 1–22. Portland,

ME, 1873 (1874). Also printed in U.S. Congress, House, *Cultivation of timber and the preservation of forests*, 43rd Congress, 1st sess., 1874, H.R. 259 (serial no. 1623).

———. 1878, 1880, 1882. *Report upon forestry.* Submitted to Congress by the Commissioner of Agriculture. Vol. 1 (1878); vol. 2 (1880); and vol. 3 (1882). Vol. 4 (1884) was prepared by Nathaniel H. Egleston. Washington, DC: GPO.

Irving, Washington. 1820. *Rip Van Winkle, and the legend of Sleepy Hollow.* 1848 ed., reproduced, Tarrytown, NY: Sleepy Hollow Restorations, 1974.

James, William. 1899. On a certain blindness in human beings. In *Talks to teachers on psychology: and to students on some of life's ideals.* New York: W.W. Norton, 1958.

Jefferson, Thomas. 1955. *Notes on the state of Virginia.* Edited by William Peden. Chapel Hill: University of North Carolina Press.

Lapham, Increase A., J.G. Knapp, and H. Croker. 1867. *Report on the disastrous effects of the destruction of forest trees now going on so rapidly in the state of Wisconsin.* Madison: Atwood and Rublee, State Printers. Reprint, Madison: State Historical Society of Wisconsin, 1967.

Marsh, George Perkins. 1848. *Address delivered before the Agricultural Society of Rutland County, Sept. 30th, 1847.* Vermont: *Rutland Herald*.

———. 1864. *Man and nature: or physical geography as modified by human action.* Another edition with introduction by David Lowenthal, Cambridge, MA: Harvard University Press (Belknap Press).

Marx, Leo. 1964. *The machine in the garden: Technology and the pastoral ideal in America.* New York: Oxford University Press.

Nash, Roderick. 1967. *Wilderness and the American mind.* New Haven: Yale University Press.

Pattison, William D. 1957. *Beginnings of the American rectangular land survey system, 1784–1800.* Department of Geography Research Papers. No 50. Chicago: University of Chicago.

Penn, William. 1663. *Some fruits of solitude.* Philadelphia.

Richardson, James D. (ed). 1897. *A compilation of the messages and papers of the presidents.* 10 vols. Washington, DC: Bureau of National Literature.

Sargent, Charles Sprague. 1884 *Report on the forests of North America (exclusive of Mexico)*, vol. 9 of *The tenth census of the United States* (1880). Accompanied by a *Folio atlas of forest trees of North America*. Washington, DC: GPO.

Starr, Frederick. 1865. "American Forests: The Destruction and Preservation." USDA. *Annual report*, 1865. Washington DC: Government Printer.

Thoreau, Henry David. 1937. "Walking," in "Nature Essays." In *The works of Thoreau*. Edited by Henry Seidel Canby. Boston: Houghton Mifflin (Riverside Press).

Tocqueville, Alexis de. 1831. A fortnight in the wilds. In *Journey to America*. Translated by G. Lawrence. Edited by L.P. Mayer. London: Faber and Faber, 1960.

———. 1838. *Democracy in America*. 3rd ed. 4 vols. Translated by Henry Reeves. London: Saundars and Otley.

Turner, Frederick Jackson. 1920. *The frontier in American history*. New York: Holt.

Turner, Orsamus. 1851. *History of the pioneer settlement of Phelps and Gorham's Purchase and Morris' Reserve*, Rochester, NY: W. Alling.

USDA. 1878. *Annual report of the Commissioner of Agriculture*. Washington, DC: GPO.

Williams, Michael. 1989. *Americans and their forests*. Cambridge and New York: Cambridge University Press.

———. 2003. *Deforesting the earth: From prehistory to present crisis*. Chicago: Chicago University Press.

Wolfe, Linnie Marsh. 1938. *John of the mountains: The unpublished journals of John Muir*. Boston: Houghton Mifflin.

ABSTRACT

Trans-Atlantic exchanges were influential in the development of U.S. forestry. Americans of the 1870s who were concerned about timber cutting read British and German texts, joined organizations that disseminated information on European forestry, and went abroad to learn from European foresters. Convinced that deforestation had caused the decline of great civilizations, they predicted a similar catastrophe on the American continent. But in the United States—where government lacked legal authority to control private landowners and was required to give away the public domain—the prospects for forestry looked bleak. France, in contrast, had governmental jurisdiction over public and private forests, administrative structures, training for foresters, and penalties for timber poaching. F. P. Baker, drawing on what he observed in France, proposed a multifaceted plan to manage the vast American landscape: schools of forestry, regional experiment stations, federal inventories of timbered lands, protection against fire and illegal cutting, and an end to low-cost public timber sales. The professional foresters who took up his challenging program were either trained in Europe or Europeans themselves. With Chief Gifford Pinchot at their head, they secured the political authority to determine national forest policy while establishing a bureaucratic apparatus, a professional society, and graduate education. Subsequent leaders of the Forest Service tacked to the timber interests and underestimated shifts in public needs, setting the agency up for a hard fall when the conservation movement was surpassed by the environmental movement.

CHAPTER 6

Le Coup d'oeil Forestier: Shifting Views of Federal Forestry in America, 1870–1945

Char Miller

Lucien Boppe, assistant director of l'École Nationale Forestière in Nancy, France, made a "tremendous impression" on Gifford Pinchot. The young student, the first American to attend classes at the venerable French forestry school, was captivated by Boppe, a man of short and stocky stature with immense vitality; Boppe had "a great contempt for mere professors," for he had "learned in the woods what he taught in the lecture room." His lectures made forestry come to life, Pinchot remembered, made it "visible," a visibility reinforced when he took his students into the woods. "We measured single trees and whole stands, marked trees to be cut in thinnings, and otherwise practiced the duties of a forester." Such work, Pinchot later recalled, "was far more valuable than any reading."[1]

Just as invaluable was another lesson he absorbed from his French mentor: "the master quality of the forester," Boppe assured Pinchot, is "*le coup d'oeil forestier*"—the forester's eye, which sees what it looks at in the woods." This form of insight was as environmental in focus as it was political in vision. Trained to acquire an in-depth understanding of the woods, foresters, to practice their craft, must put their knowledge to work, creating wealth from board-feet harvested. That's why Boppe also told Pinchot that when "you get home to America you must manage a forest and make it pay."[2]

Sound advice, no doubt, but advice that presumed a society and politics committed to intensive land management of the kind Boppe taught and on the scale that France (and other continental countries) practiced. Such did not pertain in the United States, either prior to Pinchot's departure for France in 1889 or upon his return a year later. So little did his fellow citizens understand of Pinchot's new profession that a well-meaning woman, hearing of his European studies, remarked: "So you are a forester! How very nice! Then you can tell me just what I ought to do about my roses."[3]

Her confusion of forestry with gardening, matched by Pinchot's perplexity in how to respond, is an important reminder of the complications that can come with the cross-cultural transfer of ideas. Pinchot recognized that the importation of forestry to his native ground would be difficult, and not just because of his encounter with one clueless person. Over the preceding three decades, many other Americans had confronted the same misconceptions or, worse, outright rejection of the kind of state-regulated land management that forestry entailed. He knew too that without this earlier generation of conservation activists, his prospects for establishing governmental forestry were dim. Without them, he would not be able to employ his freshly trained forester's eye.

Turning Points

One of those on whom Pinchot's future depended was German forester Bernhard E. Fernow. He had emigrated to the United States to attend the 1876 U.S. Centennial Exposition in Philadelphia, as an official Prussian observer to the celebration of American Independence, and most likely attended the concurrent sessions of the second annual meeting of the American Forestry Association. He stayed, married his American fiancée, whom earlier he had met in Germany, and sought work commensurate with his training. There was little to be found, but by 1886 he was named the third chief of the U.S. Division of Forestry. Fernow's life story neatly embodied the 1870s westward migration of European ideas that would do so much to introduce forestry to the American mind and landscape.[4]

That decade marks the beginning of a remarkable period during which the United States proved particularly receptive to a host of European notions about the proper role of the government in setting national policy.

Until the early 1940s, the "reconstruction of American social politics" was intimately bound up with "movements of politics and ideas throughout the North Atlantic world that trade and capitalism had tied together," historian Daniel T. Rodgers has argued. "This was not an abstract realization, slumbering in the recesses of consciousness. Tap into the debates that swirled throughout the United States and industrialized Europe over the problems and miseries of 'great city' life, the insecurities of wage work, the social backwardness of the countryside, or the instabilities of the market itself, and one finds oneself pulled into an intense, transnational traffic in reform ideas, policies, and legislative devices." To make this "Atlantic era in social politics" possible required the creation of a "new set of institutional connections" with European societies, "new brokers" who would facilitate the intellectual exchange, and a cultural shift that would allow Americans to suspend, perhaps for the first (and only) time, its "confidence in the peculiar dispensation of the United States from the fate of other nations."[5]

In his compelling analysis, Rodgers makes only the briefest mention of forestry, and it is a late reference at that: in 1936 "a delegation of public foresters, including the chief of the U.S. Forest Service himself," were among those whose travel to Germany the Oberlaender Trust underwrote, funneling these "social policy experts through the familiar stations of German social progress." They may have been the last of such delegations. Yet over the preceding 60 years, the number of formal and informal exchanges had been continual and of crucial importance to the development of the American forestry, none more so than the arrival in the United States in 1876 of that first, essential cultural broker, Bernhard Fernow.[6]

Fernow was not the only participant in this rich trans-Atlantic cultural process, and given that Rodgers's model depends on an *American* hunger for European knowledge, perhaps he is not the best marker of it; Fernow brought his technical knowledge of forestry with him. Other Americans of the 1870s who were concerned about the rapid devastation of the nation's wooded estate and who believed that European scientific forestry might correct this difficult problem had to seek out the relevant information. They avidly read British and German forestry texts, corresponded with their authors, joined the recently formed American Forestry Association (1875) and similar organizations that gave them access to additional information on European forestry, and traveled abroad to learn directly from European foresters. In so doing, they testified to the power of ideas to change social policy and alter political behavior.

For this cohort, there was much that needed changing. As early beneficiaries of the Industrial Revolution, they were also acutely aware of the new economy's remarkable and unrestrained consumption of wood. To generate the steam necessary to power industrial machinery and the new transportation mechanisms, as well as to construct the massive new cities and shore up innumerable mines, required the clearcutting of once-bountiful forests; wholesale harvesting spread from the Great Lakes to the South and later to the Rocky Mountains and Pacific coast. As the trees crashed down, alarmed voices rose up. The *Sacramento Daily Union* warned in 1878 that if the then-current rate of logging continued unchecked, "the exhaustion of the forest growth in the Sierra is only a question of some ten years, and…if the rate of consumption is increased the catastrophe will occur considerably sooner." Similarly vexed was Interior Secretary Carl Schurz, who a year earlier had predicted that it was only a matter of time before the nation's dire shortage of wood would jeopardize its capacity to build new homes. A timber famine seemed at hand.[7]

Even as observers fretted over this dangerous possibility and projected a catastrophe that would rival the fall of the ancient civilizations of the Mediterranean—they easily compared the future of the United States to a treeless (and impotent) Greece—these images of doom led some to ponder how best to change the status quo. One of those was Franklin B. Hough, a physician and statistician, who understood forestry to be "a composite of natural history, geology, mathematics, and physics." From his New York State home, Hough had been assessing census reports that reflected what those involved in the lumber business already knew—that extensive harvests had depleted most eastern and Great Lakes forests, and that southern and western woods were now under assault. This economic fact required a social prescription, the good doctor believed, and in 1873 he presented his findings in a paper entitled "On the Duty of Governments in the Preservation of Forests" to the American Association for the Advancement of Science. That governments had such an obligation was a radical departure for a nation wedded to laissez-faire capitalism. But Hough's reading in the scientific literature, his visits to Europe, and his long-standing correspondence with German forester Dietrich Brandis offered him access to different models of governance; these allowed him to suggest alternatives to the rapid transfer of public wooded lands to corporate interests or homesteaders.[8]

The text he most depended on was George Perkins Marsh's seminal volume, *Man and Nature: The Earth as Modified by Human Action* (1864), which the American diplomat had written while living in Italy. Marsh (and by extension, Hough) was persuaded that deforestation of the European landscape had been responsible for the decline and fall of its great civilizations; he was equally persuaded that the only way to reverse this process—which he and Hough feared the United States was in danger of replicating—was to understand the close link between human profligacy and woodland devastation. To restrain human consumption of natural resources, to abate "the restless love of change which characterizes us," Marsh proposed the adoption of a conservative land management policy to be administered through a paternal form of government reminiscent of that in France and Germany, the leaders in European forestry.[9]

The wide gap between European energy and American lethargy impelled progressives to petition Congress and succeeding presidents to pass social legislation to protect the citizenry and enhance the commonwealth. For some reformers, Rodgers notes, this led to calls for top-down initiatives focused on better housing, sewage treatment, or efficient transportation. Forest advocates sought legislation to reflect their faith in the duty and capacity of government to regulate resource exploitation. Hough was an assiduous lobbyist and prolific publicist, and his campaigning bore fruit in 1876 when he was commissioned to report on the nation's "forest supplies and conditions." The fact of his 650-page document is as critical as the conclusions he reached within it; never before had the federal government published such an in-depth finding on the subject. What Hough found was troubling, but then he expected to be disturbed by the lumber industry's terrible swift saw and the resultant scarred and battered terrain; his prior studies had paved the way for this negative reaction.[10]

Another who partly shared Hough's worries was Charles Sprague Sargent, head of the Arnold Arboretum and future publisher of *Garden & Forest*, which would become the central forum in the 1880s for sustained discussions in the United States about forestry and conservation. Encouraged by Interior Secretary Carl Schurz, who wanted a detailed analysis for the 1880 census on the nation's available timber supplies, the Smithsonian Institution hired Sargent to write what would become the *Report on the Forests of North*

America; it appeared in 1884. Based on Sargent's on-site investigations, it delineated the distribution of tree species and forest densities and assessed the economic value of the nation's varied woods. Sargent concurred with Hough's assessment that there had been rapid harvesting of timber in New England, the Mid-Atlantic, and the Midwest, but the vast forests of the South and Pacific Northwest had yet to be exploited; a timber panic was not yet in the offing.[11]

Yet Sargent's evidence, when read against the historical record of forest devastation in other sections of the United States, suggested the grim outlines of the future, which is why worries about the extent of clearcutting found their parallel in uncertainties about how to restore cutover land. In a country devoid of technical experts, lacking the kind of legal authority that enabled European governments to control private landowners, and required by federal law to give away the public domain and therefore sell significant amounts of valuable forested public lands on which it could wield power, the prospects for forestry looked bleak. Even though Congress had established the Division of Forestry in the Department of Agriculture in 1881, with Hough as its first head, the small agency had no woods under its control—and would secure none until 1905. The new division's energies were also deflected by internecine struggles that embroiled Hough, his subordinates, and superiors, and that led, in the summer of 1882, to Hough's demotion to the status of "agent" of the Department of Agriculture. "Feel very low spirited," he scribbled in his diary, "and all my ambition is gone." His depression deepened when that October he and a colleague launched the *American Journal of Forestry*, only to have to suspend its publication 11 months later. Doubtful that anyone read the voluminous reports he had written while in office and convinced that no real change had occurred in the nation's consciousness about the dangers timber devastation posed, he was left to compare unfavorably the differences between European forestry and American lumbering. Those differences once had inspired him and other reformers to enter the political arena; now they were a painful reminder of how much remained to be done.[12]

Hough was a bit hard on himself. He and others had pioneered a new profession, created its first national voluntary associations, including the American Forestry Association and the American Forest Council, which would subsequently merge. They published the initial accountings of the status of the American forests, tried to raise the nation's consciousness about the evils of uncontrolled lumbering, and had even gained some governmental recognition of their concerns through the creation of the forestry division. Without these contributions—however tentative and incomplete—the labor of those who followed would have been that much more difficult.

Among those indebted to this pathbreaking work was F.P. Baker, who, two years after Fernow traveled to the United States, set sail for Europe as one of the U.S. commissioners to the 1878 Paris Universal Exposition. In the opening to his report on the exhibition of European forestry, he doffed his cap to those on whose scholarship his scant knowledge of the subject depended; he readily acknowledged that his account was "hampered by the reflection that already the whole subject has been ably treated by writers who have brought to bear upon it the resources of immense observation and profound scholarship." Mindful of the attainments of Marsh and Hough, Baker preemptively limited the significance of his observations to an updating of "the history of forestry in Europe to a later period" and the addressing of "hitherto unreported progress." More important, he believed, would be the opportunity once more to "impress upon the American people of the United States the vital importance of the subject of forestry." If they were at all engaged by his account, he would "have accomplished all that can be reasonably expected of him."[13]

The 1878 Paris fair was itself impressive: like other international expositions of the late 19th century, it offered participating nations the opportunity to flash their commercial wares, tout their industrial might, and demonstrate how they exemplified the virtues of rationality, progress, and civilization— values that, as the event's title asserted, were presumed "universal." Emblematic of these cultural displays was the magnificent French pavilion on forestry, known as Le Chalet. Built "entirely of woods grown in France, at least 200 varieties being used in its construction," the striking edifice housed geological and entomological collections maintained at the French national school of forestry, as well as "maps, plans, photographs, and models representing the processes of reforesting mountains and of retaining the shifting surface of sand hills." Most stunning was a Disneylike model of mountain forestry that was set within Le Chalet: trudging up a "zigzag track" carved into a fabricated hillside were workers "dressed in the peculiar costume of the country"; on their backs were heavy sleds. When they reached the peak, they loaded logs on the sleds and sent them down a "timber slide," which curved past charming representations of "the natural features of mountain scenery, the yawning ravines and plunging water-courses … " A stunning example of French attention to detail, this display, and others that illustrated investment in reforestation and afforestation, convinced Baker of the great benefits to be derived from "French skill and industry," through which "every foot of earth" was carefully "preserved and patiently and laboriously cultivated." Improvident Americans had much to learn from their thrifty European compatriots.[14]

Whether Americans could replicate the successes of French conservation was another matter, for governing land management in France was a strict set of laws that granted governmental jurisdiction over the forests of the public domain, as well as communal woods, and private woodlots; its administrative structures, rigorous training for foresters in the national forest school, and stiff penalties associated with timber poaching or careless handling of the forest, amazed the American commissioner. The existence of a rational landscape and disciplined people, Baker quoted a French writer approvingly, enabled the "forest corps" to take "its place beside the great public services" whose futures would be determined "by the scientific skill and industry which characterizes our age."[15]

But neither law nor custom controlled industrialization's excesses in the United States, a point of some disappointment to Baker. Although he did not propose that his native land adopt uncritically European legal models and technical training, he was clearly fascinated by how the French, Germans, British, Scandinavians, and Swiss were able to regulate the exploitation of natural resources so as to build up their national treasuries even while protecting the land. By contrast, Americans were "famous destroyers of the forest," devastating "thousands of acres of noble forest trees … merely to rid the earth of them. The Western pioneer," Commissioner Baker concluded, "has passed his life in toilsome labor of chopping and burning trees which his descendants would gladly replace."[16]

Absent a strong central state and without cultural support for conservation, what "can we do to preserve and restore our forests, to repair the waste of the past, and provide for the needs of the future?" Baker's answer was couched in language that revealed the limitations he believed would prevent European forestry—for all its strengths and prospects—from being easily transplanted to American soil. Although acknowledging that in the United States of late "a growing sentiment has sprung up in favor of the preservation and cultivation of trees both for ornament and use," he was not yet convinced the public would support the level of governmental "interference" that gave European foresters such unrestrained authority. He doubted, moreover, that legislative mandates were "efficacious" in any event: "few statutes have been more persistently violated," Baker observed dryly, than the already extant ones prohibiting "cutting timber on government land."[17]

It is odd therefore that he put his faith in the U.S. Timber Culture Act of 1873, one of the most persistently violated of federal laws. Designed to encourage prairie homesteaders to plant trees on 40 acres of a 160-acre tract to meet residency requirements and facilitate subsequent ownership, it had no appreciable effect—except, that is, in wooded areas, where the act became a vehicle for fraudulent claims that enabled corporations to clearcut timber without charge. Still, Baker was convinced that what salvaged this act was its political acceptability: it did not promote forestry through "repressive" means, but by holding out "substantial inducements for the cultivation of trees, [it] becomes the patron and encourager of forestry, and thus fosters a popular sentiment in favor of tree-growing." Democracy could be pushed only so far.[18]

For Baker, push came to shove in 1884. That year he gave a speech to the American Forestry Congress in which he advocated the withdrawal from sale or entry of federally owned forests draped on either side of the Rocky Mountains so as to protect the headwaters of the Platte, Rio Grande, and Arkansas rivers. Drawing on his earlier observations of European forestry, he now proposed a multifaceted plan to preserve and manage this vast landscape, including the development of schools of forestry and regional forestry experiment stations, and the funding of federal surveys to establish the inventory and value of these timbered lands. He also called for protection against fire and illegal cutting and the end to low-cost sales of wood harvested on public lands. "Government timber," he advised, "should nowhere be sold at $1.25 an acre. If sold at all a price should be fixed upon it somewhere near its value." Like Hough before him, Baker had reached a point where it was no longer acceptable simply to work within current political constraints. Given relentless destruction of the nation's woods, the time had come to mount a more sustained challenge to the status quo.[19]

That challenge would be taken up, and with considerable success, by a new cohort, many of whom were professional foresters who either had emigrated from Europe (Bernard Fernow and Carl Schenck) or had been trained there (Gifford Pinchot and Henry S. Graves). These men may not have agreed on all points about the course of American forestry, but as educators, reformers, and activists, they taught a wider public about the value of forestry to the commonweal. In their capacity as civil servants, Fernow, and later Pinchot and Graves, also built the bureaucratic apparatus and secured the requisite political authority to establish a national forest policy. This crucial organizational work was reinforced through their contributions to myriad professional organizations; it is significant that one of these, the Society of American Foresters (1900), supplanted the older American Forest Association as the leading voice on forestry affairs.

That said, their achievements were predicated on the previous generation's intellectual commitments and political activism. It was Marsh, Hough, and Baker, among others, who were the first to actively engage in the fertile trans-Atlantic exchange of ideas, who sought out and popularized European forestry; it was they who discovered just how complicated it would be to introduce its principles to an industrializing America. In sparking a civic debate over the future of the nation's forests, they proved instrumental to the coming struggle to restructure the nation's government so as to develop new controls over public lands. These early "lovers of the forest," Pinchot later confirmed, "deserve far more credit than they ever got for their public-spirited efforts to save a great natural resource."[20]

The New Order

Pinchot meant what he said, but it is also true that he worked assiduously to distinguish professional foresters from earlier grassroots organizers, academics, and intellectuals. As he and his peers laid down the political foundation for the emerging national system and a forest agency designed to manage these public lands and built the educational infrastructure to educate the first generation of home-grown scientific foresters, he was at pains to distinguish these mammoth organizational efforts from his predecessors' ultimately fruitless activism. Despite the many meetings they had attended, the articles they had written, and the legislation they had promulgated, despite their impulse to "urge, beg and implore; to preach at, call upon, and beseech the American people to stop forest destruction and practice Forestry," they failed in an all-important and pragmatic respect: "in the year 1891," Pinchot observed, "there was not…a single acre of forest under Forestry anywhere in the United States."[21]

Fourteen years later, as Pinchot left government service, more than 150 million acres was under active management of the Forest Service, which employed hundreds of fire lookouts, forest rangers, clerical staff, silviculturists, hydrologists, and range specialists. Pinchot knew well what he and his conservationist colleagues had achieved and was convinced that their achievement had come as a result of a deliberate decision to choose different tactics, a new means to an old end. "Under the circumstances," he recalled, "I had to play a lone hand. I could not join the denudatics," his name for preservationists such as his friend and ally John Muir, "because they were marching up a blind alley." Neither would he cooperate with lumbermen "because forest destruction was their daily bread. There was nothing left for me but to blaze my own trail." By which he meant to put forestry "into actual practice in the woods, prove that it could be done by doing it, prove that it was practical by making it work." Experimentation, the young forester believed, would demonstrate the virtue of his work and generate essential public support and political sanction.[22]

At the time, he was less confident about his chances for success, and with good reason. He was not as thoroughly educated in forestry as he might have been: "I intended to be a practicing forester all my life, yet I thought I could spare but thirteen months to get ready for my lifework," he wrote more than a half-century after he had brought a hasty end to his studies at l'École Nationale Forestière. He had attended the school at the suggestion of famed German forester Dietrich Brandis, who became an important mentor for Pinchot, as did the school's silviculturist, Lucien Boppe. And each man cautioned the young American that his desire to return to the United States to put his ideas into practice was premature; he was not well-enough educated, they warned.

He ignored their warning, but its wisdom dawned on him when, after his return to the United States, he was hired as forester for George W. Vanderbilt's vast estate in western North Carolina. Confused as to how to implement forest management in a landscape with which he had had no experience, he wrote lengthy appeals to his European advisers seeking guidance; in one such epistle to Brandis, he confessed that "the time has come, as you foretold it would, when I begin to feel the scantiness of my preparation."[23]

Pinchot's forestry work also went slowly because of the condition of the Vanderbilt forests, which suffered from fire and overgrazing, factors that undercut his reclamation efforts. Moreover, the operation's labor force was untrained and its startup costs were high, further frustrating Pinchot's initiatives; although he would say publicly that he had made forestry pay at Biltmore, in fact he operated at a loss.[24]

Those manifold problems would lead him privately to acknowledge to having "done little in the work of my profession" during his first year there, and he was relieved to move on to New York to begin a career as a consulting forester; thereafter he monitored Biltmore's new forester, the much more thoroughly trained Carl Schenck. This shift in personnel relieved his European mentors, too: "the best thing that Pinchot has done," Brandis confided to forester William Schlich, "is that he secured Schenck for America." Pinchot's self-promotional conception of Biltmore as the cradle of American forestry was a piece of false labor.[25]

Pinchot would give birth to the real thing in succeeding years, but the reality he brought to life was not so much the development of forestry in the woods as its professional standing in society, a critical contribution to its future in a modernizing America. His timing was perfect, for it was in the late 19th century that other professions began to determine the requisite education and training, sanction the appropriate degrees, and shape the behavior required of their future practitioners. The American Bar Association (1877), like the American Chemical Society (1875), American Historical Association (1884), American Economic Association (1888), and American Sociological Association (1905), established who could call himself a lawyer or chemist, a historian or economist or sociologist—and just as important, who could not. Pinchot would seek the same ends for the forestry profession while serving as head of the USDA Division, later Bureau, of Forestry (1898–1905) and then as chief of the Forest Service (1905–1910). Between 1900 and 1905, he was a driving force behind the creation of the three major institutions without which no profession can exist: a professional society, graduate education, and employment.

In 1900, at his grand home located at 615 Rhode Island Avenue in Washington, DC, Pinchot called to order the first meeting of the Society of American Foresters, an organization that over time would bestow scientific legitimacy on, and structure the ongoing education within, the profession. Pinchot also had a hand in founding the *Journal of Forestry*, which would become the profession's leading journal; it organized and gave legitimacy to the scientific research and practical experience it disseminated with every issue.

A professional society requires members, naturally enough, and here again Pinchot was instrumental in establishing the nation's first graduate program, through his family's endowment of the Yale School of Forestry in 1900. Together with the school's first dean, Henry S. Graves, whom Pinchot had tapped for the job, he contributed to the development of an appropriate curriculum in forestry education and occasionally taught within it. Because all professions must provide hands-on experience for their neophytes, Yale students spent their summers in Milford, Pennsylvania, training outdoors on Pinchot family woodlands and indoors in classrooms the family constructed in town. Having completed their training and received their degrees, young foresters would need gainful employment, and the bulk of them would be hired by the federal agency for which Pinchot was the founding chief, the Forest Service.

The agency in turn functioned as an extension of the New Haven campus. An "enthusiastic Yale man," Pinchot sought there to recreate the fraternal life he had cherished while an undergraduate. When he became the head of the Division of Forestry, for instance, he tapped close friends to work with him. Henry Graves (Yale '92) was Pinchot's right-hand man; George Woodward and Phillip Patterson Wells, who graduated with Pinchot in 1889, provided invaluable legal advice; fellow classmate Herbert A. Smith lent his editorial skills to the agency's publications; Thomas Sherrard (Yale '97) contributed his considerable forestry acumen to the agency's daily work.[26]

Yalies were also well represented within the rank and file. In 1905, for example, nearly half of those passing the civil service exam in forestry were Yale graduates, a fact that Pinchot reported enthusiastically to his mother. Little wonder then that the Forest Service was known affectionately as the Yale Club, a club of collegiate camaraderie that bound together this first generation of American foresters.[27]

Given the significance of Pinchot's organizational activities and the speed with which they unfolded—from 1898 to 1910—and given the agency's esprit de corps and the profound sense of mission that Pinchot inspired, it is hardly surprising that he believed he *was* the profession, and the profession was him.[28]

After Pinchot

However understandable, Pinchot's intense psychological identification with forestry posed problems for his successors. It did not help, either, that his departure from the Forest Service was the result of an explosive confrontation between him and the president of the United States, William Howard Taft. They clashed shortly after Taft had replaced Theodore Roosevelt in 1908 because, in Pinchot's mind, the new president did not share Roosevelt's passion for using executive branch clout on behalf of conservationism. News of suspicious coalfield leases in Alaska led Pinchot publicly to confront the administration, forcing his dismissal. In this, Pinchot practiced as he had preached: the last advisory in chief forester's "Rules for Public Service" is, "Don't make enemies unnecessarily and for trivial reasons; if you are any good you will make plenty of them on straight honesty and public policy …"[29]

Pinchot's shrewd insight and brave words nonetheless left his successors in a bind. Henry Graves, whose European forestry education Pinchot had underwritten, and who had served as his associate forester before becoming dean at the Pinchot-funded Yale School of Forestry, became the second chief. Less provocative and charismatic, Graves understood that the agency's continued existence depended on his ability to rebuild internal morale, reknit frayed relations with the White House and Congress, and reclaim public confidence, all the while placating an occasionally nettlesome Pinchot. None of this work came easily, and yet despite being hindered by sharp budget cuts and congressional hostility, Graves managed to stabilize the agency, smoothing the way for William B. Greeley to become its third chief in 1920.

Unlike Graves, Greeley immediately picked a series of fights with Pinchot, challenging his still-profound influence on Forest Service; only in this way, Greeley reasoned, could he reform the organization in his own image. More conservative than the founder and more comfortable with the corporate Republicanism dominating that era's political arena, Greeley promoted cooperative relationships with the timber and grazing industries; he countered Pinchot's faith in rigorous regulation by advocating through the Clarke-McNary Act (1924) an accommodation with powerful interest groups. When Pinchot contended that federal regulatory controls ought to be extended to industrial forestry operations, Greeley blasted his idea as "un-American." Years earlier, Greeley had been thrilled to have "lost caste in the temple of conservation on Rhode Island Avenue," a sneering reference to Pinchot's Washington manse, and while chief he did little to repair their relationship. His perspectives on the agency's political purpose, social significance, and economic agenda so dominated professional forestry in the 1920s that an embittered Pinchot resigned from the American Forest Association and stopped attending meetings of the Society of American Foresters.[30]

Greeley was much less deft in his response to a more serious bureaucratic threat posed by an aggressive National Park Service. Founded in 1916 and headed by former advertising executive Stephen Mather, the Park Service quickly came into its own at the expense of the Forest Service. Proclaiming that its mission was to serve the recreational needs of car-crazy American culture, the Park Service moved rapidly to publicize the national parks and develop highway connections between them. As it built public (and congressional) support for its services, it used this goodwill to press its case for managing the national monuments and majestic parklands then under Forest Service control. So effective were Mather and his managers, and so flat-footed did their Forest Service counterparts appear, that they plucked one gem after another out of the national forest inventory, most notably the spectacular Grand Canyon.

In tapping "the pulse of the Jazz Age," historian Hal Rothman has observed, the Park Service sold "Americans leisure and grandeur at a time when…outdoor recreation increased"—an understanding of contemporary needs the Forest Service failed to appreciate. Although individual employees, such as Arthur Carhart, Aldo Leopold, and Bob Marshall pushed the Forest Service to delineate wilderness areas and promote backcountry recreation, in general the agency's goals in the newly competitive environment seemed "undefined and utterly up in the air." Once proactive, the Forest Service had become reactive and lost momentum.[31]

The Great Depression, ironically enough, offered the agency an opportunity to make up lost ground. Greeley had resigned in 1928 to become secretary of the West Coast Lumberman's Association—proof of his real allegiance, Pinchot averred; his replacement, Robert Y. Stuart, a Pinchot ally, was chief until he died in a fall from his Washington office window in 1933. Ferdinand Silcox, an adept administrator, then navigated the agency through the hard and harrowing times. Taking full advantage of large influx of federal dollars flowing through the Civilian Conservation Corps, among other New Deal funding mechanisms, the Silcox-led organization purchased millions of acres of abandoned and devastated lands in the South, Midwest, and Great Plains. These new forests and grasslands became employment opportunities for CCC enrollees, who planted seedlings, built shelterbelts, repaired eroded terrain, and constructed cabins and trails. The can-do agency was in its element, serving as the vigilant custodian of the nation's public lands.

Its vigilance was tested in the in the early 1930s, when Interior Secretary Harold Ickes lobbied President Franklin Roosevelt to support the creation of a new cabinet-level Department of Conservation that would absorb all federal land management agencies, with the Forest Service as its core organization. Convinced that efficiencies would result, the president approved the plan, muzzled Secretary of Agriculture Henry Wallace, and had him prevent the Forest Service from defending itself. In need of outside aid and indefatigable allies, Silcox, through Associate Chief Earl Clapp, contacted Gifford Pinchot, then 70, to crusade on the agency's behalf. He did so gladly, and between 1935 and 1940, Pinchot and Ickes engaged in one of the most bruising bureaucratic brawls in modern U.S. political history. Over the radio, in newspapers and magazines, and from one podium to another, they pounded each other while rallying their supporters. In the end, Pinchot triumphed, a remarkable testament to his infighting skills and dogged perseverance.[32]

His victory was not unalloyed. There was a personal cost for at least one high-ranking forester who had cooperated with the old chief's campaign. The president never promoted Earl Clapp beyond "acting chief," a position he had assumed following Silcox's death in 1939, because he was convinced, correctly, that Clapp had orchestrated the stout resistance to Ickes's transfer scheme. In sacrificing his career for what he perceived to be the greater good, Clapp paid a heavy professional price.

The same could be said about the Forest Service itself. In its fierce fight for survival, it may have missed an important opportunity to reconsider how conservationism had evolved and how it would be implemented in the coming years; it also failed to reflect on the governmental structure best suited to conserve the lands it managed. As it entered the war years, the agency was intact and independent, but it was also insular in orientation. Its insularity would have dire consequences in the aftermath of World War II, complicating the Forest Service's implementation of new forest management techniques, damaging its once-vaunted reputation, and hindering its ability to react to massive social changes, especially the emergence of a potent environmental movement. By the 1970s, the Forest Service and the public it served no longer always saw eye to eye.[33]

Notes

1. Gifford Pinchot, *Breaking new ground*, fourth edition (Washington, DC: Island Press, 1998), p. 11.
2. Ibid., p. 11.
3. Ibid., p. 28.
4. Andrew D. Rodgers, *Bernhard Eduard Fernow: A story of North American forestry* (Durham: Forest History Society, 1991 reprint), p. 17.
5. Daniel T. Rodgers, *Atlantic crossings: Social politics in a progressive age* (Cambridge, MA: Belknap Press of Harvard University Press, 1998), pp. 3–4.
6. Ibid., p. 421.
7. Donald J. Pisani, Forests and conservation, 1865–1890, in Char Miller (ed.), *American forests: Nature, culture, and politics* (Lawrence: University Press of Kansas), p. 18.
8. Harold K. Steen, *The U.S. Forest Service: A history* (Seattle: University of Washington Press, 1976), p. 9.
9. George Perkins Marsh, *Man and Nature: The earth as modified by human action* (New York: Charles Scribner, 1864), p. 27–80.
10. Steen, *U.S. Forest Service*, p. 13; see also Harold Clepper, *Origins of American conservation* (New York: John Wiley & Sons, 1966).
11. Michael Williams, *Americans and their forests: A historical geography* (Cambridge: Cambridge University Press, 1989), pp. 376–77; C. S. Sargent, *Report on the forests of North America (exclusive of Mexico), Vol. 9 of the Tenth Census of the United States* (Washington, DC: Government Printing Office, 1880).
12. Franklin B. Hough, *Report upon forestry* (Washington, DC: Government Printing Office, 1878), pp. 6–9; Steen, *U.S. Forest Service*, p. 17.
13. F.P. Baker, Forestry, in *Reports of the United States commissioners to the Paris Universal Exposition, 1878* (Washington, DC: Government Printing Office, 1878), vol. 3, p. 391.
14. Rodgers, *Atlantic crossings*, pp. 8–9; Baker, Forestry, pp. 393–94.
15. Baker, Forestry, p. 398.
16. Ibid., p. 423.
17. Ibid., p. 423.
18. Steen, *U.S. Forest Service*, p. 123; Rodgers, *Fernow*, p. 12; Baker, Forestry, p. 423.
19. Rodgers, *Fernow*, p. 93.
20. Pinchot, *Breaking new ground*, p. 29.
21. Ibid.
22. Ibid., pp. 29–31.
23. Ibid., pp. 16–22; Gifford Pinchot to Dietrich Brandis, August 21, 1893, Pinchot Papers, Library of Congress.
24. Gifford Pinchot, *Biltmore Forest* (Chicago: R.R. Donnelly & Sons, 1893).
25. Gifford Pinchot to Mr. Wetmore, March 21, 1893, Pinchot Papers, Library of Congress; William Schlich to Carl Schenck, January 13, 1897, Carl Schenck Papers, University Archives, North Carolina State University.

26. Pinchot, *Breaking new ground*, p. 151.
27. Miller, *Gifford Pinchot and the making of modern environmentalism*, pp. 159–61.
28. Steen, *U.S. Forest Service*, pp. 60–64.
29. This document is reproduced in Char Miller, Crisis management: Challenge and controversy in Forest Service history, in *Rangelands*, June 2005, p. 15.
30. Miller, *Gifford Pinchot*, p. 282.
31. Hal K. Rothman, "A regular ding-dong fight": The dynamics of Park-Service controversy during the 1920s and 1930s, in Char Miller (ed.), *American forests*, p. 114; David A. Clary, *Timber and the Forest Service* (Lawrence: University Press of Kansas, 1986), pp. 84–89.
32. Miller, *Gifford Pinchot*, pp. 346–49, 351–55.
33. Miller, Crisis Management, in *Rangelands*, p. 14–18; Paul Hirt, *A conspiracy of optimism: Management of the national forests since World War Two* (Lincoln: University of Nebraska Press, 1994).

ABSTRACT

The two most influential people in the development of the U.S. national forests, John Muir and Gifford Pinchot, have been unfairly stereotyped as being adversaries and having conflicting views regarding managing them: Muir advocating wilderness preservation at the expense of forest utilization, and Pinchot, vice versa. The controversy has affected the management of public lands over the past century. Surprisingly, however, Pinchot in 1900 called the destruction of the giant sequoias "as deplorable as the untouched forest is unparalleled, beautiful and worthy of preservation." And Muir understood that "wild trees had to make way for orchards and cornfields." Even the phrase "wise management" of timber resources for "the common good" came from Muir, as did the idea of opening public forests to recreation. Pinchot, as did Muir, emphasized the interrelation between trees, animals, climate, soil, and water and called the forest "a complex community with a life of its own." Both men looked to the federal government to guard America's woodlands from the insatiable demands upon them by the U.S. timber industry. Although often associated with an industry-friendly utilitarian philosophy, Pinchot even fought to regulate timber cutting on private land. This chapter seeks to correct the perceived adversarial relationship between Muir and Pinchot, showing that both men advocated sustainable forestry on public and private lands.

CHAPTER 7

Breaking Old Stereotypes: John Muir, Gifford Pinchot, and American Forestry

John Perlin

I feel greatly honored that such a prestigious gathering would ask me to present at this colloquium. I also feel a little sad, too, as John Muir's and Gifford Pinchot's ideas of sustainable forestry remain largely ignored, requiring it to be the focus of this conference.

The accepted wisdom among those in environmental and forestry circles has been that irreconcilable differences arose between John Muir and Gifford Pinchot, major icons of early American forestry, at the end of the 19th century over competing philosophies regarding the management of public forests. The timber industry and environmentalists, who rarely see eye-to-eye, agree on one issue: Pinchot favored clearcutting[1] and saw "forests as standing two-by-fours." It therefore follows, as night follows day, that he would have "founded an agency that has the same vision."[2] In contrast, environmentalists have deified—and many sympathetic to the forest industry have demonized—Muir as the progenitor of the belief "that public lands should be preserved rather than used."[3]

The historical evidence paints a different picture. Muir and Pinchot, for example, shared a common concern: America was quickly losing the best forests nature had ever planted. They saw the great change happen in their lifetimes. The year before John Muir was born, William Peabody could write, "He who would behold sylvan scenery on its most magnificent scale, should cross the Alleghenies, and visit the vast tracts into which the axe of woodsmen has never penetrated. These are covered with a cast of vegetable mould, exceeding the depth of our richest soils ... [and] we find trees at every step, of six or seven feet in diameter."[4] Less than 50 years later another journalist could write that this same area "so recently a part of the great East-American forest [has] now a greater percentage of treeless area than Austria and the North-German Empire, which have been settled and cultivated for upward of a thousand years."[5] Pinchot described the devastation as "The most rapid and extensive forest destruction ever known."[6]

Both knew the root cause of deforestation in America. Muir remarked sympathetically, "Many of nature's five hundred kinds of wild trees had to make way for orchards and cornfields. In the settlement and civilization of the country, bread more than timber was wanted."[7] In the same vein, Pinchot recognized that "[T]his gigantic and lamentable massacre of trees had a reason behind it, of course. Without wood and plenty of it, the people of the United States could never have reached the pinnacle of comfort, progress and power they occupied before this century began."[8] Although both could explain the reasons for the growing threat to America's forests, they also both agreed that the clearing of America's woodlands had, in the words of John Muir, "surely now gone far enough."[9] They feared that if the devastation continued, America would end up like much of the Old World, "barren as Palestine or Spain."[10]

The destruction of America's forests also struck a personal chord. The love for wilderness pulsed through both of their veins. "None of Nature's landscapes are ugly so long as they are wild," asserted John Muir.[11] For Pinchot, Crater Lake had all the makings of "a wonder of the world. It had "[a] great body of the clearest water miles across ... set in majestic forest, and ... without a visible sign of human occupation."[12] In his autobiography Pinchot wrote rapturously about the "overpowering sense of bigness" that emanated from the gigantic forest "of the Olympic National Reserve." He even had an enlarged photograph of these magnificent trees in his Washington office. Pinchot always expressed "wonder and delight" no matter how many times he looked at the picture.[13] But just as parents who love their children have one special child, Muir and Pinchot, who loved all forests, showed special affection for the giant sequoia, better known at the time as the big trees. Muir revealed his favoritism when stating, "Other trees may claim to be about as large or as old: Australian gums, Senegal baobabs, Mexican taxodiums, English yews, and venerable Lebanon cedars, trees of renown, some of which are from ten to thirty feet in diameter. We read of oaks that are supposed to have existed ever since the creation, yet, strange to say, I can find no definite accounts of the age of any of these trees, but only estimates based on tradition and assumed average rates of growth. No other known tree approaches the Sequoia in grandeur, height and thickness being considered, and none, as far as I know, has looked down on so many centuries, or opens such impressive and suggestive views into history."[14]

Likewise, the giant redwoods found a special place in Pinchot's heart. He actually visited the groves in the company of Muir. He called them "the grandest, the largest, the oldest, the most graceful of trees."[15] As an indication of the high regard Pinchot held for Muir, he quoted him extensively in his account of the big trees.[16] That a great proportion of these trees could possibly end up as lumber, greatly disturbed both men. Pinchot warned, "Most of the...groves are either in the process of, or in danger of, being logged."[17] Muir witnessed the consequences. "In these noble groves and forests," he reported, "the axe and saw have long been busy, and thousands of the finest Sequoias have been felled, blasted into manageable dimensions, and sawed into lumber by methods destructive almost beyond belief...."[18] Pinchot, too, had witnessed such massacres himself. Though many years had elapsed since the event occurred, he still vividly recalled, "At Millville...I ran into the gigantic and gigantically wasteful lumbering of the great Sequoias...I resented then, and I still resent, the practice of making vine stakes hardly bigger than walking sticks out of these greatest of living things."[19] Indeed, as early as 1900, Pinchot made the following judgment: "The devastation [of the big trees] is as complete and deplorable as the untouched forest is unparalleled, beautiful and worthy of preservation."[20]

Muir and Pinchot had a common solution to the encroaching devastation of the American landscape: since "every other civilized nation in the world has been compelled to care for its forests," why not the United States as well?[21] As their European brethren had done earlier, both looked to the national government as the natural guardian of America's woodlands and held up examples in France, Germany, and Switzerland as inspiration for national forest reserves in the States.[22]

When President William Harrison set aside the first forest reserves—the precursors to the present national forests—almost all of them were located in the West in mountainous areas. For such terrain, Muir and Pinchot agreed that the forest cover must be preserved, having learned from examples in France and Switzerland. Influenced by studies done mostly in Europe, they described in great detail how forests in mountainous regions act to moderate the flow of water over the year, avoiding the extremes of floods and droughts. Without forests, Muir wrote, "During heavy rainfalls... the streams would swell into destructive torrents.... Drought and barrenness would follow."[23] Pinchot added two succinct observations to Muir's discussion. First, he emphasized that in "mountain countries, where floods are most common and do most harm, the forests on the higher slopes are closely connected with the prosperity of the people in the valleys below."[24] Second, he recognized that "silt is the chief foe of irrigation and the only remedy is the forest."[25] In Muir's opinion, "Everybody on the dry side of the continent is beginning to find" just how necessary intact forests are for those downstream where most had settled to farm. For this reason, western farmers, Muir had discovered, are "growing more and more anxious for government protection"[26] and had come to realize, as had Pinchot, that "successful irrigation involves and demands the preservation of the forests."[27] Therefore, like French forestry practices in the 19th century, the American government should forbid clearing along mountainous or hilly terrain of the new reserves, since protecting the basins of streams, according to Pinchot, "is their most important use."[28]

On flatlands, where watershed protection did not pose as great an issue, the government would engage in "wise management," as Muir envisioned, by "keeping out destructive sheep, preventing fires, and selecting the trees that should be cut for lumber, and preserving the young ones and the shrubs, and sods of herbaceous vegetation ..." If loggers conducted business in this fashion, Muir concluded, America's "forests would be a never failing fountain of wealth and beauty."[29]

Muir deferred to Pinchot, who had studied forests and forestry in Europe, on how to "produce as much timber as possible without spoiling them" for "the common good of the people."[30] As Pinchot saw it, there were two mutually exclusive ways to produce timber, "Destructive Lumbering" and "Conservative Lumbering." Usually, in Muir and Pinchot's time, timber companies practiced destructive lumbering, by which an entire forest "is cut down without care for the future."[31] Pinchot called it "ordinary lumbering" or "the usual way."[32] Today, we call it clearcutting. "Its result, according to Pinchot, "is to annihilate the productive capacity of forest land for tens or scores of years to come."[33] The consequence, according to Pinchot's protégé Major George P. Ahern, told "a moving story of forest devastation, abandoned towns, abandoned farms, the closing down of hundreds of wood using industries as the centers of lumber production shift from the Northeast to the Lake States, to the South, and finally to the last stand in the Pacific Northwest."[34]

Muir and Pinchot advocated the second type of timbering, conservative lumbering, in which the yield is eventually far greater but cutting in the same area occurs over a longer time while young growth is "encouraged and protected." The overriding rule: "Draw from the forest while protecting it, the best return which it is capable of giving."[35] Following Pinchot's teachings would inevitably lead to the two basic goals of conservative lumbering, "the use and the preservation of the forest."[36]

Surprisingly, even though today everyone in the forestry field attributes to Pinchot the idea of "wise management" of our timber resources for "the common good," these phrases came from Muir's pen.[37] Equally surprising to our assumptions about the two icons, it was Muir who introduced the idea of using American public forests for recreation. Muir wholeheartedly supported the trend: "Thousands of tired, nerve-shaken people are beginning to find that going to the mountains is going home... that mountain parks and reservations are useful not only as fountains of timber and irrigating rivers, but as fountains of life."[38]

Conversely, Pinchot, to the surprise of those wishing to cling to old stereotypes, emphasized the interrelation between the trees, animals, climate, soil, and water within the forest. As Pinchot observed, the forest was "a complex community with a life of its own."[39] His strong antipathy toward the timber industry should surprise many as well. It is true that early in his career Pinchot tried to work with the lumber industry, going so far as to state, in 1903, that the U.S. Bureau of Forestry, which would later become the U.S. Forest Service, should "serve the lumber interests."[40] As time went by, though, Pinchot began to see things differently. He came to believe that unless the ax wielded by industry were controlled by government, "there can be no solution... of the problems of forest devastation in America."[41] In fact, Pinchot added, "The lumber industry is spending millions of dollars ...trying to fool the American people into believing that the industry is regulating itself...."[42]

The relationship between Muir and Pinchot did become contentious, but not over forests and forestry. They had a parting of the ways over water—namely, the Hetch Hetchy Dam in Yosemite. Still, Pinchot's affection for

John Muir endured. In his autobiography, *Breaking New Ground*, Pinchot could be very acerbic toward people he had worked with and did not like. Yet a feeling of delight comes up in every passage regarding Muir. Pinchot writes, for example, that in Montana on July 16, 1896, when he met with the National Forest Commission, set up by the National Academy of Sciences to study the importance of forests and the role the government might play in their protection, "[T]o my great delight, John Muir was with them.... I took to him at once."[43] Again, on a trip from Pelican Bay, California, to Crater Lake in that same year, Pinchot credits John Muir for making "the journey short with talk that was worth crossing the continent to hear."[44] Three years later, he, Muir, and Hart Merriam, head of the Biological Survey, "made a memorable trip to the Calaveras Grove," where many of the giant sequoias grew. Pinchot gave the time spent with these two men a rave review. "This little journey was for me in the nature of a liberal education. Never were there two more delightful talkers than Muir and Merriam, or with a richer fund of experience to talk about."[45]

Their finest hour came when the National Forest Commission arrived at the Grand Canyon. While the other members drove through the woods, John Muir and Gifford Pinchot did not join them. Instead, Pinchot recalled 50 years later as if it were just yesterday, they "spent an unforgettable day [together] on the rim of the prodigious chasm, letting it soak in. I remember at first we mistook for rocks the waves of rapids in the mud-laden Colorado, a mile below us. And when we came across a tarantula he wouldn't let me kill it. He said it had as much right there as we did."

Wishing that such a marvelous day would never end, Pinchot hoped they could spend the night together. "We had left from our lunches a hard-boiled egg and one small sandwich apiece, and water enough in our canteens. Why go back to the hotel?" Pinchot asked Muir. The feeling seemed mutual, as the idea "suited Muir as much as it did me. So we made our beds out of Cedar boughs in a thick stand that kept the wind away, and there he talked until midnight. It was such an evening as I have never had before or after."[46]

I recently had the opportunity to tell Pinchot's grandson, Peter Pinchot, this story. He thought a statue of the two eating sandwiches by the campfire should go up in Milford near Grey Towers. What a wonderful statement it would make for friendship and for forestry!

Notes

1. See for example, National Products Association, March 9, 1976, The Monongahela issue: A spreading economic malady, p. 5.
2. Ed Marston, March 18, 2002, Will the real Gifford Pinchot please stand up?, www.hcn.org/servlets/hcn.Article?article_id=11090.
3. http://www.fs.fed.us/gobal/lzone/student/tropical.htm.
4. W. Peabody, October 1832, American forest trees, *North American Review* 35, p. 338.
5. F. Oswald, 1877, Climatic influence of vegetation—A plea for our forests, *Popular Science Monthly* 11, p. 388.
6. G. Pinchot, 1972, *Breaking new ground*, p. 23.
7. J. Muir, 1897, The American forests, *Atlantic Monthly* 80, p. 147.
8. Pinchot, 1972, p. 23.
9. Muir, 1897, p. 147
10. Muir, 1897, p, 147; see also G. Pinchot, 1914, *The training of a forester*, pp. 22–23.
11. J. Muir, 1897, The wild parks and forest reservations of the West, *Atlantic Monthly* 81, p. 17.
12. G. Pinchot, 1972, pp. 100–101.
13. Pinchot, 1972, pp. 126–27.
14. J. Muir, 1901, Hunting big redwoods, *Atlantic Monthly* 88, p. 308.
15. G. Pinchot, 1900, *A short Account of the Big Trees of California*, p. 8.
16. John Muir quoted in Pinchot, 1900, pp. 18–19.
17. Pinchot, 1900, p. 7.
18. J. Muir, 1920, Save the redwoods," in W. Cronon (ed.), 1997, *John Muir: Nature writings*, p. 829.
19. Pinchot, 1972, pp. 102–103.
20. Pinchot, 1900, p. 30.
21. J. Muir, 1897, p. 147.
22. See G. Pinchot, 1891, Government forestry abroad, *Publications of the American Economic Association* 6(3) (May), pp. 7–54.
23. Muir, 1897, p. 155.
24. G. Pinchot, 1905, A primer of forestry. Part II: Practical forestry, U.S. Department of Agriculture, *Farmer's Bulletin* 358, p. 69.
25. G. Pinchot, 1901, Trees and civilization, *The World's Work* 2, p. 992.
26. Muir, 1897, p. 155.
27. Pinchot, 1901, p. 994. R. Zon, 1927, *Forests and water in the light of scientific investigation*; probably the most exhaustive study on forests and their role in watersheds, this publication had strong support from Pinchot.
28. G. Pinchot, 1909, A primer of forestry. Part I: The forest, U.S. Department of Agriculture, *Farmer's Bulletin* 173, p. 34.
29. Muir, 1897, p. 155.
30. Muir, 1897, p. 147.
31. Pinchot, 1909, Part I, p. 34.

32. Pinchot, 1909, Part II, p. 22.
33. Pinchot, 1909, Part I, p. 34.
34. G. Ahern, 1929, *Deforested America: Statement of the present forest situation in the United States*, p. vii.
35. Pinchot, 1909, Part I, p. 34.
36. Pinchot, 1909, Part II, p. 3.
37. Muir, 1897, pp. 155, 147.
38. Muir, 1898, p. 15.
39. Pinchot, 1914, p. 14.
40. G. Pinchot, 1903, The lumberman and the forester, *Forestry and the timber supply*, Bureau of Forestry Circular 25, p. 14.
41. G. Pinchot, 1929, Foreword, in Ahern, 1929, p. vi.
42. Pinchot, 1929, p. vi.
43. Pinchot, 1972, p. 100.
44. Pinchot, 1972, p. 101.
45. Pinchot, 1972, pp. 171–72.
46. G. Pinchot, 1972, p. 103.

ABSTRACT

Silvicultural methods for sustainable forestry exemplify the circular course of the U.S. Forest Service through the 20th century. The agency's first chief, Gifford Pinchot, and likeminded progressive reformers were leaders in conservation movement; to end the destructive cut-and-run practices of the timber industry, they advocated selective cutting as the best method of harvesting to preserve the important ecological functions of the forest and its watersheds. At midcentury, however, the federal agency and the forestry profession had more in common with the timber industry, and profitable, industrial clearcutting— efficient and inexpensive—was promoted. Now, after three decades of conflict between environmentalists and the Forest Service, the pendulum has begun to swing back, and the agency again embraces harvesting methods that preserve the integrity of the forest. The reasons for these shifts are complex and reveal that forestry is as much a product of history and culture, responding to social, economic, and political currents, as it is of science.

CHAPTER 8

Back to the Future: The Rise, Decline, and Possible Return of the U.S. Forest Service as a Leading Voice for Conservation in America, 1900–2000*

Paul Hirt

In early June 2005, the Second U.S. Circuit Court of Appeals blocked a timber sale on a national forest in Vermont, giving what environmentalists and the news media referred to as "a solid victory for conservationists."[1] News reports like this have been common for decades in the United States, reflecting a deeply entrenched antipathy between conservation organizations and the U.S. Forest Service over forest management in general and timber harvest practices in particular. In popular understanding today, forestry and conservation appear to be in conflict. From a historical perspective, this is an ironic, surprising, and fascinating turn of events. One hundred years ago in America, the terms forestry and conservation were inseparably linked; in fact, one of the founding fathers of American forestry, Gifford Pinchot, helped invent the word *conservation*. As the first chief of the U.S. Forest Service, Pinchot made conservation the guiding land management philosophy of that agency. How odd that a defeat for the Forest Service is now seen as a victory for conservation.

During Pinchot's tenure as Forest Service chief, his main allies were progressive social reformers at the forefront of the newborn conservation movement, including many leaders in the forest protection movement.[2]

Pinchot's staunchest opponents, on the other hand, came from the timber and livestock industries and their political allies. Around the middle of the 20th century, however, this configuration of Forest Service allies and opponents reversed, with the federal agency and the forestry profession finding more common ground with the timber industry and growing increasingly alienated from conservation organizations. Lawsuits like the one mentioned above are a legacy of this era of antagonism. Interestingly, after three decades of bitter conflict between environmentalists and the Forest Service, from the 1970s through the 1990s, the pendulum began to swing again. Future analysts may identify the first decade of the 21st century as the point at which the Forest Service and the forestry profession in America began to recover their former roles as conservation leaders.[3] The reasons for these shifts are complex and reveal much about the inextricable social embeddedness of Forest Service culture, the forestry profession, forest management science, and conservation policy.

In this chapter, I explore the causes and conditions behind the Forest Service's and the forestry profession's rise to leadership in the conservation movement a hundred years ago, their subsequent fall from that position after midcentury, and the changes taking place at present. Since the full dimensions of this story cannot be covered in one essay, I focus on one particular piece of the puzzle: changing scientific orthodoxy regarding proper silvicultural methods for sustainable forestry, especially on public lands, as it evolved during the 20th century. I will use this historical survey of forestry prescriptions as a lens through which to understand how specific historical contexts have shaped forest policies and practices, U.S. Forest Service culture, and the changing relationships between the timber industry, foresters, and conservationists.

Clearcutting and Deforestation

The science of forestry in America grew slowly and hesitantly from its infancy in the 1870s until the second decade of the 20th century. Europeans had been refining the art and science of silviculture for generations by then, but the wild, ancient, and highly diverse forests of North America required adjustments to Old World practices, which had been designed for the highly managed and ecologically simplified forests of western Europe. European "textbook forestry" did not transfer well to America, so the first generations of American foresters had to work out, through trial and error, new systems compatible with the conditions they confronted. A workable silvicultural program had to factor in biological, climatic, topographic, and soil conditions, as well as the unique economic, legal, and social conditions in the United States.[4]

At the turn of the century, foresters, who were the backbone of the conservation movement at that time, had the overriding goal of stopping the destruction of America's forests and transforming timber exploitation into sustainable timber production.[5] They were an essential element of a larger forest protection movement that included a diverse array of interests with many different but complementary concerns and objectives. Some people in the forest protection movement were concerned about a possible timber famine, others about stream flow, flooding, and sedimentation in the nation's rivers; others feared the disappearance of wildlife; others worried about monopoly and its threat to democracy; and still others hoped to ameliorate the 19th-century pattern of boom-and-bust economics. All these interests coalesced into a forest conservation movement, with Gifford Pinchot one of its leading evangelists. Pinchot's acerbic commentary about this time period was right on the mark: "When the Gay Nineties began, the common word for our forests was 'inexhaustible.' To waste timber was a virtue and not a crime.... The lumbermen, whose industry was then the third greatest in this country, regarded forest devastation as normal and second growth as a delusion of fools, whom they cursed on the rare occasions when they happened to think of them. And as for sustained yield, no such idea had ever entered their heads."[6]

Reversing forest destruction and building a sustainable timber economy obviously would be a time-consuming and difficult task. The reform options available to conservationists involved essentially three strategies: 1) regulate private forest practices; 2) find incentives to get voluntary cooperation from industry; or 3) reserve important publicly owned forest lands from private ownership and manage them in the public interest. All three paths were pursued, with varying degrees of success.

Proposals for the first option, government regulation of private forest practices, found little support in the laissez-faire climate of 19th-century American political economy. Calls for regulation surfaced perennially, but nothing ever came of those proposals at the federal level. Some states initiated forest practices regulations in the 20th century to fill the vacuum at the national level, but such reforms were usually weak, inadequately enforced, and rarely in place when they were most needed. The second option, cooperation and incentives to the private sector, was a one-way street, with a few intrepid federal foresters in the Department of Agriculture advocating sustained-yield forestry to a mostly disinterested industry. Government foresters did make some progress promoting cooperative fire suppression, but this campaign had only limited success in reducing fires and had virtually no impact on the problems associated with destructive logging activities and cut-and-abandon business practices.

Option 3, however, did become a significant and lasting public policy. A movement in the 1880s to have the federal government retain some of its lands in the West, rather than turn them all over to private or state ownership, succeeded in 1891, with the Forest Reserve Act authorizing the president to set aside forest reservations from the existing federal land base.[7] Over the next two decades, a succession of presidents reserved tens of millions of acres, mostly in the western United States, which formed the basis for the national forests of today. Those "forest reserves," as they were called at that time, however, represented only a small fraction of the lands in the West holding valuable timber—and the bulk of the federal reserves were in remote, inaccessible areas and did not contribute much to the lumber economy. The majority of productive and accessible timberland passed out of government ownership in the 19th century, mostly ending up in the hands of large railroad and lumber corporations.[8] A source of some frustration for government foresters, private timberlands supplied 95 percent of commercial lumber for the U.S. market until World War II, leaving federal forests marginal to the timber economy in the first half of the 20th century. Instead of being able to showcase sustainable timber harvest practices, federal foresters spent most of their time and energy fighting fires, tracking down timber poachers, inventorying resources, building roads and trails, and promoting the idea of conservation. As a consequence, to have any salutary effect on logging practices or the lumber economy, foresters would have to continue seeking the cooperation of private industry—option 2 above. And to accomplish this, they would have to be creative.

Cut-and-abandon logging practices typical of the time often resulted in conditions that worked against the reestablishment of adequate forest cover. Abandoned cutover forests full of logging debris easily caught fire in subsequent years, killing new growth, while the bared soil washed away in rainstorms, clogging streams and rivers with silt and leaving behind less fertile ground for forest regeneration. Foresters and the general public viewed these scenes as frightful, ugly, economically wasteful, and socially irresponsible. Conservationists in government employ took on the task of converting the lumber industry from destructive exploitation to responsible forest management. Foresters' initial task, then, was to clarify the difference between exploitation and responsible management. The less technical aspects of defining responsible management were the easiest: good forestry looked toward the future; every action should contribute to the maintenance or improvement of the forest. America's first professional forester, Bernhard Fernow, wrote in 1902 of the distinction between the "lumberman" and the "forester." The former takes all that is valuable from a forest and leaves the rest behind. The least useful species then reproduce themselves—if subsequent fires in the logging debris do not kill everything—and the area loses much of its value (timber value, that is) for the coming generation. The forester, in contrast, first culls the *undesirable* species to give the desirable ones an advantage, Fernow said, and thus "improve the composition of his crop." Subsequent logging practice is oriented toward encouraging reproduction of desired species, not just skimming the cream from the crop. Fernow admitted that ideal forestry practices, such as culling unmarketable species or leaving valuable trees standing to provide a seed source for the next generation, sometimes did not pay for themselves right away, but the improved forest crop would eventually profit the landowner.[9]

The technical aspects of responsible forestry were more difficult to define. How was one to actually log a forest without destroying it? Answering this question led to the establishment of the first round of scientific orthodoxy in American forestry. Although conditions (biological, physical, and economic) varied widely by region, some generalized conventional wisdom based on specific historical circumstances did emerge. In Europe, especially Germany, clearcutting in small blocks was a common practice. But those forests were intensively managed, highly accessible, and composed of even-aged stands of timber that would mature in uniform blocks.[10] The forests of America were generally unmanaged, with primitive or nonexistent transportation systems, and composed of a wide diversity of species growing in uneven-aged stands. These conditions, along with poor markets for many kinds of trees, demanded a different approach.

Selective Cutting for Forest Preservation

The approach that first achieved dominance was "selective" harvesting. In fact, the 1897 Pettigrew Act (also referred to as the National Forest "Organic Act") enshrined this silvicultural practice into law, legally mandating its application to the forest reserves until it was repealed in 1976. Forestry leaders in the United States played a major role in drafting the Pettigrew Act. According to the legislation, only "dead, matured, or large growth" trees could be sold and removed from the forest reserves (renamed national forests in 1907), and only after the trees were individually marked for sale and appraised. This selective style of harvesting was designed "for the purpose of preserving the living and growing timber and promoting the younger growth on forest reservations."[11]

Fernow's 1902 textbook, *Economics of Forestry*, reiterated this approach in subsequent years with only a mild variation. Although a variety of silvicultural practices were available, he said, one approach was preferable: selective cutting. In his words, "… a gradual removal of [desirable trees] takes place, either of single individuals here and there, or of groups of them, making larger or smaller openings; or else more or less broad strips are cleared, on which seed falling from the remaining neighboring growth can find lodgment, and sprout; and, as the young seedlings require more light for their development, gradually more of the older timber is removed, or the openings are enlarged for new crops of young growth, and thus the reproduction is secured gradually, while harvesting the old crop."[12] Notice that he used "gradual" to describe the logging process three times in this one sentence, and that openings were to be no bigger than what neighboring trees could naturally and readily reseed.

Gifford Pinchot, as head of the Division of Forestry in the Department of Agriculture starting in 1898, also endorsed conservative selective cutting. In his 1906 *Use Book*, an instruction manual for forest rangers, he stated, "Wherever possible a stand of young, thrifty trees should be left to form the basis for a second crop."[13] Reiterating Fernow's gradual approach, Pinchot repeatedly emphasized that timber cutting should be exercised with deliberate care in order to preserve the essential integrity of the forest. In 1905 Pinchot penned a letter for Agriculture Secretary James Wilson's signature that established policy for management of the forest reserves; in part, the letter stated, "The permanence of the resources of the reserves is … indispensable to continued prosperity, and the policy of this Department for their protection and use will invariably be guided by this fact, always bearing in mind that the *conservative use* of these resources in no way conflicts with their permanent value"[14] (emphasis in the original). Similarly, one of Pinchot's early forestry manuals called for "conservative lumbering to maintain and increase the productivity and the capital value of forest land."[15]

Before offering timber for sale from the forest reserves, Pinchot instructed his rangers to consider possible damage from logging, the likelihood of successful reforestation, and whether other land values might be more important than the market value of the timber. When choosing a harvest system, rangers were instructed not to place profitability above reforestation and watershed protection considerations. "The object of a sale is not solely to realize the greatest possible money return from the forest. The improvement and future value of the stand both for forest cover and for the production of timber must always be considered. In many cases, the need of preserving an unbroken forest cover for the protection of watersheds will influence the method of cutting recommended."[16] Pinchot's deference to watershed values was an important feature of early public forestry. Watershed considerations would come to play only a small role in management decisions by midcentury, however, and would not make a significant comeback until the tenure of Chief Michael Dombeck in the 1990s. This midcentury decline in the attention foresters paid to the broader ecological and social functions of forests (watershed, wildlife, recreation) in large part accounts for the growing alienation of the Forest Service from the conservation community (more on this below).

Echoing Pinchot's concern for preserving "forest cover," Fernow made an interesting and important distinction between the "protection forest" and the "supply forest." The former had values other than timber production to consider. For such lands, a very slow removal of the timber crop was indicated, Fernow said, "even if financial and silvicultural results would make other methods desirable." For the "supply forests," the greatest continuous growth of desired species at the lowest cost was the objective, according to Fernow, and whatever system worked best to accomplish that was acceptable, so long as it did not impair the productivity of the land.[17] This distinction between protection and supply forests was just as important a determinant of logging prescriptions as other factors, such as markets, species, and growing conditions.

In the following years, while attempting on-the-ground applications of silvicultural theory, foresters confronted a variety of conditions that required increasing flexibility in their prescriptions. Henry S. Graves, first dean of the Yale School of Forestry and second chief of the Forest Service, wrote a silvicultural textbook in 1911 that extended many of Fernow's and Pinchot's ideas. He acknowledged that different economic, physical, and biological conditions called for different silvicultural prescriptions. Selection harvests, Graves said, were best adapted to "a newly developed country" where trees of all age classes exist in mixed stands. Selective cutting was also preferred where a continuous forest cover is desired, where public recreation values are prominent, and where markets are poorly developed. Clearcutting *in small blocks* worked best in even-aged stands, in stands composed of very large trees that destroy smaller growth when felled, and in instances where reproduction of shade-intolerant trees was desired. While approving of clearcutting under these restricted conditions, Graves added the following caution: "The disadvantages of clear-cuttings are in direct ratio to the size of the clearings. Many of the evils of clear-cutting can be obviated by reducing the size of the area clear-cut. Intensive forestry avoids large clearings." He noted that in Europe, the ideal size of clearcut openings (he called them "group selection" harvests) ranged from 50 to several hundred feet across.[18] This contrasted sharply to the wholesale liquidation of hundreds or thousands of acres of timber in valleys and mountainsides then common on industrial forest lands in North America.

Another endorsement of the early preference for selective cutting came in a series of silvicultural textbooks written by Yale forestry professor Ralph C. Hawley beginning in 1921 and updated periodically over the next several decades. His descriptions of the various logging methods contained a discussion of each method's advantages and disadvantages. In his original 1921 edition he lists eight "advantages" to the selection method, including the following three: "[It] affords a high degree of protection to the site and to reproduction and minimizes the danger of snow-slides and land-slides"; it is "the method that best satisfies the aesthetic purpose"; and "there is less danger of disastrous fire." In the 1946 edition Hawley added one more advantage to the list: it is "superior... for the conservation of wildlife." All of the disadvantages to the selection system pertained to practical matters related to logging efficiency and getting the greatest dollar value out of the timber stand. Conversely, clearcutting's disadvantages included temporary loss of forest cover, desiccation, erosion, changes in stream flow, and aesthetic impacts. All its advantages pertained to the practical business of profit-oriented lumbering.[19] Thus, even as late as the 1940s, leading forestry academics judged clearcutting as physically disruptive but economically efficient for lands dedicated to log production, while prescribing the selection system as most appropriate for forests dedicated to a broader spectrum of values. As caretakers of the national forests, the Forest Service held to these general distinctions and preferences, strongly favoring selective cutting from the time of Pinchot until shortly after World War II. Although this ideal made eminent sense for the national forests, industry never warmed to the vision. As long as the forestry profession remained dominated by Pinchot's generation of social reformers, industry remained largely skeptical, even hostile, to forestry and conservation, and conservation organizations remained forestry's allies.

Clearcutting for Efficiency

A crucial shift occurred around midcentury. National production of timber on private lands had reached its peak in the 1920s, and from then on private supplies were in decline. Pressure on public lands to take up the slack might have begun in the 1930s if not for the Great Depression and the subsequent housing market slump. World War II–related production, however, suddenly caused an explosion in lumber demand in 1940, and industry could not meet that market demand solely from private commercial timberlands. Justified because of the "war emergency," timber harvests and clearcutting practices consequently accelerated rapidly on the national forests during the war. Instead of being a temporary deviation, however, national forest timber sale volumes continued their rapid climb. The unprecedented economic boom in the postwar era greatly boosted lumber demand while private timber inventories continued to decline. With private lands no longer able to meet market demand, government and industry turned to public forests to make up the difference.[20]

Foresters and policy makers concerned about lumber supply began to consider protection forests an expendable idea and to view selective cutting as unacceptably inefficient. Consequently, they argued that the objectives and practices used on supply forests should be applied to all areas of the national forests with commercial timber. By the late 1950s, little difference could be discerned between harvest practices on public versus private forests, as both were dedicated to a management philosophy increasingly focused on aggressive wood production. This shift in national forest management represented in part a change in how national forest managers viewed the public properties under their charge, but it also reflected a revolution in foresters' confidence in their ability to remake natural forests into timber factories. Most professional foresters in the postwar era simply adopted the industrial forestry maximum-production ethos wholesale.

The Forest Service held onto the selective harvest orthodoxy longer than the forestry profession as a whole, questioning if not actually resisting the shift to large-scale clearcutting until the late 1950s. In a *Journal of Forestry* article in 1949, former Forest Service Chief Earle Clapp argued that protecting the forests for the future required "the greatest emphasis on selective cutting, which removes only the larger and more mature trees and leaves the smaller trees for future growth." "Clearcutting," he continued, "and clearcutting in its modified form of leaving only a bare minimum of seed trees, must be reduced to an absolute minimum."[21] Within a decade, however, clearcutting replaced selective cutting as the harvest method of choice for most timber sales on the national forests, in part because of an unprecedented, unmanageable flood of demand for national forest timber that the Forest Service felt obligated to meet and for which clearcutting provided the most practicable means for accelerating harvests.[22] To meet the timber demand, Congress increased the Forest Service's personnel budget, and young professionals streamed into the agency from the swelling ranks of forestry schools around the country. The new recruits were virtually all assigned to timber management.

Studies touting the benefits of clearcutting rapidly proliferated. Foresters had always known that the commercially valuable Douglas-fir regenerated best in sunlight, but by the late 1950s they were using this "sunlight" argument to rationalize ever-larger clearcuts and had added a whole host of other trees to the list of species supposedly dependent on clearcuts for regeneration. In 1958, for example, the chief of the Management Division of the Forest Service's Intermountain Forest and Range Experiment Station, George Craddock, substantially broadened the list of trees judged fit for clearcutting at a meeting of the Society of American Foresters: "Clearcutting is essential to the regeneration of the densely growing forests of western white pine, spruce, fir, and lodgepole pine in Idaho and Montana."[23]

Ironically, Craddock's presentation was on watershed problems in the region caused by destructive logging and overgrazing.

Some foresters who had lived through this transition period with the Forest Service continued to lament the transformation in the agency's silvicultural approach. Edward Crafts, who served as Forest Service deputy chief for Programs and Legislation in the 1950s, recalled in an oral interview in the early 1960s, "Frankly I was quite dissatisfied with Forest Service timber management policies... the chopping up of the hillsides with too many roads, and resulting erosion, excessive cutting... use of clear-cuts instead of selective cuts, excessively large blocks of clearcut timber, failure to require adequate regeneration measures, failure to require adequate slash disposal and erosion control measures. This is a whole list of things that I thought the Forest Service timber management people were far too easy on with the timber industry."[24]

A Profession Geared to Industry

Those midcentury changes in the forestry profession reflected new conditions in the political economy of the timber business. In light of the increased demand for wood during and after World War II, forestry schools began pumping out record numbers of graduates trained for a job market keyed to getting the most timber out of a forest.[25] This employment emphasis accelerated as a result of a growing trend by industry to hire trained foresters. Several specific historical conditions dictated this new trend. First, with U.S. lumber companies running out of unclaimed, uncut forests to migrate to, permanent landownership and timber farming became one of the few options available for those who wished to stay in business. Second, increased mechanization and corporate consolidation in the timber industry and associated pressures for operations to run on larger scales to enhance profits led to a greater need for professionally trained land managers. Third, rising lumber prices after the long depression led to the kinds of revenues that made timber farming—as opposed to cut-and-abandon timber mining—economically feasible. The long-term capital investment required to maintain a timber plantation had a greater chance of paying off as lumber prices repeatedly hit new record highs in the postwar years.

Finally, the timber industry's sporadic and belated conversion to an agricultural paradigm in the 1940s received added impetus from revitalized government threats to regulate private forest practices. Public welfare legislation and vast new regulatory agencies had proliferated during the New Deal of the 1930s. The profound failures of unregulated capitalism evidenced during the Depression toppled laissez-faire philosophy from its political pedestal, giving Gifford Pinchot and his allies and acolytes in the Forest Service license for increasingly bold calls for public regulation of the timber industry. This made lumber leaders nervous: they stepped up their promotion of public-private *cooperative* forestry programs, especially cooperative fire control, government-subsidized reforestation, and forestry research; and they initiated an industry-sponsored experimental "tree farm" program to improve their public image and deflect Forest Service criticism.[26] Both tactics paid off, as federal regulation never materialized.

For long-term success in these campaigns, industry had to make some visible changes, such as committing to reforestation, adopting more responsible logging practices, and hiring qualified foresters to lend a credible aura of professionalism to its operations. Higher lumber and pulp prices allowed industry to expand its activities in each of these areas. Thus, private sector employment opportunities for foresters ballooned, until by the 1950s most forestry school graduates went into private practice rather than Forest Service employment. Indicatively, in 1947 the Society of American Foresters for the first time elected its president from the ranks of industry rather than government or academia.[27] The end result was that forestry schools increasingly geared their curriculum to industry needs, and the Society of American Foresters (which defines orthodoxy and accredits forestry schools) tipped increasingly toward forest industry perspectives and approaches to forest management that emphasized maximum wood fiber production and economic efficiency.[28] As Forest Service old-timers like Earl Clapp retired, advocates of selective cutting and the broader philosophy of conservation receded into the background of the agency and the profession as a whole.

Timber company executives had an influential hand in accelerating forestry school responsiveness to their agenda. Besides hiring graduates, industry provided significant funding for silvicultural research; industry leaders endowed forestry schools, served on their boards of directors, established experimental forests, and wielded considerable clout in state legislatures, which hold the purse strings of many universities. Whereas in the past, most foresters had advocated the broader conservation vision and criticized the timber industry for its irresponsible practices, in the postwar era a new generation of private sector–oriented technological optimists took the reins of the profession, as well as its research estate, and focused forest science on productivity and profits.

An exemplary proponent of this new wave in forestry was Yale professor Herman H. Chapman, who initially made a name for himself in the 1920s as a critic of the Forest Service. During the middle decades of the century, he trained a dozen academic generations of foresters in his southern industry-oriented management style. "[F]orest management," he stated in a 1950 treatise, "aims to secure 1) continuity of production of the forest crop, and 2) benefits in excess of the cost of production." Silvicultural decisions must be guided by the "ultimate goal" of establishing a forest "in which the maximum possible volume of timber can be grown and cut annually." This was the essence of the new forestry: a simplified, technologically based, commodity output-oriented ethos of maximum productivity. The 19th-century foresters' interest in maintaining forest cover for watershed protection and preserving natural forest functions, botanical diversity, wildlife, recreation, and aesthetics all faded into the distant background, seeming quaint or irrelevant to the no-nonsense production foresters.

Inevitably, the Forest Service would reflect these changes, too, since its ranks were swelling with young graduates trained by people like Chapman. Besides, as industry clamored for more access to national forest timber, Congress obligingly pumped huge amounts of capital into the Forest Service's timber sales and road construction budgets. The agency not surprisingly turned wholeheartedly to the business of maximizing timber production rather than maintaining forest cover in part because that's where the money was.[29] Its traditional concern for resource values other than timber remained in place rhetorically, but in practice, timber production priorities overwhelmed everything else. In the process, the old distinction between protection and supply forests essentially disappeared. A "conspiracy of optimism" among silviculturists and policy makers held that timber production could be maximized at the same time that other forest values were protected. Selective cutting was considered inefficient and unnecessary; clearcutting to facilitate even-aged management became not only desirable but, in the words of many foresters, *necessary*.[30] Foresters touted this new orientation as wholly "scientific," labeling dissidents as uninformed or behind the times.

All of these new developments led to a deep and lasting split between the Forest Service and conservation organizations, as well as a vigorous debate over the meaning of conservation and the purpose of the national forests. The timber industry and the forestry profession as a whole adhered to a utilitarian definition of conservation as wise use, which was consistent with Pinchot's original vision, but in the postwar years many leaned increasingly toward the view that wise use actually meant intensive management for "full utilization," with little room for conservative low-intensity uses and no room for nonuse. As the annual *Report of the Chief* of the Forest Service in 1948 stated, "In handling the national-forest timber resources the Forest Service is working toward intensive management for maximum continuous production."[31] During most of the next decade, the Forest Service worked to produce a monumental assessment of the nation's timber supply outlook, published in 1958 as *Timber Resources for America's Future.* Premised on the assumption of a looming timber famine and carving out a role for the Forest Service in that environment, this document set the stage for national forest management for the next three decades. In the foreword, Forest Service Chief Richard McArdle wrote, "To meet future timber demands will take earnest effort. Meeting those needs will require not only early action but an intensity of forestry practices that will startle many of us." Accordingly, the long-term goal for the management of the national forests was a "prompt and very substantial expansion and intensification of forestry...."[32]

This was not the direction that many conservation groups favored for the management of the nation's public forest lands. Leaders within the Wildlife Management Institute, Izaak Walton League, National Wildlife Federation, Sierra Club, Wilderness Society, Audubon Society, Federation of Western Outdoors Clubs, and many other groups interested in resources other than timber became nervous, then appalled, then downright hostile in the face of this intensive-forestry, full-utilization rhetoric.[33] Attitudes toward old-growth forests and wilderness provide good examples of this divergence of opinion. As logging roads and activities extended into the backcountry, large tracts of wild, ancient forests grew ever scarcer and more precious to Americans. Consequently, many conservation group leaders focused considerable energy on protecting these remaining old majestic forests from harvest. At the very same time, ironically, the forestry profession ramped up its campaign to eliminate these wild areas from what it considered to be the working landscape of the national forests.

The Society of American Foresters sponsored a technical conference called "Converting the Old-Growth Forest" at its national convention in 1955. The underlying assumption behind the conference was that old-growth was wasteful and inefficient and had to be replaced by thrifty, young commercial timber stands as rapidly as possible. At the conference, both industry and Forest Service leaders all agreed on this premise, even though most conservation organizations saw these old-growth forests as a priceless national heritage deserving protection. In an opening speech at the conference, E. P. Stamm of Crown Zellerbach Corporation stated the typical view of industry: "There is no argument but that complete utilization of timber resources is necessary." To get complete utilization, all old-growth had to be liquidated, Stamm contended. And since most remaining old-growth forests were on federal lands in the West, America needed to focus intensively on those lands: "The large volumes of big old growth, rugged topography, and long distances from woods to market [on western federal lands] all spell ROADS, MORE ROADS, AND BETTER ROADS in big, red capital letters." Underscoring his contempt for unroaded, unharvested forestlands, Stamm concluded, "Lack of management is synonymous with mismanagement."[34] Forest Service leaders at the 1955 conference, such as Regional Forester A.W. Greeley of Alaska, echoed these sentiments, though in more restrained language.[35]

In contrast to Stamm, many conservation leaders offered an equally vigorous dissent against the Forest Service's growing enthusiasm for road construction and timber sales. One of the most articulate, Supreme Court Chief Justice William O. Douglas, gave vent in 1961: "There is hardly any place left in the country (outside Alaska) where one can get more than ten miles from a road. Roads, roads, roads—they seem to be the compulsion of state and federal officials.... The Forest Service is the main offender. In some states it is hardly more than the voice of the logging interests.... Multiple-use in actual practice is a high-sounding term that means loggers and automobiles take over."[36]

Seeing uses other than timber becoming increasingly marginalized in the new regime, conservationists promoted a multiple-use bill and a wilderness bill in Congress in the late 1950s to remind the agency that it had a broader mandate than lumber production. The maximum-utilization advocates countered that wilderness was a nonuse and therefore contrary to the utilitarian definition of conservation that they proffered. Siding with those who advocated a broader conservation vision, President Kennedy's Secretary of Interior Stuart Udall boldly stated in 1962—two years before the passage of the National Wilderness Preservation Act—that "wilderness preservation is the first element in a sound national conservation policy."[37] But dominated by an industrial forestry perspective, the Society of American Foresters and the Forest Service ignored these objections, considering them misguided or naïve, and continued on their chosen path, which they vigorously defended as the true path of conservation. On a quite divergent trajectory, the great majority of old-line conservation organizations—and quite a few new ones that surfaced in the 1960s and 1970s—came to view the Forest Service as an enemy of conservation rather than its leader.[38]

Clearcutting Disavowed

With multiple competing definitions of conservation on the table, the public was understandably confused at times. Eventually, a new label, "environmentalism," was coined to distinguish the values and concerns of the new conservationists from the utilitarian production forestry advocates who also claimed the mantle of conservation. Use of the word *conservation* was never fully abandoned by environmentalists, however. Still today, the term is invoked by widely divergent interests and employed to signify substantially different land management visions. Today, too, the Forest Service and the forestry profession remain largely alienated from the conservation-environmental community, although that is starting to change.

The growing postwar antagonism between production forestry advocates and the new conservationists led to several showdowns in the late 1960s and early 1970s. Most of the controversy centered on the national forests. Critics of industrial-style even-aged management argued that although intensive even-aged silviculture might be acceptable for private timber plantations, it was unacceptable on the public forestlands that served so many other purposes. As usual, when the debate entered the public arena, it degenerated into a simple dichotomy: conservation groups said that clearcutting was bad, while the Forest Service insisted that clearcutting was good. Dramatic photos of mountainsides in the West stripped of their forest cover with poorly designed and hastily built roads washing into rivers galvanized public outrage.[39] The agency responded with its own flurry of defensive propaganda, and the fight was on. From approximately 1969 until 1976 a "clearcutting controversy" raged in America's newspapers and policy circles. A series of congressional hearings sponsored by Idaho Senator Frank Church rallied the affected parties, while the Forest Service, universities, public interest groups, the Nixon administration, and industry all launched and published studies of the problem.[40]

Bringing the controversy to a head was a lawsuit by the Izaak Walton League challenging the legality of clearcutting based on the 1897 Organic Act's requirement that trees to be cut from the national forests be mature or of large growth and individually marked and appraised—a selective harvest system. To the shock of the forestry establishment, the conservation group won its lawsuit in federal court in 1975. This threatened to force a significant reduction in the rate of logging on the national forests. Those with a stake in the higher harvest levels succeeded in getting a bill passed through Congress the next year legalizing clearcutting. This law, the National Forest Management Act, actually restructured the entire statutory authority of the Forest Service and prescribed goals, procedures, and management practices in some detail. Although it legalized clearcutting, it also empowered the previously marginalized dissidents by mandating "interdisciplinary planning" for national forest management. Critics had complained that narrowly trained, narrowly focused timber managers wielded too much authority over management decisions, so the act's writers forcibly opened the decision arena to a wider array of professionals and to the public at large. Over time, more natural resource professionals trained in disciplines other than timber management were hired, and thus the seeds were planted for a second wave of controversy in the 1980s—one that led to both a revolt of dissidents within the Forest Service and a dramatic revision of forestry orthodoxy.[41]

Recent events are perhaps familiar to many readers, so I will summarize here. The 40-year period of old-growth liquidation and high timber harvest levels on the national forests came to a crashing halt in the early 1990s. The harvest level on national forest lands dropped by an average of 70 to 80 percent across the nation during that decade, including the timber-rich Pacific Northwest, where the annual harvest went from five billion board feet in 1987 to less than one billion board feet by the early 1990s. The reasons are complex and have been treated in detail elsewhere.[42] The timber industry blamed environmental regulations and the Endangered Species Act, which did play a role, but the bust was actually a long-term result of many other factors, including decades of unsustainable harvest levels, skyrocketing costs accompanied by declining benefits, extensive resource damage, changing markets, and changing public values. As these historical forces all came to a head in the 1980s, the forestry profession underwent changes, too. The marginalized minority among foresters who remained critical of large-scale clearcutting, old-growth liquidation, and even-aged monocultural timber farming slowly came back into the mainstream by the late 1980s, led by academic researchers and by conscientious objectors within the Forest Service.

These dissidents reemphasized the value of small clearcuts, more selective harvests, retention of natural biological diversity, and watershed and wildlife protection. This "New Forestry" rose to some contentious prominence in the forestry profession by 1993. Perhaps the most recognized exponent of New Forestry at the time (there were many) was Professor Jerry Franklin. He argued that selection cuts should be used more often, that clearcuts should be reduced to smaller sizes, and that in logged areas a percentage of older trees should be left standing to provide "vertical diversity" as well as a seed source for the second-growth forest. Streamsides should also be protected from disturbance to reduce sedimentation of watercourses and to protect these crucial fish and wildlife habitat corridors. So far, other than "vertical diversity," New Forestry simply reiterated the prescriptions of Fernow, Pinchot, and Graves from the turn of the previous century. What makes Franklin's forestry modern (or "postmodern," as some have called it)[43] are the following additional prescriptions: "Snags," dead trees that old-school foresters considered insect and disease bait, should be retained rather than removed or burned so that they can perform their important tasks of contributing to nutrient recycling, organic soil litter, wildlife habitat, and erosion control; and large, interconnected blocks of wilderness should be set aside as native genetic diversity reserves and as habitat for species that cannot tolerate human disturbance.[44]

Although some of the scientific language describing New Forestry is novel, its thrust is in many ways a return to a conservative style of silviculture that aims to retain natural forest conditions as much as possible while removing some of the timber for commercial use. It recognizes the importance of forest cover for watershed, wildlife, and aesthetic purposes and seeks to balance all of these values rather than maximize one of them. American forestry had come full circle to its roots.

The Forest Service, as usual, reflected this trend by adopting in 1989 a program called "New Perspectives," followed a few years later by its current management paradigm, "Ecosystem Management."[45] As with the shift that occurred in forestry orthodoxy at midcentury, ecosystem management in the 1990s was not simply a product of scientific progress but instead resulted from a set of specific historical conditions. First, the abundance of economically valuable old-growth timber present on the national forests of the West at midcentury had for the most part been either liquidated or placed off-limits—mostly liquidated—by the 1990s. The sector of the timber industry dependent on national forest old-growth for its timber supply then turned back to second- or third-growth forests on both public and private lands or migrated to new sources of supply or simply invested in other businesses.

At the same time, the exploitable timber that remained on national forests was of such marginal value relative to the costs required to get at it that the will to exploit these remaining remote areas declined. Other than a few exceptionally profitable national forests in the Northwest and Southeast, most national forest timber sales from the 1960s through the 1980s—the era of intensive timber production—cost the government more money than they returned to the Treasury in timber sale revenue. Since World War II the government had willingly supported a high level of national forest timber sales. The Forest Service took the opportunity to use timber values to pay the cost of constructing an extensive road system that the agency felt was needed for administrative and firefighting purposes. But the escalating costs and maintenance problems associated with building roads over remote and often rugged terrain to reach increasingly marginal stands of timber eventually eroded the political support of fiscal conservatives who otherwise were quite probusiness. Thus, the lack of trees and lack of will to subsidize marginal logging lifted some of the pressure from the Forest Service and allowed the agency to shift back toward the "custodial" role it held prior to World War II, when timber demand was lower.[46]

This shift away from the emphasis on timber production is also due in part to increased workforce diversity in the Forest Service, as mentioned earlier.[47] Natural resource professionals from disciplines other than timber management have incrementally gained positions of influence in the agency and even achieved some top leadership positions in the past 15 years, including two recent chiefs of the agency, Jack Ward Thomas and Michael Dombeck. Also as mentioned earlier, change is partly attributable to the increased political clout of environmental organizations since the 1970s—a significant social development that many scholars have assessed.[48] Both of these new developments reflect macrolevel changes in society that have affected essentially all aspects of agricultural and natural resource politics. The forestry profession responded to this pluralization of the political environment and the broadening of employment opportunities by adopting new curricula at forestry schools that emphasize the whole forest, rather than just the timber crop, and an ecosystem approach to management. Whether these ideological shifts will control actual forest management for more than a brief transition period remains to be judged by future historians. Many of the incentives for a commodity orientation in forest management remain in place.

∎∎∎

In what Alan McQuillan has called a "circular century,"[49] foresters entered the 20th century at the vanguard of the forest protection and sustainable use movement; then at midcentury they shifted to facilitating rapid and environmentally damaging commercial exploitation of forests in a classic boom-and-bust pattern; and finally they exited the 20th century adopting an ecosystem management paradigm emphasizing forest health. Similarly, national forest management cycled from a custodial phase focused on conserving resources and suppressing fires at the start of the century; to a gung-ho intensive management phase focused on maximizing the provision of goods and services (motto: "Land of Many Uses") in midcentury; to the current posttimber bust and custodial phase focused once again on resource protection and fire suppression (motto: "Caring for the Land, Serving the People"). The social role and reputation of foresters and the Forest Service followed the same trajectory during the 20th century: from progressive reformers viewed as underdog heroes fighting corruption and greed; to bureaucrats in a politicized production machine serving special interests; to …perhaps servants of the public interest caring for the land. Time will tell.

Gifford Pinchot himself evolved through similar phases in his thinking about forestry and conservation in response to changing political and social contexts. In 1899, immediately after he was appointed head of the Division of Forestry, he wrote the following passage in his book *A Primer of Forestry*:

The forest is the most highly organized portion of the vegetable world. It takes its importance less from the individual trees which help to form it than from the qualities which belong to it as a whole. Although it is composed of trees, the forest is far more than a collection of trees standing in one place. It has a population of animals and plants peculiar to itself, a soil largely of its own making, and a climate different in many ways from that of open country. Its influence upon the streams alone makes farming possible in many regions, and everywhere it tends to prevent floods and drought. It supplies fuel, one of the first necessities of life, and lumber, the raw material, without which cities, railroads, and all the great achievements of material progress would have been either long delayed or wholly impossible.... The forest is as beautiful as it is useful.... From every point of view it is one of the most helpful friends of man. Perhaps no other natural agent has done so much for the human race and has been so recklessly used and so little understood.[50]

Pinchot wrote that while he was still young, idealistic, and fresh to the political arena. He and John Muir were still friends, and Muir was still enthusiastically promoting Pinchot. Teddy Roosevelt, Pinchot's great friend and ally, was not yet president, the forest reserves were still out of Pinchot's reach in the Interior Department, and western political rebellion over the establishment of the reserves was still at a manageable level. Pinchot's rhetoric would change rather markedly, however, when he was ensconced in a much more politicized environment. Only six years later, in 1905, upon the transfer of the reserves to Pinchot's direct control in the Department of Agriculture, after Roosevelt had added tens of millions of additional acres to the forest reserves, to the considerable consternation of western politicians who felt these lands were being "locked up," Pinchot took a much more practical and politically ameliorative position on forests and what should be done with them:

In the administration of the forest reserves it must be clearly borne in mind that all land is to be devoted to its most productive use for the permanent good of the whole people.... All the resources of the forest reserves are for use, and this use must be brought about in a thoroughly prompt and business-like manner, under such restrictions only as will ensure the permanence of these resources.[51]

Unquestionably, the changed political environment of 1905, in which Pinchot was holding the proverbial hot potato, altered the way he framed his pitch for conservation forestry. As Pinchot's investment in the political arena deepened, his utilitarianism correspondingly grew to dominate his public discourse. To fend off his critics and opponents in the resource development industries, he stressed efficiency, outputs, and profits as the primary aim of forestry. By the 1910s, he and Muir had parted ways over whether to dam the Hetch Hetchy Valley in Yosemite National Park and Pinchot had earned a reputation for opposing national parks and even advocating timber harvesting in Adirondack Park.[52] This utilitarian materialist orientation stayed with Pinchot throughout his career as a politician (at the national level and in Pennsylvania). But later in life, especially after his considerable political clout had begun to fade in the 1940s, Pinchot's rhetoric began to change again. His early support for preservation resurfaced in his advocacy for the protection of California redwoods; he grew increasingly critical of the forestry profession, which he felt had become entirely too cozy with the timber industry; his diatribes against corporate power grew less and less restrained; and in his 1940s revisions to his textbook on forestry, he added sections on ecology and specifically advocated ecological forestry of the type being pioneered by Aldo Leopold at that time.[53]

Pinchot's evolution in many ways presaged the evolution of the Forest Service itself. When little was at stake, the agency was free to take a broad vision of its mission, advocate reform, and challenge the status quo. When tremendous market and political pressures came to bear on the national forests at midcentury, the agency adopted a management program replete with justifying ideologies that catered to those pressures and strengthened its position and resources in that new production-oriented environment. By the 1990s, when the good timber was mostly gone, the maximum production ethos had been discredited, and environmental interests were gaining considerable clout in the political arena, the Forest Service adapted again. Likewise, the science of silviculture cycled through a parallel suite of paradigm shifts in response to similar historical conditions and sociopolitical pressures. Historical context powerfully molds both political rhetoric *and* scientific orthodoxy. Forestry is, has always been, and always will be as much a product of *culture* as it is of science.

The same is true of conservation. Throughout the century, conservation organizations embodied a broad set of values and hopes, ranging from responsible utilitarianism to mystical nature worship and everything in between. Conservation group leaders usually tried to reflect and balance this variety of orientations in a diverse agenda, advocating a mix of conservation practices that ranged from park and wilderness preservation to fish and game enhancement to campground development to responsible grazing and sustainable logging. The Forest Service *thought* it was balancing multiple uses and blending preservation with responsible development during its midcentury go-go years, but its dominance by a professional forestry culture that had succumbed to a timber-production obsession blinded the agency to its own biases, and its postwar technological optimism overwhelmed dissenting voices with a promise of cornucopia for all. Those days seem to be over now. But the future is not settled. We can hope that the two conservation visions—nature protection and sustainable development—will come together again, as they did in the late 19th century, in a new Forest Service culture founded on a forest science that accounts for whole ecosystems, a public that understands and values complex systems, and a political economy that resists demanding too much from America's ecosystems and their stewards.

Notes

* Portions of this paper are drawn from the author's book *A conspiracy of optimism: National forest management since World War Two* (Lincoln: University of Nebraska Press, 1994).

1. Susan Smallheer, Court blocks logging plans, *Rutland Herald*, June 10, 2005; http://www.rutlandherald.com/.
2. Still the best historical analysis of Progressive Era conservation is Samuel Hays, *Conservation and the gospel of efficiency* (Cambridge: Harvard University Press, 1959).
3. In 1992, the Forest Service officially adopted a new management paradigm, termed Ecosystem Management, to replace the postwar timber production model of forestry. For about a decade, observers have debated whether this is window dressing or a true and lasting reform of agency culture. One guardedly optimistic assessment of this organizational shift is found in James J. Kennedy and Thomas M. Quigley, Evolution of USDA Forest Service organizational culture and adaptation issues in embracing an ecosystem management paradigm, *Landscape and Urban Planning* 40 (1998), pp. 113–22. For a hopeful discussion of how ecosystem management paradigms are penetrating the U.S. forestry profession as a whole, see V. Alaric Sample, *Land stewardship in the next era of conservation* (Milford, PA: Grey Towers Press, 1991).
4. For good summaries of early American forestry and its relationship with European forestry, see Char Miller, *Gifford Pinchot and the making of modern environmentalism* (Washington, DC: Island Press, 2001), chs. 4–5; and David A. Clary, *Timber and the Forest Service* (Lawrence: University Press of Kansas, 1986), ch. 1.
5. One of the best studies of U.S. forest exploitation in the 19th century is Michael Williams, *Americans and their forests: A historical geography* (Cambridge: Cambridge University Press, 1989). On the role of foresters in the reform of government land laws and forest protection policy in the 1880s and 1890s, see Harold K. Steen's classic *The*

U.S. Forest Service: A history (Seattle: University of Washington Press, 1977), pp. 8–46. For a similar discussion covering a somewhat broader time period, see Clary, pp. 1–28. See also Hays, Conservation and the gospel of efficiency.
6. Gifford Pinchot, Breaking new ground, commemorative ed. (Washington, DC: Island Press, 1947, 1998), p. 27.
7. General Revision Act of March 3, 1891, ch. 561, 26 Stat. 1095, 1103.
8. Williams, pp. 309–15.
9. Bernhard E. Fernow, Economics of forestry (New York: Thomas Y. Crowell Co., 1902), pp. 166–72.
10. Fernow.
11. Act of June 4, 1897, ch. 2, 30 Stat. 34. For analysis and commentary on this act, see Charles F. Wilkinson and H. Michael Anderson, Land and resource planning in the national forests, Oregon Law Review 64(1,2), pp. 46–52; and Harold K. Steen, The origins of the National Forest System (Washington, DC: USDA Forest Service, 1991).
12. Fernow, pp. 169–70.
13. USDA Forest Service, The use book, 1906, p. 43.
14. Letter of Secretary of Agriculture James Wilson to Chief Forester Gifford Pinchot, Feb. 1, 1905. Reprinted in full in Pinchot's Breaking new ground (Harcourt, Brace and Co., 1947), pp. 261–62.
15. The second Pinchot quote is cited in Roy Keene, Salvage logging: Health or hoax, Inner Voice 5(2) (March–April 1993), p. 1.
16. Use book, p. 43.
17. Fernow, pp. 171–72.
18. Henry Solon Graves, Principles of handling woodlands (New York: John Wiley and Sons, 1911), pp. 69–70, 83–87.
19. Hawley, The practice of silviculture, 1926, pp. 40–41, 102–03; Hawley, The Practice of Silviculture, 1946, pp. 92–94, 163.
20. Paul W. Hirt, A conspiracy of optimism: National forest management since World War Two (Lincoln: University of Nebraska Press, 1994), chs. 2–6. See also Clary, pp. 110–25.
21. Earl H. Clapp, Public forest regulation, Journal of Forestry 47(7) (July 1949), pp. 528–29.
22. For a discussion of the Forest Service's shift to clearcutting in the 1950s and the resulting controversy over it, see Clary, pp. 180–88; Hirt, pp. 245–51, 260–63.
23. George W. Craddock, Watershed management problems of the Intermountain West, Proceedings: Society of American Foresters Meeting, 1958 (Washington, DC: SAF, 1959), p. 32.
24. Edward C. Crafts, Congress and the Forest Service, 1950–1962, an oral history conducted in 1965 by Amelia Fry, Regional Oral History Office, The Bancroft Library, University of California, Berkeley, 1975, pp. 43–44. Courtesy of the Bancroft Library.
25. In 1903–1904, 19 students were enrolled in undergraduate programs in forestry at U.S. academic institutions. Ten years later, 1914–1915, there were 904 undergraduates enrolled. By 1929–1930, the number had risen to 2,123, and enrollment stayed generally at that level for the next four years. Partly because of demand for foresters from the Civilian Conservation Corps, enrollment jumped to 6,067 between 1934 and

1938. Enrollment crashed during World War II to a low of 503 in 1943–1944. During the four academic years following World War II, 1945–1949, enrollments soared from 1,473 to 7,010 to 7,454 to 8,212. Ralph S. Hosmer, Education in professional forestry, ch. 17 in Robert K. Winters, ed., *Fifty years of forestry in the U.S.A.* (Washington, DC: SAF, 1950), pp. 313–14. The enrollment data were actually compiled and contributed by Cedric H. Guise. Beginning in 1934, Guise contributed annual articles to the *Journal of Forestry* titled, Statistics from schools of forestry: Degrees granted and enrollments.

26. William Hagenstein, a prominent national spokesman for the lumber industry from the 1940s through the 1970s, has said, "Those of us hired by industry used the tree farm program as a vehicle to get public support for good [fire and pest] protection, for reasonable taxation. At the time there was a drive on in the United States—a political drive by the Roosevelt Administration—to allege that a long-term crop like timber couldn't be handled by anyone except the government. The government either would have to grow timber on its own lands alone, or would have to regulate the private owners. And there was nobody in our industry who looked with favor upon the idea of the federal government coming in and telling us how to do it. So the tree farm program was in part a vehicle to build up some public confidence that here was an industry prepared to do the job of managing these lands." William Hagenstein, oral interview with Ray Raphael, *Tree talk: The people and politics of timber* (Covelo, CA: Island Press, 1981), p. 33. On the issue of public regulation of private forest practices as it was debated by the Society of American Foresters, see Shirley W. Allen, ch. 15 in Winters, pp. 276–78. On the Forest Service's support for public regulation of private forest practices, see Clary, pp. 104–07, 147–49.

27. Writing in 1950, a prominent Yale forestry professor said, "Until the last decade American foresters were mostly federally employed …" Herman H. Chapman, Forest management, ch. 4 in Winters, p. 79. Also, a long-tenured secretary of SAF noted in 1950 that the organization's membership before World War II consisted mainly of federal and state employees, but by 1950 their predominance had dwindled. Shirley W. Allen, ch. 15 in Winters, pp. 272–75.

28. In a 1993 oral interview, retired forest ecology professor Robert Zahner recalled this important transition period. When he got his master's degree from Duke University in 1950, he had dozens of job offers from industry, he said, while in contrast only a small number of jobs were available with the Forest Service. Industry also paid better wages, so the cream of the academic crop went into the private sector. He distinctly remembered a transition in silvicultural orthodoxy occurring in the late 1940s during the last two years of his bachelor's degree program: His professors and recent textbooks were beginning to emphasize "even-aged" management over selective cutting. New research supposedly showed the benefits of that system. Personal interview with Robert Zahner, November 14, 1993, Prescott, Arizona.

29. See Hirt, chs. 4, 6, and 9 for a discussion of the "maximum production" ethos of the postwar era, the dramatic increases in timber production budgets for the Forest Service, and the effect of both of these on national forest management and on

resources other than timber. For a survey of these critical postwar developments as they played out in the inland Northwest forests, see Nancy Langston, *Forest dreams, forest nightmares* (Seattle: University of Washington Press, 1995), pp. 264–74.

30. A history of federal forest management in the southern Appalachians written by Shelly Smith Mastran and Nan Lowerre, *Mountaineers and rangers*, FS-380 (Washington, DC: USDA Forest Service, 1983) offers these telling comments about the switch to clearcutting: "In the early 1960s—under policy directives to increase National Forest timber production, with the support of long-awaited new silvicultural research findings, a more stringent need for economy and efficiency in harvesting, and with demand increasing from the region's pulpwood industry—clearcutting in patches (called even-aged management by foresters) became a more prominent practice ..." Clearcutting accounted for 50 percent of timber harvest volume in the East in 1969, more in the West; p. 144.

31. USDA Forest Service, *Report of the Chief*, 1948, p. 5.

32. USDA Forest Service, *Timber resources for America's future*, Forest Resource Report 14 (Washington, DC: GPO, 1958), pp. iii, 108.

33. On the growing split between the Forest Service and conservation groups in the 1950s–1970s, see Samuel Hays, *Beauty, health, and permanence* (Cambridge: Cambridge University Press, 1987), ch. 1, From conservation to environment, pp. 13–39; Grant McConnell, The conservation movement—past and present, *Western Political Quarterly* 7 (September 1954), pp. 463–78; and Hirt, chs. 7, 10, and 11.

34. E.P. Stamm, Converting the old growth forest: Utilization and road problems, SAF *Proceedings* (1955), pp. 5–7.

35. A.W. Greeley, Protecting the public's interest in converting the old forest to new, SAF *Proceedings* (1955), pp. 2–3.

36. Justice William O. Douglas, editorial in *Mazama* 43(13) (December 1961).

37. Stewart Udall, Wilderness, *Living Wilderness* 80 (Spring–Summer 1962), p. 4.

38. On the debate over wilderness within the Forest Service, see Clary, pp. 169–76; and Dennis Roth, *The wilderness movement and the national forests* (College Station, TX: Intaglio Press, 1995).

39. Clary, pp. 177–94.

40. See, for example, *"Clearcutting" practices on the national timberlands*, hearings before the Senate Committee on Interior and Insular Affairs, 92nd Cong., 1st Sess., May 7, 1971; A university view of the Forest Service, prepared for the Committee on Interior and Insular Affairs, U.S. Senate, by a select committee of the University of Montana, 91st Cong, 2nd Sess., Dec. 1, 1970, Senate Document No. 91–115 (often called "The Bolle Report"); USDA Forest Service, Wyoming Forest Study Team, *Forest management in Wyoming*, reprinted in hearings on *"Clearcutting" practices on the national timberlands*, pp. 1116–201; Daniel R. Barney, *The last stand: Ralph Nader's study group report on the national forests* (New York: Grossman Publishers, 1974); and *Report of the president's advisory panel on timber and the environment* (Washington, DC: GPO, April 1973). Dennis C. Le Master discusses these clearcutting studies and related events in chapter 2 of his excellent study, *Decade of change: The remaking of Forest*

Service statutory authority during the 1970s (Westport, CT: Greenwood Press, sponsored by the Forest History Society, 1984). A more recent critical reflection on this controversy from an eyewitness at the Montana School of Forestry is found in Richard W. Behan, *Plundered promise: Capitalism, politics, and the fate of the federal lands* (Washington, DC: Island Press, 2001), pp. 151–58, 192–94.

41. On the revolt within the Forest Service, see Forest Service, *Sunbird Proceedings: 2nd National Forest Supervisors' Conference, Tucson, Arizona, November 13–16, 1989*; James J. Kennedy, Richard S. Krannich, Thomas M. Quigley, and Lori A. Cramer, *How employees view the USDA-Forest Service value and reward system* (Logan, UT: College of Natural Resources, March 1992); Paul Mohai and Phyllis Stillman, *Are we heading in the right direction? A survey of USDA Forest Service employees, executive summary* (Ann Arbor: School of Natural Resources and Environment, University of Michigan, March 1993); Greg Brown and Charles C. Harris, The U.S. Forest Service: Toward the new resource management paradigm? *Society and Natural Resources* 5 (July–September 1992), pp. 231–45. On the recent revisions in forestry orthodoxy, see Sample, *Land stewardship in the next era*; also Richard W. Behan, Multiresource forest management: A paradigmatic challenge to professional forestry, *Journal of Forestry* 88(4) (April 1990), pp. 12–18; Jerry Franklin, Toward a new forestry, *American Forests* 95(11–12) (November–December 1989), pp. 37–44; and Hal Salwasser, Gaining perspective: Forestry for the future, *Journal of Forestry* 88(11) (November 1990), pp. 32–38.

42. Hirt, ch. 12. For three excellent accounts by deeply embedded journalists, see William Dietrich, *The final forest: The battle for the last great trees of the Pacific Northwest* (New York: Simon and Schuster, 1992); Kathie Durbin, *Tree huggers: Victory, defeat, and renewal in the Northwest ancient forest campaign* (Seattle: Mountaineers Books, 1996); Richard Manning, *Last stand: Logging, journalism, and the case for humility* (Salt Lake City: Peregrine Smith Books, 1991).

43. Alan McQuillan, From Pinchot to post-modern—The circular century of U.S. forestry, *Inner Voice* 4(6) (November–December 1992), pp. 10–11.

44. Jerry Franklin, Toward a new forestry, pp. 37–44.

45. On the changing orthodoxy in the Forest Service and the forestry profession as a whole, see Hal Salwasser, Gaining perspective: Forestry for the future, *Journal of Forestry* 88(11) (November 1990), pp. 32–38; Jeff DeBonis, New perspectives, *Inner Voice* 2(4) (Summer–Fall 1990), pp. 8–9; Memo of F. Dale Robertson to Regional Foresters and Station Directors, re: Ecosystem management of the National Forests and Grasslands, June 4, 1992; Hanna J. Cortner and Margaret Ann Moote, Ecosystem management: It's not only about getting the science right, *Inner Voice* 5(1) (January–February 1993), pp. 1, 6; Sample, *Land stewardship in the next era*; Elliot A. Norse, *Ancient forests of the Pacific Northwest* (Washington, DC: Island Press, 1990), ch. 8: Sustainable forestry for the Pacific Northwest, pp. 243–70.

46. The issue of below-cost timber sales has an extensive literature. The Natural Resources Defense Council initiated public attention to this issue in the 1980s with a report by Thomas J. Barlow, Gloria E. Helfland, Trent W. Orr, and Thomas B. Stoel, Jr., *Giving away the national forests: An analysis of Forest Service timber sales below cost* (New York:

NRDC, June 1980). A U.S. General Accounting Office investigation followed up on the NRDC report: Report to the Congress: Congress needs better information on Forest Service's below-cost timber sales, GAO/RCED-84-96, June 28, 1984. Contemporary with the GAO report was a strongly critical analysis by the Wilderness Society written by economist V. Alaric Sample, *Below-cost timber sales on the National Forests* (Washington, DC: The Wilderness Society, 1984). A rather defensive Forest Service response to the controversy is found in Ervin G. Schuster and J. Greg Jones, Below-cost timber sales: Analysis of a forest policy issue, *General Technical Report INT-183* (Ogden, UT: USDA Forest Service, Intermountain Research Station, May 1985). Responding to a request from Congress, the Forest Service came up with a Timber Sale Program Information Reporting System (TSPIRS) in 1987 to compare costs and benefits. The best critical analyses of TSPIRS and timber program economics in general have been written by natural resource economists Robert Wolf and Randal O'Toole. See O'Toole's *A critique of TSPIRS*, cowritten with Randy Selig, CHEC Research Paper 20 (Oak Grove, OR: CHEC, November 1989) and *Growing timber deficits: Review of the Forest Service's 1990 budget and timber sale program*, CHEC Research Paper 23 (Oak Grove, OR: CHEC, April 1991). Robert Wolf offers an equally detailed critique and a historical evaluation of the below-cost timber sales issue in National forest timber sales and the legacy of Gifford Pinchot: Managing a forest and making it pay, *University of Colorado Law Review* 60(4) (1989), pp. 1037–78.
47. Jennifer C. Thomas and Max Weintraub, Change in the U.S. Forest Service's workforce: A quantitative analysis of workforce trends in the past ten years; and Paul Sabatier, John Loomis, and Catherine McCarthy, Factors affecting output levels in NFMA plans; both in *Proceedings: Society of American Foresters 1993 National Convention* (Washington, DC: SAF, 1994). Also, see the September–October 1994 issue of *Inner Voice* 6(5). *Inner Voice* is a bimonthly journal on Forest Service reform, and this issue is dedicated to workforce diversity within the agency.
48. One interesting though dated look at the influence of environmentalists on public land management is Paul Culhane, *Public lands politics: Interest group influence on the Forest Service and the Bureau of Land Management* (Baltimore: Johns Hopkins Press, 1981). For a more recent assessment, see Sabatier et al., 1994. Other excellent studies include Hays, *Beauty, health, and permanence*, and Roth.
49. McQuillan.
50. Gifford Pinchot, *A primer of forestry* (1899), pp. 7–8; excerpted in Harold K. Steen, ed., *The conservation diaries of Gifford Pinchot* (Durham: Forest History Society, 2001), p. 96.
51. Letter of Secretary of Agriculture James Wilson to Chief Forester Gifford Pinchot, February 1, 1905 (written by Pinchot), reprinted in full in Pinchot's *Breaking new ground*, pp. 261–62.
52. On Pinchot's utilitarianism, see Hays, *Conservation and the gospel of efficiency*, ch. 3; Michael Frome, *The Forest Service* 2nd ed. (Boulder, CO: Westview Press, 1984), pp. 21–23; and Patricia Limerick, *The legacy of conquest* (New York: W.W. Norton, 1987), chapter titled Mankind the manager, especially pp. 297–300. Most recently,

Char Miller's biography of Pinchot recognizes the complexity of the man, but also traces his philosophical evolution and "political two-step" that led to his alienation from Muir: *Gifford Pinchot*, ch. 6, pp. 119–44.

53. Char Miller traces much of Pinchot's intellectual evolution near the end of his life in the final chapters of *Gifford Pinchot*. In very recent work on Pinchot and Muir, John Perlin has traced a significant ecological impulse in Pinchot's early and later life work. See his essay in this volume.

ABSTRACT

At the end of the 19th century, as Gifford Pinchot and the European foresters were articulating their concept of the "wise use" of natural resources, another, more contemplative idea was being developed—an ethical approach to man's responsibility toward nature. When this approach was confronted with the rapidly expanding science of ecology between 1930 and 1970, it divided itself into two branches, each claiming very different philosophical foundations but more or less sharing the same view of forestry, both quite different from that of Pinchot. Public opinion has shaped this debate of ideas and is dominated by an urban perspective, with ideas about nature that borrow, in some way, from the Romantic period. Today, perceptions of forestry are influenced by the powerful image of tropical rainforests, which have become the archetype of an ideal forest, with utilitarian, ethical, and aesthetic values, raised to the rank of a world heritage site. In confronting this, the managers of temperate forests must choose between two alternatives: either advocating a cultivated forest model that protects so-called natural forests, or creating a new social and cultural consensus around managed forests that can represent, according to each person's desire, shared values of utility, ethics, and aesthetics.

CHAPTER 9

Man, Nature, and Forest: The Great Debates of Ideas

Christian Barthod

When it comes to forest management, foresters are responsible not only for the growth and development of trees, but also for ideas, dreams, and a vision of man, nature, and the forest. A journey into the history of these ideas and visions over the past century will illuminate the questions and concerns of today that must be answered tomorrow.

At the end of his training period in Europe, Gifford Pinchot[1] shared with the great majority of French and German foresters a steadfast faith in technique, progress, and sustained economic growth. He was convinced that solutions could be found for environmental problems and resource scarcity without necessarily having to significantly alter the status quo. Pinchot also believed that success relied on competent forestry consultants backed by enlightened governmental policies. With their utilitarian approach and vision of social welfare, forest engineers focused on maximizing for society a broad expanse of goods and services, thus exalting the civilizing effect of technology on a fundamentally perfectible nature that was neither immutable nor optimal.

In the second half of the 19th century, foresters of western Europe were faced with forests that had been degraded by almost 6,000 years of human pressure; since that time, this degradation has slowed, largely because of the implementation of ingenious modes of forestry management. Even before Pinchot tried to adopt the dynamic methods applied by the European foresters in his own country, other Americans were raising their voices in protest of mankind's negative effects on nature. Henry David Thoreau and John Muir[2] had called for a strict respect for nature based on the aesthetic values and the religious experience that came largely from their "discovery" of the forest. This reawakening stirred in both Thoreau and Muir an extreme sensitivity to the grandeur of nature and to individual and collective responsibility: nature should be affected little or not at all by human activity. (Contrary to the European situation, Thoreau and Muir could admire a nature without long and great human influence.)

It was the colonial experience in equatorial Africa that gave French foresters their first encounter with forests where human activity could be considered negligible and nature considered impressive. From this shock of the *tremendum et fascinans*, strong experiences emerged, of two perspectives: respect for the awe-inspiring unknown and attempting to accommodate to it through new forestry techniques, or a desire to brutally harvest it, wiping the slate clean and preparing the ground for the familiar plantation. But it was also through contact with the tropical humid forest and its fauna that Albert Schweitzer,[3] an Alsatian familiar with European forests, felt an emotional shock leading him to enlarge the ring of ethics beyond interpersonal relations to include respect for life, in all its forms, and for all the living beings met in nature. The legacy of Schweitzer's ethic was immense, especially in Anglo-Saxon countries.

For Schweitzer, as for Thoreau and Muir, initial wonder was followed by a broadening of vision in discovering something beautiful that had existed prior to man. A strong link of aesthetics and ethics connected these three men through the mediation of a personal and profound religious experience—one commonly associated with the Judeo-Christian ethic: God as creator, man His creature, nature His creation. The logical conclusion of this view is an expression of man's duty toward nature.

In 1939, a North American forester named Walter Lowdermilk[4] extended this line of thinking by announcing the "11th commandment," adding the duties of man toward nature to those toward the creator and fellow man. This early concept of "stewardship" called for defining the relationship between man and nature. The term insists not only on the moral duty of man toward nature, but additionally associates with it the idea of service, the connotation of legal guardianship of a defenseless minor, as well as a religious background with the image of the "good shepherd."

But Aldo Leopold,[5] another North American forester who had observed forests, deer, and wolves, came to new and audacious conclusions entirely opposed to those of Schweitzer: first, that respect was accorded the upper levels of organization and integration—what could be considered the "whole," which was more than the sum of the parts—and not the individual parts, which may die; and second, that the system's integrity (and not only the living components) was far more important than the well-being of individual organisms. The ecological processes behind the species and the environment are the keystone of an equilibrium in which human beings, much like other species, are only beneficiaries and form a component. The maximum respect must therefore be given to "virgin nature," which is the "perfect norm."

The land ethic departed from the Judeo-Christian tradition and linked with some profound and ancient beliefs still alive in the tradition of North America's Indian communities. This approach has grown and developed during the past 30 years, in various and at times radical forms, with reflections at the interface between ecological sciences and philosophical and ethical views, and contributions to national and international public debates. The outcome is expressed in calls for the rights of nature.

The rupture caused by Leopold is considerable, since it resulted in the following dictum with both ethical and operational implications: "A thing is right when it tends to preserve the integrity, stability, and beauty of the biotic community. It is wrong when it tends otherwise." The land ethic represents an important departure on two accounts: it is ecofocused and no longer anthropocentric, and its logic has identified new rights that are not tied to obligations, as they are in the Judeo-Christian tradition. From a legal point of view, nature can then be construed as an entity endowed with rights but placed under the responsibility of a guardian, who is authorized to defend it in court and to demand, in its name, restitutions in case of harm. The North American old-growth forests, which were endangered by forest harvesting, were an appropriate and highly symbolic field of application for such an approach. It is not by chance that the first article taking a stand on this question was titled "Should trees have standing? Towards legal rights for natural objects."[6]

The classic anthropocentric approach of the wise use of natural resources is then replaced by a desire to characterize the intrinsic value of nature. For the first time in modern history, an approach that is meant to be scientific implies moral action, as in the ancient Greek quest for the articulation of *ethos* and *physis*. Nature is seen or can be seen both as a model and as a norm and can be thus the object of true religious fervor. It is difficult at this stage to predict whether the mystique of wilderness, already so developed in North America with its attendant considerations of man's moral regeneration through contact with a virgin and wild nature, will assert itself in Europe on a long-term basis. All that can be underlined is that some elements of it exist, notably in the Scandinavian and German cultures (shared in France in Alsace and in a part of Lorraine), less in Latin cultures.

Parallel to the scientific and cultural departure that put the whole above the sum of the parts was another scientific and cultural revolution, a corollary to the first: in the 1970s it became evident that one could hurt a living being without attacking it directly—by attacking its habitat. There was no longer an irreducible conflict between those who favored the individual living beings and those who favored the integrative view of the ecosystem and its ecological processes and functions.

René Dubos,[7] a French agronomic engineer and world-renowned microbiologist, moved to the United States and during the 1960s and 1970s attempted to synthesize the tradition embodied by Schweitzer and Lowdermilk with Leopold's ideas about the primacy of ecological processes and functions. He accepted Leopold's view and regarded disruptions and diseases as normal, even essential processes of the functioning biosphere, in this way discounting the traditional concept of the balance of nature. At the same time, he tried hard to defend an enlightened anthropocentrism. In so doing, he accepted the principle of manipulating nature in the name of mankind's best interests, on the basis of the "stewardship" concept espoused by Lowdermilk regarding man's duties toward nature, a view widely accepted by traditional Judeo-Christian societies.

Michel Badré, a contemporary French forester, notes that European foresters, who essentially pursued the "good use" of forests until the 1980s, were then confronted by the sudden emergence of what has become a goal as well a yardstick: the "good state" of forest. The simultaneous attainment of these two sensibilities, good use and good state, is called for in the criteria of the second Ministerial Conference for the Protection of Forests in Europe (Helsinki 1993). But the subsequent E.U. conservation laws based around the "Natura 2000" network of natural heritage sites (notably the 1992 E.U. directive on habitats, fauna, and flora and the E.U. framework directive on water) turns only on the logic of the good state, which is as yet not clearly defined. Badré hypothesizes a link between this new good state and raised awareness of quickening global changes since the early 1980s, whereas previously, the overall environment was considered invariable over the course of forest rotations.

The divergences are no longer between what forests are or what they should be or whether they are seen as an image of nature, but rather, to what extent it is legitimate to manipulate forest species and their habitats and how to evaluate the impact on the ecological processes and functions (while integrating disruptions and crises). Now, the forest is seen as a community of interdependent living beings, with no a priori distinction among useful, useless, or harmful species. And an essential task for foresters is to guarantee the biological stability, integrity, and resilience of forest ecosystems, which should be manipulated only with the greatest care.

It is interesting to note that the basis of the most famous French forest maxim of the 19th century, Imiter la nature, hâter son oeuvre ("Imitate nature, accelerate its work," Louis Parade, 1862), is not very different from Leopold's view of nature, but it places more emphasis on the model than on the norm. This approach to forest management is deeply rooted in the concept of "good use" of the forest and the forest models that have been developed in accordance with it.

Urbanite Nature and Country Nature

Nature has been domesticated since Neolithic times in western Europe. In modern urban societies, nature is no longer something most people experience, but instead a tame object observed from afar or on summer Sunday afternoons. Tropical rain forests have become familiar but in an artificial way—on the television screen. From this distance emerges an urban confusion that makes it difficult to perceive the range of possible interactions, negative but also positive, between man and nature.

Nevertheless, a distinction between "countryside" (land highly manipulated by agricultural activities) and "nature" is still alive in the European urban imagination. In the West, the Romantic period of the first half of the 19th century saw a prevailing culture whose people sought out the senses, the imaginary, the irreplaceable personal experience, and nostalgia. The Romantic quest sometimes confused beautiful and good, embraced oriental cultures and mystics, and dissolved the difference between dream and reality. Modern culture—essentially an urban Romantic culture—favors individualism and perceives nature both as a unique whole and as a source of happiness that is now otherwise elusive. This view is very distant from the traditions of rural people, who mainly (but not exclusively) see in nature a source of truly concrete and directly useful goods and services, as essential to their lives as air or water.

Between the two main traditional views of nature, modern man, as a latter-day romantic, tends to choose the organic view and renounce the mechanistic view that prevailed in forest engineering for so long a time. In the same romantic way, he cultivates the nostalgia of a wild and mystic nature, incarnating "the world's soul." Just as Romanticism emerged from the Enlightenment of the 18th century and from the supreme reign of reason, modern man is the offspring of the triumph of science from 1945 to 2000; he doesn't deny science and reason, but he seeks an antidote to a promethean view of man and world, soul-destroying and terrifying. There exists a need to "re-enchant" the world through nature.

Sociologists declare that wild nature has become the most shared reference in public opinion in the West. The current debates on sustainable management and biodiversity are regularly put in perspective with cultural and technical models that claim wild nature as reference, not historically "humanized" nature.

The preferred European forest scenery gives the impression of not having evolved; it is "natural," unchanging, and out of time. Since the 1970s, major conflicts in France have been rooted in the quick evolution of landscapes (a consequence of clearcuts), artificial reforestation, and natural fallow. This evolution has been rejected by both the city dwellers and country people, possibly because it drastically simplifies what should be complex and not betray any visible human impact. In this way, the most famous forest model in Europe has become the selection forest, because it gives the untrained eye the illusion of constancy.

Managers of natural ecosystems must not underestimate the prevailing urban cultural universe, without ignoring still-strong resistance in the rural world. These two types of cultures coexist— sometimes in the same person— to different degrees and logic, depending on the moment and the degree of personal involvement. Rural residents have access to modern channels of information that are creating a new concept of nature, at the same time that city dwellers cultivate, probably more in France than elsewhere, the nostalgia of their fading rural roots. Furthermore, the accelerating geographic mobility in Europe can only further destabilize what is still a relatively coherent and specific rural culture. The fact that retired people who have once worked in town become decision makers in small local jurisdictions is already an indicator of a probably irreversible change.

But the attachment to rural culture, claimed by the urbanites even when their understanding of nature is completely different from that of the rural population, could explain why forest debates have not yet experienced the paroxysms of the North American spotted owl crisis. If the progressive obliteration of this old, rural cultural heart continues inexorably, will we no longer anticipate radicalization of the conflicts in the media and the law? Are the "ecowarriors" of Fontainebleau or the "green panthers" of Saint-Germain en Laye mere epiphenomena, copies of North American movements that will disappear as quickly as they appeared? Or are they the first manifestation of a new underground force, representing a part of the urban public opinion that has not yet found its expression in Europe, as was the case in 1850–1860 in the forest of Fontainebleau near Paris?

The New Paradigm: The Tropical Rain Forest

Humid, tropical ecosystems abound with a luxury of forms, behaviors, colors, and sounds—a profusion of diversity. The biologist Portman says, "nature doesn't only offer the image of structures in the service of their own conservation and of the species, but a display of richness, and this is called liberty." In these forests, at least as portrayed on television, beauty is associated with laughter and joy; the forest frees men from time and work, finally restoring freedom. It is something of a lost paradise; and the presence of a small number of famous specie that have somehow escaped the mercenary and frighteningly destructive forces of the world, only reinforces the idealization of an universe in which man and nature would live in harmony.

The great floral and faunal richness of tropical rainforest ecosystems, with species yet to be discovered and identified and biodiversity's role in the functioning of these ecosystems still not understood, has given tropical rain forests a renewed utilitarian defense. Calls for saving the rainforest are founded on the precautionary principle: plants with compounds that may be one day necessary for man may exist in tropical rainforests and are just waiting to be found. This approach reduces the value of these forests, and at the least relativizes the most common uses—wood, game, minor products, and land area to be cleared for the food-producing agriculture.

Unlike the climate change debate, in which scientists have succeeded in shaping public opinion through reason and simplified predictive models, the biological diversity debate has not succeeded in swaying the public by offering an "icon"—the tropical rainforest—to the world. The argument confuses people by calling the rainforest crucial for the biosphere, and therefore for humanity, without identifying how to maneuver for action and by admitting that the goal is, at best, only to slow down or stop a degradation that is already well along. This type of forest presumes an ethical duty.

The facts that tropical rainforests are under constant assault and that forest harvesting is limited to half of the cases of deforestation or degradation diffuses culpability in western minds, since intervention in these forests is based on economic reasons. In modern references to nature we also see a rejection of the economic ideology, which is perceived as fundamentally contrary to ethics, a denial of the primacy of the individual, who is essentially only a producer. On the other hand, the finite extent of natural resources confirms the interdependence of man and nature.

What seems to me to be most important is that the western, urbanized perception of tropical rainforests reconciles the three values underlying all human judgment and sensible action: usefulness, ethics, and aesthetics. This perfect synthesis can elevate an object to the rank of a world heritage site not only in the eyes of experts and "enlightened" citizens but also for public opinion as a whole. Thus, to many people, the tropical rainforest has been internalized as the forest archetype, and as such it changes their view of all other forest types, notably temperate forests.

What About Temperate Forests for Tomorrow?

We have a choice:
- either temperate forests are managed for one part as nature reserve resembling the ideal of the tropical rainforest and for the other part as "cultivated forest" becoming acceptable to the public as preserving natural forest from harvesting; or
- new types of management and silviculture must create an alternative synthesis, acceptable to the public, of utilitarian, ethical, and aesthetic values different from those recognized for tropical rainforests.

The correct proportion between the two types of options is itself an important societal choice that must be taken into account in every coherent and socially acceptable forest policy.

In the current public international debate, temperate forests—at least those in Europe and surely those in the eastern and southern parts of the United States—are indeed caught between the image of tropical rainforests (as sold by the media) and the less clearly focused picture of boreal forests. Boreal forests and rainforests are the two big types of forest where man has been harvesting and destroying ecosystems and where positive human action has been less pronounced.

The first alternative has only two approaches that seem to be worth trying: a priori zoning of a region, with intensive plantations to meet legitimate human needs and lessen the pressure on "natural" forests. Its ideology is agronomic, and it mobilizes the most recent agronomic research to create a system of high productivity in an almost artificial environment. Pinchot's comparison—when he likened silviculture to growing crops of corn (maize)—could be acceptable at this price for our modern urban societies. The more productive the plantations are, the better their effectiveness in protecting natural forests from harvesting.

Such an option necessitates, nonetheless, a social consensus on the a priori zoning of the land, and thus its practical feasibility turns on the landownership structure and the forest owners. This consensus depends also on the deintensification of forest management or even the classification of vast areas as strict reserves, in exchange for the intensive management of small areas without any thought to biodiversity. In this model, such an option is conceivable only for North America and certain limited areas of Europe, such as the maritime pine forest area of Aquitaine.

A derivative of that model could rely on "degraded" forms: less intensive plantations that are nonetheless managed, zoned a posteriori, ending in a "leopard pelt" landscape. This landscape would reflect the diversity of owners' choices. The great number of private forest owners with an ancestral link to their family land, as well as their eventual capacity to generate products of value to the country, would create a very positive view of the "cultivated forest" in the landscape seen by travelers and tourists. But in the framework of modern landownership structure and law in France, such an outcome is very difficult to realize.

The second alternative presupposes that the public can share with foresters a similar vision of maintaining the utilitarian, ethical, and aesthetic value of each of type of temperate forest.

As far as utility is concerned, the problem is getting beyond concerns about only rotations and employment while developing the temperate forest's contribution to socially desirable goals, such as human health and water quality, and restoring wood's noble image by distinguishing it from the quickly perishable and disposable products that are emblematic of waste.

As far as ethics is concerned, we must voluntarily accept the idea that certain areas should be designated as strict reserves, intended for restoration of a "virgin nature," and then engage in a continual debate—public and transparent, at both national and local levels—about the appropriateness of manipulating different types of forest ecosystems and the consequences on living beings and ecological processes and functions. These two preconditions will be difficult to satisfy but are probably not out of reach for forest partners, given that the stakes—avoiding the paralysis of forest decision making—are so high.

By far the most complex job will be translating an aesthetic perception of the forest into the terms of forest planning and management. Foresters generally feel ill at ease when the aesthetic argument intervenes in a decision-making process. They entrench themselves behind rational arguments that hide their own emotional attachment to the forest. The aesthetic dimension of forestry is almost always present, even when hidden or relativized. We must emphasize the significant cultural dimension of aesthetics, rooted in the diversity of people's histories and the physical and biological environments that shape our outlook and experience—the diversity that enriches Europe. We must also appreciate the diversity of opinions in every region, and often the different appreciation of urban and rural dwellers. Each has a particular view about what constitutes a "beautiful forest." But all remember their emotions when standing in a forest landscape and want to reconnect with those feelings.

The first aesthetic challenge requires us to stand *outside* the forest. This is the challenge of seeing the landscape from beyond the forest, from roads and edges and tourist vistas, and most of all, identifying the landscapes that are changing too quickly. Landscape has become both a goal and a symptom. It is the element of nature most easily seen and most often shared by urbanites. Hence the challenge of landscape planning: a long-term, prospective vision for the landscapes structured by forests. Following the pioneering British work, forest managers are grasping this problem. But foresters must accept an approach that integrates their action in a wider geographic scope before they can return to manipulating the intraforest landscape.

The second aesthetic challenge involves the feelings of liberty and appreciation for the complexity and beauty aroused by forest stands. The question of free access to forests is a social, technical, and legal issue, at least for a culturally Latin country. But in a contrasted and somewhat provocative way, the expansive accessible areas of the Landes forest inspire more aesthetic appreciation than the large enclosed areas of Sologne, even if the latter are surely more diversified. Furthermore, the liberty and the beauty of both have something to do with being pleasantly surprised; this agreeable surprise comes from the lack of monotony and a certain harmony in the elements of the perceived landscapes.

The third aesthetic challenge is reconciling silviculture and the "natural" forest—the perception that man lives alongside but does not dominate nature. Some planted high forests, those that are no longer monocultures, inspire this feeling because of the irregular spatial distribution of trees, but to a lesser extent than high forests that developed from natural regeneration. Selection forests, in all their rarely respected orthodoxy, are forests managed through silviculture that is "close to nature": they can satisfy the requirements. The complexity of stands formerly managed as coppice-with-standards, with cuts that have been relatively erratic for several decades, can also be satisfying. The diversity met at three levels—in the stands, in the understory, and in both the visible and the hidden fauna—is incontestably an indicator of "naturalness" for our fellow citizens. But naturalness also presupposes the presence of overmature trees or stands and senescent trees or stands as well— the spectacular, colossal growth that indicates longevity. It remains difficult to tell which silvicultural regimes are the most compatible with public perceptions of naturalness, but each of us can identify a great number of options that are *in*compatible.

Conclusion

Nature is both a scientific and a cultural object. During the past century, scientific progress has been significant, but western foresters may underestimate the cultural evolutions that have touched the societies in which they work. We can no longer ignore the movement of ideas and sensibilities that affect public opinion, which from now on will be predominantly urban. The public view of forests is more cultural; that of foresters is more scientific. As Leopold advocated, the manager of natural environments must therefore develop an ecological engineering strategy that integrates the cultural dimension. This engineering must necessarily encompass three dimensions—utilitarian, ethical, but also aesthetic—to make managed forest types, in public opinion, more comparable to the "nature" symbolized by the tropical rainforest.

Forests, as well as seas, are the great symbols of nature for industrialized and urban societies. The debate that emerged in the 1970s over the actions of man in the forest and has continued ever since was therefore inevitable. This debate, which has questioned, challenged, and sometimes even attacked foresters, is part of a debate far more vast, both ancient and very modern, on the relationship and interdependence between man and nature. It mobilizes the latest scientific experience in service of a quite different thesis, something between the Judeo-Christian view and the resurgence of ancient and profound trends that are difficult to characterize but could provisionally be termed shamanistic.

Given the different forest contexts in old Europe and young North America, the debate has not been, and could not be, the same. During the past century, creativity has been more North American than European. But old Europe has caught up in globalization and scientific and cultural strength. The questions now are much the same on both shores of the Atlantic. There is no solution except to work on a synthesis of the best part of the European forest tradition—the "good use" we associate with Pinchot—and the "good state" that Leopold formulated on the basis of scientific ecology. But this synthesis will doubtless not be the same for the European and North American foresters, considering the types of forests and certain cultural legacies. The nature of western Europe will never quite be the same as that of North America.

American problems and questions are European problems and questions, or will be soon, but solutions will probably not be exactly the same. Nevertheless, the challenges we both face are a good basis for a stronger cooperation between American and European foresters.

Notes

1. Gifford Pinchot (1865–1946): responsible for the American federal forests from 1898 to 1910, close friend to President Theodore Roosevelt, founder of the U.S. Forest Service in 1905, influenced by his long journey of forest training in Europe, notably in Nancy in 1890, before his shorter stays in Switzerland and Germany.
2. Henry David Thoreau (1817–1862): American pacifist influenced by Hindi mysticism and German idealism, theoretician of nonviolence and civil disobedience, famous writer who lived in the vicinity of Appalachian forests already affected by human influence and celebrated nature and its benefits; John Muir (1838–1914): American pacifist, advocate of the beauties of wild nature in the Sierra chain and old-growth forests of the Pacific coast, promoter of the idea of national parks to preserve still-inaccessible regions, proponent of the first California park (Yosemite) and founding father of the Sierra Club, which today has more than 400,000 members.
3. Albert Schweitzer (1875–1965): French clergyman, philosopher, and theologian, musicologist and organist, doctor who founded the Lambaréné hospital (Gabon), winner of the Nobel Peace Prize in 1952.
4. Walter Clay Lowdermilk (1888–1974): American forester who studied in England (Oxford) and Germany before he was assigned to Montana, then specialized in soil erosion control in China and California, and finally joined the Soil Conservation Service of the U.S. Department of Agriculture.
5. Aldo Leopold (1887–1948): U.S. Forest Service forester, then professor of flora and wildlife management at the University of Wisconsin in Madison.
6. Should Trees Have Standing? Towards Legal Rights for Natural Objects, by Christopher Stone (1972), in *Southern California Law Review* 45: 450–57.
7. René Dubos (1901–1982): became a naturalized American in 1938, was inventor in 1938 of the first antibiotic commercialized worldwide (gramicidin), coauthor of the basis report ("We only have one earth") of the first international conference on human environment (Stockholm, 1972), at the establishment of the United Nations Environment Program (UNEP).

ABSTRACT

European forests play a prominent role in timber production, nature protection, water conservation, erosion control, and recreation. For centuries, temperate forests in Europe have been affected by forest devastation and soil degradation. During the past 150 years, regeneration efforts have been undertaken to eliminate severe wood shortages. Today's high growth rates and growing stocks indicate that that goal has been met, but at the cost of a shift to non-site-adapted tree species, an increasing average stand age, and a reduction in resistance to storm, snow, ice, droughts, insects, and fungi. Society now demands sustainable forestry, emphasizing biodiversity and close-to-nature forest management. Forest managers must find new ways to cope with uncertainty and risk, using adaptive and flexible strategies. Diversity in site conditions, ownership, economic, and sociocultural conditions requires strategies adapted to local and regional needs. Pursuing sustainable forestry will ensure higher resilience of forests, which in turn will increase the economic and social benefits of forests and reduce the risks.

CHAPTER 10

The Evolution of Forest Management in Europe

Heinrich Spiecker

Forests in Europe cover a large bioclimatical range, from the mild Atlantic zone in the west to the rough continental zone in the east, from the Mediterranean zone in the south to the boreal zone, as well as from floodplain to timberline. The number of tree species, however, is relatively small. Genetic diversity was reduced during glacial times, and immigration during the warmer interglacials could not compensate for the extinctions during the cold periods. Human land use also had a major effect on tree species composition. Beginning at the end of the 13th century and especially during the past 200 years, conifers have been introduced on sites naturally dominated by broad-leaved species. European temperate forests today are the result of centuries of vast human activities. Because site conditions, ownership, and cultural, economic, and social conditions vary at short distances across Europe, forest management is likewise different. Three phases of forest management can be distinguished: forest exploitation, forest reestablishment, and forest conversion.

Forest Exploitation

Most of Europe was once naturally forested, but forest area in central Europe decreased drastically starting at about 1000 B.C. and continued to decrease until the end of the 18th or even the end of the 19th century. Forests offered many resources: forage and litter for livestock, land for agroforestry, shifting and swidden cultivation, and forest pasture, as well as nuts and

Figure 1. Pig feeding. Manuscript illumination from the Grimani Breviary, early 16th century.

nonwoody plants and fruits. Forests were also a source of industrial raw materials, such as woody and nonwoody biomass for fuel, potassium carbonate from ash for glass production, shrubby and herbaceous vegetation and bark for tanning and fiber, resin and tare for various purposes, and timbers for mining. Wood was an important source of energy. Fuelwood

Figure 2. Collection of fuelwood. Wood engraving by Albrecht Dürer, 1483.

was obtained by coppicing or by collecting dead woody biomass, branches, and often stumps. Charcoal was produced in the forest or at landings from woody biomass, often including branches and shrubs. Finally, wood was an important raw material for construction.

As wood became scarce, it was extracted from deeper in the forest by horses or rafted over long distances. Because of this exploitation, many forests in Europe were overused, devastated, or destroyed. Dense and unbalanced game populations, stimulated by the hunting ambition of nobility, worsened forest conditions.

Forest Reestablishment

Reestablishment of exploited and devastated forests. During the 19th and 20th centuries, the urgent need to reestablish forests became apparent. To eliminate the severe wood shortages of those days (Kirby and Watkins 1998), successful countermeasures were taken to regenerate and tend highly productive forests that would cover larger areas and supply more goods and services. Forest management plans were established with special emphasis on safeguarding a sustainable supply of wood.

The area of European forests has been increasing up to the present day. From 1980 to 1995, forest area increased by 4.1 percent; in developing countries it decreased by 9.1 percent (FRA 2000). Increases in forest area can be found in almost all countries in Europe. In some countries, the area of forest available for wood supply has decreased but the area not harvested has increased (TBFRA 2000). In the European part of the Russian Federation, the area of forests increased by 1.2 million hectares during the period 1966 to 1998 (Pisarenko et al. 2000). In Denmark, the United Kingdom, Ireland, and the Netherlands, forest cover accounts for only about 10 percent of the total land area, but in Austria, Estonia, Latvia, Liechtenstein, Slovakia, and Slovenia, forest cover exceeds 40 percent. In Austria, Ireland, Poland, and the United Kingdom, more than 60 percent of the forest area is primarily coniferous; in Bulgaria, Croatia, France, Hungary, Luxemburg, Romania, and the Republic of Moldova, more than 60 percent of the forest area is dominated by broad-leaved species. In most countries, "high forest" is by far the most common silvicultural category, but in Bulgaria, France, Liechtenstein, the Republic of Moldova, and Ukraine, coppice forests and coppice with standards account for more than 40 percent of the forest area. Most high forests are even-aged. Only in some countries—for example, Croatia, Germany, Liechtenstein, Slovakia, Slovenia, and Switzerland— does the area of uneven-aged forests amount up to about 15 percent (TBFRA 2000). Currently, about 26 percent of the total land area of Europe is covered with forest, much of it fragmented.

Coniferous species, primarily Scots pine (*Pinus sylvestris* L.) and Norway spruce (*Picea abies* [L.] Karst.), were often favored because they were easy to establish and manage and were expected to yield high volumes of timber. The advantages of coniferous species included their low cost of planting and relatively low susceptibility to deer browsing, and foresters understood the management of these forests. With faster-growing species substituted for less productive species, not only did wood production increase, but revenues increased even more because softwood commanded higher prices than hardwoods.

Today, the area covered by coniferous species expands far beyond the limits of their natural range. Consequently, the current species composition of temperate forests in Europe is mainly determined by former management rather than by natural factors (Ellenberg 1986). European forests of today are the result of centuries of human activities. Most of the forests have been planted or seeded, and species composition has been subject to drastic changes. In central Europe broad-leaved species dominated the forest, covering

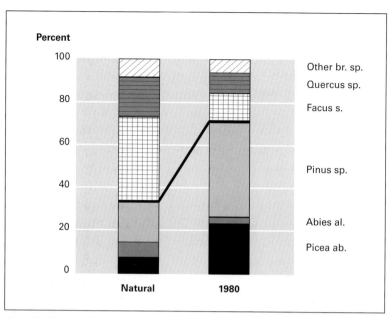

Figure 3. Changes in tree species composition in central European forests. The area of broad-leaved species has been reduced while the area of conifers has increased (Mayer 1984).

approximately two-thirds of the area, until the mix was reversed on many sites by human activities. The area of European beech (*Fagus sylvatica* L.), oaks (*Quercus* sp.), and other broad-leaved species has been reduced while the area of Scots pine and Norway spruce has increased substantially. Douglas-fir (*Pseudotsuga menziesii* (Mirb.) Franco) has also been introduced.

Norway spruce and Scots pine are today's economically most important tree species in the temperate forests of Europe. The highest coverage of Norway spruce in European temperate forests is found in Austria, where 350,000 hectares, more than 25 percent of the total land area, is covered with Norway spruce and Scots pine forests outside their natural ranges (Prskawetz 2000). A rather high coverage of Norway spruce can be found in the Czech Republic and Slovakia, with 15 to 25 percent of the total area and more than 25 percent of the forest area. In Switzerland and Germany, Norway spruce covers 10 to 15 percent of the total land and more than 30 percent of the forest land. In these two countries spruce has been often planted within its natural range in the mountains but also outside, especially in hilly regions and lowlands. The western range of Norway spruce has been expanded considerably, and now there are substantial areas of Norway spruce beyond its natural range. In Belgium, Luxembourg, the Netherlands, and Denmark, where only 10 percent of the land is forested, Norway spruce accounts for 50 percent of the forest area. Norway spruce is also grown in Great Britain, Ireland, and parts of France, again beyond its natural range. In fact, the best growth is often found outside its natural extent (Schmidt-Vogt 1987). Although Norway spruce is said to be adapted to continental climate, warm winters seem not to have a detrimental impact; long growing seasons may even increase growth rates. Scots pine has become popular on dry and temperate continental sites.

According to the national forest statistics of several European countries (e.g., Kuusela 1994), annual removals—thinnings and final cuttings—have remained at the same level during the past few decades, and the net annual increment is 40 to 50 percent higher than the fellings. Only in exceptional cases have removals reached the level of net annual increment. In some central and east European countries such as Estonia, Latvia, Lithuania, Slovakia, Rumania and Poland removals have recently increased substantially and even surpassed the allowable cut (a figure based on the sustained yield). Today, in several countries in central Europe, average growing stock exceeds 300 m^3 per hectare. In Austria, average growing stock per hectare amounts to approximately 250 m^3 (FBVA 1995), and in Switzerland, approximately 330 m^3 (Schweizerisches Forstinventar 1988). In 1952, for forests in the former West Germany, Mantel (1952) suggested a sustainable long-term growing stock of approximately 170 m^3 per hectare. From 1950 to 1990, average growing stock increased in West Germany from approximately 105 m^3 to about 300 m^3 per hectare (Wiebecke 1955; BML 1993). Even though annual removals increased considerably in the 20th century, average growing stock has continued to increase. Net annual increment has been substantially underestimated—partly because of conservative forest management planning but also because of unexpected changes in forest productivity. The productivity of European forests has increased substantially.

Average growing stock in exploitable European forests has increased during the period 1950–2000 from 98 m^3 per hectare to 143 m^3 per hectare (Kuusela 1994; TBFRA 2000). European wood resources are greater now than at any time during the past 200 to 300 years.

The increase in the area planted to conifers in the past contributes to today's observed increase in wood volume growth of European forests. The

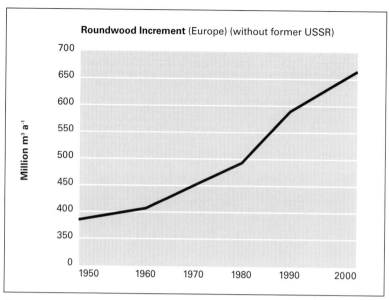

Figure 4. Annual wood increment in European forests. Annual wood increment has increased considerably during the past 50 years (Kuusela 1994; TBFRA 2000).

increased average growing stock and the unexpectedly high growth rate can be explained by several factors: inconsistencies in former inventory methods, changes in forest area, different tree species composition and age composition, and changes in site productivity and management intensity (Päivinen et al. 1999).

Current status of European forests. A forester from a hundred years ago would be excited about the state of today's forests. These forests prove that his goal—high wood supply—has been achieved. It took appropriate tree species selection, fertilization, active regeneration, intensive tending, and an increase in forest area, all contributing to a substantial rise in forest productivity as well as an increase in growing stock.

There are also, however, some developments that are not in line with expectations. Our forester from 1900 did not realize that older trees might be more susceptible to damage and disease than younger trees. In 2004 the Ministry of Consumer Protection, Nutrition and Agriculture reported that in the years from 1987 to 2004, more than 30 percent of Norway spruce older than 60 years showed considerable damage, while only less than 10 percent of Norway spruce younger than 60 years were classified as considerably damaged (BMVEL 2004). The high percentage of mature coniferous trees means a higher susceptibility to fungi and storms as well as a reduced adaptability to environmental change.

Changes in ecological conditions. Improved measurement methods and a higher awareness of environmental issues have focused greater attention on changes in environmental conditions. The global increase in atmospheric concentrations of carbon dioxide (CO_2) is one of the best-documented environmental changes. The drastic increase to about 350 ppm and its possible impacts are widely discussed. The predictions of some scientists indicate that this increase in atmospheric CO_2 will continue, and in the year 2100, CO_2 concentrations could exceed 500 and even reach 1,000 ppm.

Surface air temperatures, which have undergone wide fluctuations in the past, have globally increased in recent decades, as seen from Antarctic ice cores and atmospheric measurements (IPCC 2001). Scenarios for the future generally assume that air temperature will continue to increase globally by one to six degrees Celsius until the year 2100 (IPCC 2001). There are regional differences, however. When we compare the global air temperatures and precipitation in Europe over time, we see not only a rising air temperature and a relatively small long-term change in precipitation, but also considerable periodic variations leading to a higher frequency of extreme conditions. These extreme events do not always occur all over Europe at the same time but show a large spatial variation, with sequences of years of extremely dry and hot weather as well as extremely cold and wet conditions. The summers of 1947 and 1949 were dry and hot, as were the summers of 1976 and 1993 to 1998. One of the hottest and driest summers in recent climate history—June to September 2003—affected vast parts of Europe from Lisbon to Moscow. On the other hand, unusually cold and wet summers were observed in 1941 and from 1955 to 1958, as well as from 1966 to 1970, and 1987 and 1988. These sequences suggest that today's extreme events may tend to persist for a longer period.

Changing ecological conditions alter the growing conditions of forests. Trees do not grow as expected based on yield tables developed when the climate was different. Growth changes with changing environmental conditions. Trees may not show the well-known age trend with volume growth that declines with increasing age. In some cases we even may observe an increasing volume growth.

When subsequent tree generations at the same location were compared, case studies detected accelerated growth (Kenk et al. 1991). An increasing

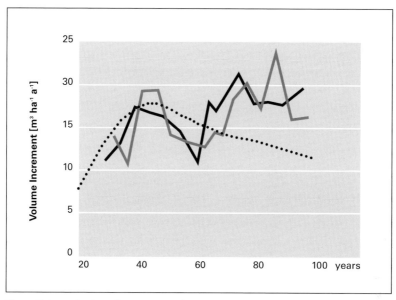

Figure 5. Annual volume increment per hectare over tree age on two permanent research plots. Annual periodic wood volume increment increased with increasing age, but yield tables would suggest a different age trend (dotted line) (Spiecker 2001).

site productivity was also evident when trees that germinated at different times showed a time-related change in height growth (Untheim 1996). Studies across Europe have shown that growing conditions have changed globally. The growth of forests in Europe has accelerated considerably in recent decades. This growth acceleration can not be explained solely by different inventory methods, increased forest area, changing species or age composition of the forests, or intensified forest management (Spiecker et al. 1996). The increase in site productivity may be explained by various causes, such as nitrogen deposition, CO_2 increase, or higher air temperature and therefore longer growing season. In addition, changes in land use may have contributed to the increase. Litter raking, for example, was once widely practiced and caused nutrient depletion but has not been much used for 50 years. Although site productivity may change naturally, the potential causes of increased site productivity are related to human activities.

Besides the long-term changes in sites, extreme events such as storms, heavy snowfall, or drought and heat may stimulate insect and fungi attacks and eventually kill trees. The heavy storm "Lothar" at Christmas 1999 destroyed many forests in central Europe, and Norway spruce was more affected than other species. Statistics describing damages from storms showed an irregular pattern, but the total volume of salvage cuttings increased overall. In some regions, salvage cuttings after one event were more than two times the annual allowable cut. Besides local variations in the severity of storms, high salvage harvests may be partly caused by changes in species composition, increased growing stock, higher average age, and the greater height of the trees. Ice and snow can also necessitate salvage cuttings of up to 40 percent of the annual allowable cut. In recent years salvage cuttings due to snow and ice were relatively small. Compared with salvage cuttings after damaging storms, salvage cuttings of desiccated trees and trees killed by fungi and insects were relatively low. Even when "forest decline" was a concern, salvage cuttings in, for example, the public forests of the Black Forest remained below 10 percent of the annual allowable cut. After 1990, when forest decline ceased to be an issue, the salvage cut volume rose to 20 percent, prompted by a sequence of dry and hot summers. The most severe event was summer 2003, after which the number of trees classified as severely damaged increased drastically. Increased susceptibility to desiccation and fungi and insect attacks is clearly related to the drought and heat of 2003, but after-effects of recent storms, such as unplanned openings and reduced management intensity outside damaged areas, have to be taken into account as well.

Numerous studies prove the strong relationship between air temperature, precipitation, and forest growth. Thus, in the Black Forest, growth was reduced substantially during and after the dry summers of the 1940s and the 1970s. When soil moisture levels improved again in cold and wet summers, growth rate increased (Spiecker 2001).

The growing conditions of the forests in Europe are continuously affected by changing site conditions and changing species composition. The extent of the changes and the increasing impact of human activities on these changes are new, however. Increased globalization and our populations' high mobility increase the danger of introducing invasive species that spread uncontrolled. Already, some exotic species have had a substantial impact on the dynamics of forest ecosystems.

The changing context. It is not just ecological conditions that are changing. Economic, social, cultural, and political conditions are subject to change as well. Changing markets, labor costs, and production techniques have an impact on forest management. Although the market value of wood has stagnated at a remarkably low level, labor costs increase continually. This development has spurred a drastic cutback in labor input into forestry, a process accelerated by technological progress. Through production of higher-valued timber and lower management costs, the profitability of forestry has improved. Environmental services, such as wildlife habitat, water, and climate change mitigation, are increasingly appreciated not only by the public but also by forest owners. Recreational and aesthetic values, as well as cultural values, are gaining in importance.

Globalization of markets affects the economics of the forest sector in Europe. One indication of this development is the increase in forest area due to the expansion of the European Union—by 53 million hectares in 1995, by 24 million hectares in 2004, and by another 10 million hectares expected in 2007. Worldwide destruction of forests, especially in developing countries, continues while the forest area in Europe is increasing. Economically less attractive agricultural land is transformed into forest. Conversion of coniferous monocultures on sites once dominated by broadleaves into mixed, site-adapted forests is under discussion—and not only for public forests. The global dimension and the increased speed of changes underline the increasing relevance of these processes for decision makers.

Social conditions are subject to change as well. Recently, forest management has been affected by the relatively decreasing importance of forests' commodity functions as energy sources, production and transport technologies, and wood utilization all change. In addition, with rising living standards, conservation, environmental, and social and cultural services have become more important. Demographic changes have an impact as well. Changing social demands today require a wider scope for forest management, and principles of sustainable forest management, forest laws, directions, and regulations are being reviewed (e.g., resolutions H1 and H2 of the Helsinki Ministerial Conference in 1993). Society wants to be more involved in forest decision-making processes. Interest groups with varying influence on these processes are getting involved. The groups and their values are changing, and the values of the society are changing as well. As a result, the objectives of forest management are changing.

Forest Restoration and Future Management

Forest conversion. Society is demanding sustainable forestry that emphasizes biodiversity and close-to-nature forest management. The discussion about the future of the forest sector has never been more intensive and politically oriented than it is today. Historically, tree species composition has been subject to drastic changes. Now the question is whether the forests of Europe should be converted to their original cover, broad-leaves. Motivating the reverse change in tree species composition is evidence of the disadvantages associated with pure coniferous forests under certain conditions: the risk of increased mortality in droughts, the relatively high susceptibility to storm damage, necessitating more salvage cuttings, and the higher susceptibility to disease and insects. In addition, a shift in tree species composition closer to that indicated by current climatic and edaphic conditions is expected to increase the capacity of forests to adapt to changing climatic conditions. Increased hardwood prices relative to softwood and the popularity of close-to-nature forestry are stimulating this shift as well. Volume growth is until recently judged less important, and wood quality is receiving more attention. The higher ecological and economic value of slower-growing species may at least partly compensate for losses in volume growth (Ripken and Spellmann 1979; Brandl 1988).

Since conversion affects almost every aspect of forest management, there is an urgent need for well-founded information regarding the conversion of secondary coniferous forests (Spiecker et al. 2004). In several conferences at the international level (Hasenauer 2000; Klimo et al. 2000; Hansen et al. 2003), the question of conversion has been discussed extensively, and national research initiatives are investigating the consequences of conversion. Research aims include a better understanding of the functioning of pure coniferous forests versus mixed forests with varying proportions of tree species and of the ecological and economic implications during the conversion phase, which in most instances will last many decades. The potential of forests to provide sustained productivity and yet remain ecologically stable—in biodiversity, water regimes, and nutrient cycles—has to be considered. Mixed stands may use resources differently and may be more resistant to physical and biotic stress factors (Cannel et al. 1992; Kelty et al. 1992). Some species combinations can increase wood production (Mitscherlich 1978). Possible reasons include a change in canopy structure and light interception, the reduction in crown and root competition, increased litter decomposition rates, and enhanced mineralization rates (Kreutzer et al. 1986; Rothe et al. 2001). Aspects of soil degradation need to be considered, including soil acidification and the processes of decomposition, as well as microbial and fungal populations and the effect on carbon budgets and future timber supply.

The vitality and stability of secondary coniferous forests is affected by a complex set of stress factors that can be categorized as anthropogenic, biotic, and abiotic. All put pressure on secondary spruce stands, as well as on new mixed forests and forests undergoing conversion, and may entail forest management problems, economic losses, and changes in social functions. At present, the forest community has no integrated forest protection strategies to prevent impairment by such factors during each stage of conversion. The risks inherent in the conversion process need to be considered, as do measures to increase the resilience of coniferous stands in withstanding the complex array of stress factors, such as the increase in windthrow and bark beetle attacks when the canopy of old Norway spruce stands is opened.

Although the economics of temperate tree species under different scenarios are well known, the economics of conversion itself have not been investigated adequately on the European scale. This applies in particular to the conversion of secondary coniferous forests into mixed broad-leaved stands (Hanewinkel 1998; Knoke and Plusczyk 2001; Knoke et al. 2001). The conversion of even-aged coniferous forests not only requires investments but will also lead to a significant change in cash flow. However, the direction and dimension of this change will vary with the silvicultural target and the silvicultural conversion strategy. The political aspects of conversion have not been thoroughly addressed in surveys (Schraml et al. 2000; Stipp and Schraml 2001), although several studies about opinions and attitudes toward forests and forestry have been made in individual European countries (Lafitte 1993; Pelissié 1992; Schmithüsen and Wild-Eck 2000; Sondergaard and Koch 2000; Kramer 1998; Wiersum 1998; Terrasson 1998; Koch and Rasmussen 1998; Rametsteiner 1999; Finnish Forest Association 2000).

Change in species composition is generally rather slow because of the long lifespan of forest trees. Only in exceptional occasions, such as after storms or serious insect attacks, can this process be accelerated. A good indicator of change in the minds of managers is future plans for species competition. The species composition planning in the public forest of Baden-Württemberg in 1965 and 1993 is an example of such changes for the long term. Whereas in 1965 the goal was to have about 40 percent of forest area in Norway spruce, in 1993, this species had been reduced to about 25 percent of forest area (Riedl 1978; Moosmayer 1993). Clearly, the human-induced changes of the past will be at least partly reversed in the future. The full impact of the planned changes will not be visible in the overall species composition for some decades, however, because the percentage of conifers in existing younger stands is relatively high.

Managing the forest for the future. Today's forests are the result of past management. Because the conditions under which forest are managed are subject to continuous, often unpredictable changes, the optimal forest of tomorrow cannot be determined today. Likely it will differ from today's ideal. Long lived and changing but slowly, forests must meet the changing needs of the public, forest owners, and the wood industry, as well as other interest groups. Today, people seek close-to-nature forests. Forest managers must consider not only the economic outcome but also biodiversity, water quality, recreation values, and other ecological, social, and cultural services provided by forests. Some aims can be achieved simultaneously, others may conflict. The tradeoffs need to be made transparent and evaluated. Changes in natural, economic, social, cultural, and political conditions require new challenges for managing the forest of tomorrow.

One general concept may not apply to all forests. Site conditions, past management and its impacts, the structure and goals of the society, ownership, economic conditions, and the political and cultural framework will all be reflected in the management concept. Local and regional differences are inevitable. Since the natural, economic, and social conditions change with time and these changes are difficult to predict, it is not possible for us to define the forest that will serve best future generations. Forest management has to adapt to the uncertain future. Plans for forest management need to be reconsidered often. The resulting adaptations to plans will help redirect the development of a forest as it grows. Although the aims of forest management may change fast, absent natural hazards or massive human intervention, forests change slowly.

Decision makers dealing with the forest of tomorrow confront a multidimensional decision space. Much information is needed to explore this space, and not all of the required information is available. Therefore, purely rational decision making in a deterministic environment is not applicable. The longer the planning horizon, the greater are the risk and uncertainty. Accurately predicting the development of forest ecosystems using today's knowledge is not possible. Nevertheless, long-term planning is essential when dealing with such a long-lasting and slow-growing object as a forest, where today's management activities may be visible many decades later. Some goods and services may be provided by the forest in the near future but others will be available only in the long run. Wrong decisions may cause irreversible consequences. To deal with risk and uncertainty, we must be guided by the best available information and ethical values and use planning tools that allow the development of alternative scenarios, prevention, adaptation, and diversification.

Scenarios are useful for analyzing management options to show how well the various options meet future needs. Tomorrow's needs are not yet well defined, however, and in the meantime, decisions must be made despite the gaps in science-based information. Scientific analysis has to be combined with the perceptions and values of society. Models and scenarios may have to be based on assumptions when the facts are not known. Scenarios describing potential future conditions as well as future options and their consequences have to be created. The best forest management options will be those that in both short run and long term can react to future changes and adapt quickly. Thus, shaping and reshaping the forest is a continual challenge for meeting the needs of today's and future generations. Ultimately, the question is how such a slow-moving and complex system like a forest can adapt to new conditions and challenges.

Especially attractive are those scenarios that allow for new options whenever new information is available or values have changed. Those management systems should be preferred that have the best capacity to adjust to new decision spaces. The resulting goods and services have to be seen in the light of market conditions and values at the time when they are provided—which today cannot be predicted with certainty. Mechanisms for adaptation and self-regulation need to be established. Good adaptation can be facilitated by sensitive reaction to indicators of changes and unexpected events. We therefore need indicators of ecological resilience and early warning signs so that we can respond fast and flexibly to small indications of change and thereby head off irreversible changes. Although some components of the system may fail to adapt, others will adapt to future conditions and help the system survive. Automatic systems that monitor current conditions and describe ongoing changes are needed, since early countermeasures may with little extra cost prevent serious damage.

In addition, increased tolerance against errors and management failures improves the robustness of the system. One approach to improving the robustness and adaptability of the system is careful tree species selection. Species covering a large range of sites are most robust when the site characteristics provide optimal growing conditions. Where these specifications are not met—where tree species are growing at their limits and show little adaptability to changes or low resistance to extreme events—forests should be converted. Sites can change with time, and tree species growing on sites close to their optimum may be less affected by changes.

Risk can be reduced through diversity as well. The wide diversity of sites, forest types, forest ownership, and culture contributes to form a robust, adaptive forest sector. Robustness and resilience are also a function of the system's ability to self-organize and its capacity for feedback, adaptation, and learning. Continuous conversion may be too costly. Long-term vision and careful consideration are required. The multidimensional decision space has to be illuminated and explored, and options selected that do not result in negative consequences but are acceptable under various potential future conditions. Once selected, strategies should not be immutable: they should be continuously reshaped. Forest design is not only the task of today but a continual process of preparing the forest to satisfy future needs in anticipation of future challenges. Those management strategies are preferable that do not rely on permanence but instead take advantage of biological processes. Homogeneous, dense forests, for example, are not able to self-regulate, whereas mixed, differentiated forests may possess this ability. A reduced stocking density can increase drought resistance and increase tree stability. Site-specific and ownership-specific adaptations of the aims, type, and intensity of management will result in high diversity, since site conditions and ownership structure vary considerably in Europe, and people tend to prefer the type of forest they are used to (Schraml and Volz 2001).

Some principles may guide the decision maker in formulating management plans. The resilience of forests against storm, snow, drought, and pathogen attacks has to be improved. Risk reduction and maintaining site productivity are critical elements of management for the future, requiring a better understanding of the ecological interactions between forests and site conditions, including ecological services of forests. What is the added value of close-to-nature forestry? Do we just want to mimic processes in untouched forests? Today's forests are far from "natural" in their species composition, age structure, spatial structure, and site conditions. Artificial forests require scientifically sound management. Habitats for plants and animals need to be preserved and improved on various spatial scales without incurring high costs, including opportunity costs.

Because the results of past management may not be appreciated until much later, the question of conversion will be an ongoing issue. However, for efficient management, some vision for the future must be combined with a thorough analysis of scenarios describing future conditions and options and trends in future values.

Adaptability to future changes is the key to sound management. Forest managers have to adapt to changing demands for goods and services through proactive thinking, through marketing innovative products, and identifying market niches in cooperation with market partners. Alternative products include nature protection services. Our knowledge in this area is growing, and new demands for services come up. The carbon issue is an example of how forests can provide new services. Wood as a renewable resource may enter a renaissance as fuel becomes scarce. Environmental services such as water are already held in high estimation. However, forests in Europe provide a livelihood for many private forest owners, who hope to continue making their living from forestry even if subsidies are reduced. The cost of management must therefore be affordable, which means revenue must be higher than cost. The value-added chain from the forest to the consumer can be increased through better cost structures, new machinery, improved information technology, and more efficient logistics and product delivery. An adjustment of management aims and management intensity to the local conditions is essential as well.

Improved adaptability to social changes, including globalization of markets and demographic development, is required. A balance between use and protection of forests is essential. Where is the appropriate balance? Which forest satisfies these functions in the best way? Is the multifunctional forest, which has proved useful in the past in Europe, outdated? Will specialized forests provide more valuable goods and services than multiple-use forests? More information is needed before we can answer such questions. Efficient research and communication may help to solve this problem. National and international collaboration as well as interdisciplinary cooperation may improve the use of the existing capacities. We need to create arenas for wide, flexible collaboration for learning, feedback, and the building of adaptive management capacity. Transfer of knowledge helps various interest groups participate in the decision-making process.

Forest management in Europe has evolved over the centuries. Management systems have developed and improved, and also given us the opportunity to learn from past mistakes. It is time to share our experience with others, especially with those in countries where forest devastation compromises their citizens' future.

References

BML. 1993. Bundeswaldinventur 1986–1990: Volume I: Inventurbericht und Übersichtstabellen für das Bundesgebiet nach dem Gebietsstand vor dem 03.10.1990 einschließlich Berlin (West); Volume II: Grundtabellen für das Bundesgebiet nach dem Gebietsstand vor dem 03.10.1990 einschließlich Berlin (West).

BMVEL. 2004. Bericht über den Zustand des Waldes 2004—Ergebnisse des forstlichen Umweltmonitorings. Bundesministerium für Verbraucherschutz, Ernährung und Landwirtschaft.

Brandl, H. 1988. Entwicklung der Ertragslage der vier Baumarten Fichte, Kiefer, Buche und Eiche im Staatsforstbetrieb von Baden-Württemberg und ihr Einfluß auf die waldbauliche Planung. *AFJZ*, 195. Jahrgang, Heft 8, S. 164–170.

Cannel, M.G.R., D.C. Malcolm, and P.A. Robertson. 1992. The ecology of mixed-species stands of trees. Forest dynamics. Oxford University Press.

Ellenberg, H. 1986. Vegetation Mitteleuropas mit den Alpen. Ulmer, Stuttgart, Germany, 4th ed.

FBVA. 1995. *Österreichische Forstinventur. Ergebnisse 1986/90.* Forstliche Bundesversuchsanstalt—Waldforschungszentrum, Wien, Nr. 92.

Finnish Forest Association. 2000. Nordic forestry: The public opinion in Holland, Germany and Great Britain. http://www.smy.fi/tiedotteet/Demoskop99.pdf, 25.09.2001.

FRA. 2000: Global Forest Resource Assessment 2000—main report. *FAO Forestry Paper* 140.

Hanewinkel, M. 1998. Der Einsatz von Betriebsmodellen und Decision Support Systems für Überführungsvorhaben. In *Überführung von Altersklassenwäldern in Dauerwälder. Berichte Freiburger Forstliche Forschung* 8, 92–106.

Hansen, J., H. Spiecker, and K. von Teuffel. 2003. The question of conversion of coniferous forests. Abstracts of an international conference. September 27– October 2, Freiburg im Breisgau, Germany. *Freiburger Forstliche Forschung.* Berichte 47.

Hasenauer H. (ed.). 2000. Forest ecosystem restoration. Ecological and economic impacts of restoration process in secondary coniferous forests. Proceedings of the international conference in Vienna, April 10–12.

IPCC. 2001. Climate change 2001: Synthesis report, Intergovernmental Panel on Climate Change. http://www.grida.no/climate/ipcc_tar/vol4/english/index.htm.

Kelty, M.J., B.C. Larson, and C.D. Oliver. 1992. *The ecology and silviculture of mixed species forests.* Kluwer Academic Publishers.

Kenk, G., H. Spiecker, and G. Diener. 1991. Referenzdaten zum Waldwachstum. *Kernforschungszentrum Karlsruhe Projekt Europäisches Forschungszentrum für Maßnahmen zur Reinhaltung der Luft* 82.

Kirby, K.J., and C. Watkins. 1998. The ecological history of European forests. Oxford: CAB International.

Klimo, E., H. Hager, and J. Kulhavy. 2000. Spruce monocultures in central Europe—problems and prospects. *European Forest Institute Proceedings* 33.

Knoke, T., and N. Plusczyk. 2001. On economic consequences of transformation of a spruce (*Picea abies* (L.) Karst.) dominated stand from regular into irregular age structure. *Forest Ecology and Management* 15: 163–79.

Knoke, T., M. Moog, and N. Plusczyk. 2001. On the effect of volatile stumpage prices on the economic attractiveness of a silvicultural transformation strategy. *Forest Policy and Economics* 2: 229–40.

Koch, N.E., and J.N. Rasmussen (eds.). 1998. Forestry in the context of rural development. Final report of COST Action E 3. Horsholm: Danish Forest and Landscape Research Institute.

Kramer, C. 1998. Umweltbewußtsein in Europa—ähnliche Probleme, unterschiedliches Verhalten. *Informationsdienst Soziale Indikatoren* 20: 10–14.

Kreutzer, K., E. Deschu, and G. Hoesl. 1986. Vergleichende Untersuchungen über den Einfluß der Fichte (*Picea abies*) und Buche (*Fagus sylvatica*) auf die Sickerwasserqualität. *Centralblatt für das gesamte Forstwesen* 105: 346–71.

Kuusela, K. 1994. *Forest resources in Europe*. Research Report 1. European Forest Institute.

Lafitte, J.-J. 1993. Sondage d'opinion sur les forêts périurbaines. *Revue Forestière Francaise* 35(4): 483–92.

Mantel, K. 1952. Entwicklungstendenzen der westdeutschen Holzerzeugung und -versorgung. *Holz-Zentralblatt*: 2016–18, 2033, 2034.

Mitscherlich, G. 1978. Wald, Wachstum und Umwelt I: Form und Wachstum von Baum und Bestand. J.D. Sauerländer's Verlag, Frankfurt a. M.

Mayer, H. 1984. *Wälder Europas*. Stuttgart and New York: Gustav Fischer.

Moosmayer H.-U. 1993. Stand und Zukunft des Waldes in Baden-Württemberg. *Allgemeine Forstzeitung* 18: 950–56.

Päivinen, R., A. Schuck, and L. Lin. 1999. Growth trends in European forests—what can be found in international forestry statistics. In *European Forest Institute Proceedings*, edited by T. Karjalainen, H. Spiecker, and O. Laroussinie, 27: 125–37.

Pelissié, D. 1992. Les Francais et leur foret. *Arborescences* 38: 30–36.

Pisarenko, A.I., V.V. Strakhov, R. Päivinen, K. Kuusela, F.A. Dyakun, and V.V. Sdobnova. 2000. Development of forest resources in the European part of the Russian Federation. *European Forest Institute-Research Report* 11. Leiden, Boston, and Köln: Brill.

Prskawetz, M. 2000. Conditions for forest restoration in Austria—analysis based on forest inventory data. In *Forest ecosystem restoration: Ecological and economic impacts of restoration process in secondary coniferous forests*, edited by H. Hasenauer. Proceedings of the international conference in Vienna, April 10–12, 223–28.

Rametsteiner, E. (eds.) 1999. Potential markets for certified forest products in Europe. Brussels, March 13, 1998, *European Forest Institute Proceedings* 25: 57–73.

Riedl, W. 1978. Forsteinrichtungsstatistik 1961–1970 für die öffentlichen Waldungen in Baden-Württemberg. Teil II Auswertungen. *Schriftenreihe der Landesforstverwaltung Baden-Württemberg* Bd. 52.

Ripken, H., and H. Spellmann. 1979. Modellberechnungen der Reinerträge der wichtigsten Baumarten sowie der gesamten Holzproduktion in den Niedersächsischen Landesforsten. Aus dem Walde, *Mitteilung der Nds. Landesforstverwaltung,* Heft 30, S. 346 ff.

Rothe, A., C. Huber, K. Kreutzer, and W. Weis. 2001. Deposition and soil leaching in stands of Norway spruce and European beech: Results from the Höglwald research in comparison with other European case studies. *Plant and Soil,* submitted.

Schmidt-Vogt, H. 1987. Die Fichte, Band I, 2. Aufl., Paul Parey.

Schmithüsen, F., and S. Wild-Eck. 2000. Uses and perceptions of forests by people living in urban areas—findings from selected empirical studies. *Forstwiss. Cbl.* 119 (6): 395–408.

Schraml, U., and K.-R. Volz. 2001. Einflüsse auf die Situation und Entwicklung von Laubwäldern aus der Sicht der Waldeigentümer und forstpolitischen Akteure. *Institut für Forstpolitik, Untersuchungsbericht* 2001, Freiburg.

Schraml, U., J. James, and K.-R. Volz. 2000. Laubwälder und Gesellschaft: Ergebnisse einer Bevölkerungsbefragung, *Posterbeitrag für die Forstwissenschaftliche Tagung 2000,* October 11–15, Freiburg, Germany.

Schweizerisches Forstinventar. 1988. Ergebnisse der Erstaufnahme 1982–1986. *Berichte Nr. 305 der Eidg. Anstalt für das forstliche Versuchswesen.*

Sondergaard, J., and N.E. Koch. 2000. Measuring forest preferences of the population—a Danish approach. *Schweizer Z. Forstwes.* 151(1): 11–16.

Spiecker, H. 2001. Changes in wood resources in Europe with special emphasis on Germany. In *World forests, markets and policies,* edited by M. Palo, J. Uusivuori, and G. Mery. World Forests Volume III, Kluwer Academic Publishers, 425–36.

Spiecker, H., K. Mielikäinen, M. Köhl, and J.P. Skovsgaard (eds.). 1996. Growth trends in European forests. *European Forest Institute Research Report* 5, Springer-Verlag.

Spiecker, H., J. Hansen, E. Klimo, J.P. Skovsgaard, H. Sterba, and K. von Teuffel. 2004. Norway spruce conversion: Options and consequences. *European Forest Institute Research Report* 18. Leiden and Boston: Brill.

Stipp, F., and U. Schraml. 2001. Zur Existenz einer "öffentlichen" Meinung zum Wald. *AFZ/Der Wald* 56: 1154–55.

Temperate and Boreal Forest Resource Assessment (TBFRA). 2000. Forest Resources of Europe, CIS, North America, Australia, Japan and New Zealand. UN-ECE/FAO, Contribution to the Global Forest Resources Assessment 2000. *Geneva Timber and Forest Study Papers* 17. Geneva.

Terrasson, D. (ed.). 1998. Public perception and attitudes of forest owners towards forest in Europe. *Editions CEMAGREF,* Antony cedex, France.

Untheim, H. 1996. Zur Veränderung der Produktivität von Waldstandorten. Untersuchungen zum Höhen- und Volumenwachstum von Fichte (*Picea abies* [L.] Karst.) und Buche (*Fagus sylvatica* L.) auf Standorteinheiten der Ostalb und des Flächenschwarzwaldes. *Mitteilungen der Forstlichen Versuchs- und Forschungsanstalt Baden-Württemberg*, Freiburg i. Br., 198.

Wiebecke, C. 1955. Zum Stand der deutschen Forststatistik, Überblick über die gegenwärtigen forststatistischen Unterlagen Deutschlands. *Forstarchiv* 26: 1–8.

Wiersum, K.F. (ed.). 1998. Public perceptions and attitudes of forest owner towards forest and forestry in Europe. From enlightenment to application. *Hinkeloord Report* 24, Agricultural University Wageningen.

ABSTRACT

Forestry in Europe and the United States, after following separate courses of development over much of the 20th century, is now converging toward a common set of generally accepted principles of sustainable forest management, aimed at achieving common goals in forest conservation. Shared concerns for sustaining wood production, conserving wildlife habitat, protecting biological diversity, ensuring water quality, and mitigating climate change are bringing about important changes in the way society views forests, and the ways in which forestry professionals manage them. Citizens of the industrialized nations on both sides of the Atlantic are becoming more aware of the effects they are having on forests in other parts of the world through consumer demand and the net import of wood products. Sustainable forestry is not only about local issues and domestic policy making; it has become a global issue at the center of international trade negotiations and environmental policy. Government regulation to promote responsible forestry is being augmented by voluntary, market-based mechanisms that allow forestry enterprises in both the public and the private sectors to demonstrate their commitment to widely supported principles of sustainable forest management.

CHAPTER 11

The Emerging Consensus on Principles of Sustainable Forest Management: Common Goals for the Next Century of Conservation

V. Alaric Sample

At no time in history has there been greater interest in—and concern over—the future of the world's forests. Images from satellites orbiting the earth convey to us how finite our forests are, and how small a proportion of our planet is green. From this satellite imagery we can also see smoke plumes from fires that annually deforest areas of the tropics equal in size to some European countries.

Some of the world's most respected scientists report that plant and animal species, many of them dependent upon forest ecosystems, are going extinct faster than at any time since the cataclysmic extinction of the dinosaurs at the end of the Cretaceous period. Biologist E.O. Wilson (1992) has estimated that we are incurring a loss of approximately 27,000 species each year. Most of these are in tropical and subtropical forests, but a significant number are in the world's temperate and boreal forests.

With 20,000-year-old glaciers in Greenland and Antarctica melting at an extraordinary rate, scientists tell us that the long-term survival of human society as we know it today may be in jeopardy. The buildup of greenhouse gases in the atmosphere—ironically a product of the Industrial Revolution, which brought such prosperity to Europe and America—is leading to climate changes that include rising sea levels, prolonged droughts in major agricultural regions, and more frequent destructive storms. Fossil fuel use is not the only human-related source of greenhouse gases. It is estimated that between 23 and 27 percent of all greenhouse gas emissions stemming from human activities has come from the deforestation in tropical forests during the past two decades (Fearnside 2000; Houghton 1999).

Changes will be needed if we are to follow a different course. Approaches to forest conservation and sustainable use that worked well a half-century ago may be less well suited to circumstances a half-century in the future. World population will have more than doubled in that time, and with it the needs for all the goods and services that forests provide. Just as importantly, there is likely to be a similar increase in the need for agricultural production, living space, and other human activities that displace forests and lead to continued net decreases in forest area.

The recent evolution of forest policy in the United States has in many ways been a prelude to the policy debates now taking place at the global level. A new environmental awareness arose in the United States beginning in the 1960s. Emblematic of this period was the 1962 publication of Rachel Carson's *Silent Spring*, which documented the lethal impacts of pesticides and other chemical toxins on songbirds and other species. Carson's book also helped usher in a new era of social activism on behalf of the natural environment, with individual citizens taking to the streets and to the halls of government, forcing industries of all kinds to reduce pollution and environmental degradation.

This activism produced a number of landmark U.S. laws and government policies during the 1970s—such as the National Environmental Policy Act, the Clean Water Act, and the Clean Air Act—that provided an entirely new framework for reducing the negative impacts of human activities on the air, water, and land upon which we all depend. At times, it was government itself that became the target of this new environmental activism. Federal and state agencies created to implement environmental laws were challenged in court when they were seen to be lax in their enforcement. Other government agencies, particularly those responsible for managing public lands and natural resources, were challenged as never before to demonstrate that they themselves were not contributing to the degradation of important environmental values.

Forests and the policies governing them became an especially high-profile focus of this environmental activism. Concerns over clearcutting and other aspects of commercial timber production prompted a series of federal and state laws aimed at limiting or prohibiting controversial forest practices, protecting wilderness areas, and conserving habitat for wildlife, especially threatened and endangered species.

Although many of us tend to think of these issues as they relate to our own locale and experiences, they are but a reflection of the broader concern over these very same issues worldwide. For example, the creation of protected forest areas to conserve biological diversity is an issue in every country with significant areas of forest, whether tropical, temperate, or boreal. The Convention on Biological Diversity, an agreement developed at the United Nations Conference on Environment and Development—UNCED, the 1992 "Earth Summit" in Rio de Janeiro—was one of the first major efforts by nations around the world to reach a common understanding and develop a common strategy to conserving biological diversity in the world's forests.

Likewise, our struggle to develop solutions to these problems is also a reflection of similar efforts worldwide. Issues in the United States in the 1970s over clearcutting and the protection of environmental values in forests were but a precursor to the current global debate over "sustainable forest management." The global debate has been like a pendulum, swinging the focus of forestry from commercial uses to environmental values and finally to a recognition that in the long run, forests can be conserved only through a strategy that protects forest productivity and environmental values, maintains economic values, and deals squarely with perennial concerns over the fair and equitable use of these important resources.

Conceptual Foundations of Sustainable Forestry

Sustainability has always been a primary focus of forestry. Indeed, it was concern that forest resources would be inadvertently depleted, leading to unacceptable social and economic impacts, that first gave rise to the systematic study of forests and a scientific approach to the long-term management of these resources.

Sustainability in forest management began as both a biological and a socioeconomic concept. Foresters developed an understanding of natural forestry productivity—and how it might be enhanced through silviculture—to maintain a continuous supply of wood, game, and other products for human use and consumption. The concept was fundamentally driven by the desire to avoid the social and economic disruption associated with shortages of timber, whether for local use or as the basis for a community export economy. Forest products clearly held the potential of being a perpetually renewable resource, and foresters undertook the responsibility of making this so.

Introduction of sustainable forestry to the United States from Europe. The origins of "sustained-yield forest management" can be found in 18th-century Europe (Heske 1938). The lack of well-developed systems for transportation and communication at this time resulted in a system of small, independent political units with high customs barriers that prevented any significant degree of regional trade (Waggener 1977). Local consumption was almost entirely dependent on local production, and communities had to be largely self-sufficient. There was a distinct possibility of exhausting local timber resources unless collective use was strictly controlled, and the production and consumption of forest products became highly regulated. Perhaps because of the opportunities it afforded for employment and income in rural communities, this approach to sustained-yield forestry persisted long after improved transportation and communication systems had reduced the need for local self-sufficiency and turned timber into an ordinary economic commodity.

It was in this context that the concept of the "regulated forest" came into being: forests were to be managed to yield periodic, regular, and sustainable timber harvest volumes. In theory, the objective of maximizing the volume of timber that can be sustainably harvested from the forest is achieved by harvesting when the annual increment of growth in the stand reaches a maximum—the culmination of mean annual increment. This approach recognizes that trees begin their growth slowly, with only modest increases in annual volume, then increase their growth at an increasing rate, and finally, beyond some age, experience a decline in annual additions to tree volume. The culmination of mean annual increment determines the rotation age at which the sustainable harvest volume is maximized. The harvest level is determined using Hanzlik's Formula, which divides the net growth over the entire area of the economic enterprise by the rotation length and indicates the average annual volume of timber that can be removed on a sustainable basis.

In the mid-19th century, it was shown that forestry enterprises seeking to maximize *financial* returns would, in most cases, harvest younger trees on shorter rotations (Faustmann 1849). Although harvesting at the culmination of mean annual increment maximized the physical volume of the harvest, it could result in lower financial returns from the forest because it did not account for the cost of capital tied up in land and forest management expenses. This concept was controversial at the time and remains so today. In Europe, where centuries-old forest enterprises have endured through wars, currency devaluations, and other events that have put other forms of capital investment at great risk, forests have served as a stable, reliable, tangible asset. Increasing long-term financial security can be as much a concern as near-term financial returns. Building value in relatively low-yielding forest assets (compared with investments in stocks or bonds) has been shown to reduce risk in a diversified portfolio of investments and thus increase overall return (Binkley et al. 1996).

The concepts of sustained-yield forest management were transplanted to the United States at a time of growing concern over the possibility of a timber famine—nationally as well as locally. Forests in the United States had been regarded as both an inexhaustible resource and an obstacle to the westward expansion of agriculture. By the late 19th century, wood was still the major building material and the predominant source of fuel. Vast areas of forest had been cleared but not reforested, and there was a very real concern that a timber shortage would begin to seriously limit the prospects for future economic growth.

As introduced in the United States by Bernhard Fernow, Gifford Pinchot, and others, forestry was largely a technical undertaking. It was broadly assumed that by maintaining a continuous supply of timber and protecting the basic productivity of soils and watersheds, a broader set of forest uses and values would automatically be protected for the American people as a whole. Federal forestry reserves were established by the Forest Reserve Act in 1891. The purpose of the federal forest reserves, later called national forests, was to provide "the greatest good for the greatest number in the long run" (Pinchot 1947).

Custodial forest management. Management of the national forests was largely custodial until the mid-1940s. Preventing theft and wildfire was the major activity on the national forests of the western United States. In the East, large areas of cutover forest and abandoned farmland were acquired by the U.S. Forest Service and gradually restored through huge replanting, erosion control, and land stabilization efforts. Conversion of forest to other land uses was generally prohibited. Little timber was cut on the national forests during this period, in part because of political pressure from timber companies seeking to minimize competition in the private wood products industry and maintain favorable prices for private timber. Management of public forests emphasized maintaining the land in its native forest cover and relying upon natural regeneration following disturbances. The underlying biological and ecological systems were not well understood, however, as evidenced by the way wildfire was viewed at the time. Rather than recognizing that wildfire was part of a natural disturbance regime integral to the functioning of the forest ecosystem, the policy of the day was to suppress wildfires whenever and wherever they occurred. Thus, even custodial management requires a thorough understanding of natural disturbance regimes and other complexities of forest ecosystem functions.

By the 1940s, many private timber companies had also come to accept the idea of sustained-yield forest management. Previously, the standard practice had been to acquire forest land, liquidate the timber assets, and abandon it—an approach known as "cut out and get out," or more succinctly, "cut and run." With the leadership of corporate pioneers such as Frederick Weyerhaeuser, private timber companies began to recognize the benefit of holding land, reforesting it, and harvesting timber on a renewable basis. Today, the management of many private forestlands in the United States reflects the sustained-yield forestry of 19th-century Europe while utilizing modern technology and adhering to certain environmental constraints. With private timber supplies drawn down by the war effort, the federal forest reserves in the late 1940s and 1950s became a major supplier of timber for economic expansion and the suburban housing boom. Greater leisure time and improved transportation systems brought more Americans in contact with the national forests, increasing demand for recreation, wildlife, and other noncommodity resource values. With growing frequency, large-scale timber harvesting activities came into conflict with these other uses, challenging the operational utility of the traditional concept of sustained yield as the maximization of timber yield constrained only by the biophysical limits of the land itself.

Multiple-use forestry. The Multiple-Use Sustained-Yield Act (1960) was an important turning point in foresters' interpretation of their responsibility for sustainable forest management. It defined sustained yield as "the achievement and maintenance in perpetuity of a high-level annual or regular periodic output of the *various renewable resources* of the national forests without impairment of the productivity of the land" (16 U.S.C. 528; emphasis added). It has long been recognized that forests generate a host of goods and services simultaneously. Medieval forests were commonly valued for their game and forest foods as well as wood for both fuel and construction (Westoby 1989). Even when forests are managed for timber, other values are commonly produced as by-products. Fish and wildlife, recreation, water and water quality, range, and other outputs are commonly generated incidentally to the production of timber. It was this notion of producing multiple benefits for multiple uses that inspired the Multiple-Use Sustained-Yield Act.

The Multiple-Use Sustained-Yield Act provided the statutory basis for the application of this approach to U.S. public forests. Public controversies over Forest Service implementation of multiple-use forestry have led to additional statutory direction for sustainable management of the national forests. The Forest and Rangelands Renewable Resources Planning Act of 1974 required the development of periodic national assessments of the supply and demand for a large array of resource uses and values, plus a strategic plan detailing how the Forest Service intended to address all demands simultaneously (16 U.S.C. 1600). The agency's answer, in a word, was *money*; with significant increases in funding, investments in intensive resource management would allow all the renewable resources of the national forests—several of which competed and conflicted with one another—to be sustained indefinitely (Sample 1990).

In the decade following the passage of the Multiple-Use Sustained-Yield Act, the public grew increasingly dissatisfied with the balance the Forest Service had struck in addressing these competing sustainability goals. The predominant focus on timber production that had developed in the agency during the 1950s persisted, in large part because of budget allocations from Congress to sustain high harvest levels over other forest management goals. Public criticism suggested that such high levels of timber removal not only imposed unacceptable impacts on the nontimber resources but threatened the long-term sustainability of timber production as well. The agency's optimism and rising estimates of the sustainable level of timber harvesting were based on technical assumptions and overlooked the fact that the needed investments in intensive forest management were simply not being funded by Congress, and thus were not being made (Hirt 1994).

In 1976, the National Forest Management Act placed numerous additional statutory limits on timber production on the national forests and required the development of detailed management plans with ample opportunity for public involvement in national forest management decision making (16 U.S.C. 1600, note). Many of these limitations were aimed at reducing the impacts of timber harvesting on nontimber resources. But concern over the sustainability of timber production itself led Congress to add a new wrinkle to its definition of sustained yield, specifying that the sale of timber from each national forest be limited to "a quantity equal to or less than a quantity which can be removed from such forest annually in perpetuity" (16 U.S.C. 1600, §13). This "nondeclining even flow" constraint was criticized by some economists as inherently inefficient for managing the extensive areas of native forest old-growth on many western national forests (Clawson 1983). Trees in old-growth forests, especially in the Pacific Northwest, can reach 800 to 1,000 years old and thus far exceed the point of culmination of mean annual increment—the age at which harvest is typically set for maximizing timber production.

Incorporating social norms into sustainable forestry. Taken together, those laws represented an important shift in the concept of sustainability in forest management, from the maximization of a single objective subject to ecological and environmental constraints, to the simultaneous pursuit of multiple objectives. Whereas sustained-yield forestry has as its sole objective maximum timber volume production (or maximum financial returns), the intent of multiple-use forestry was to satisfy numerous resource output objectives simultaneously. In developing national forest plans to implement the laws, the Forest Service took a highly technical approach, using complex mathematical models to determine which mix of products and "nonmarket values" could be expected to maximize "net public benefits." In practice, this mechanistic approach simply did not work, as evidenced by the political and legal gridlock that came to characterize national forest policy and management. Additional concerns over endangered species brought sudden and immediate court-imposed reductions in timber supply in some areas, leading to a political impasse and a fundamental reexamination of what forest managers were to sustain, for whom, and to what purpose.

This reexamination has led to a further evolution in the definition of sustainability in forest management, one that explicitly rather than implicitly includes social and economic as well as biological objectives. A crucial tenet of sustainable forest management is that it must be not only ecologically sound but also economically viable and socially responsible (Aplet et al. 1993). If it is lacking in any one of these three areas, the system will sooner or later collapse.

Most conservation interests now acknowledge that it is not possible to accomplish long-term protection of forest ecosystems without incorporating the economic and social needs of local communities into conservation strategies. Economic development and commercial interests are recognizing that the ecological soundness of their activities not only helps ensure raw material supplies for the future, but also helps maintain essential social and political support (Schmidheiny 1992). Communities are no longer willing to accept the social disruptions and family dislocations that have always accompanied a boom-and-bust approach. They are recognizing that government policy makers alone cannot lead the way toward stable, resilient, and economically diverse communities—that there is an important role for partnerships among federal and state governments, business interests, and the communities themselves in finding a new basis for sustainable resource use and sustainable communities.

Development of Principles for Sustainable Forest Management

The debate over sustainable forest management in U.S. forest policy is not taking place in isolation but is part of a larger international effort to define, articulate, and encourage the practice of sustainable forestry. At the 1992 United Nations Conference on Environment and Development, consensus was reached on a set of nonbinding "forest principles." This first global consensus on forests addresses the need to protect forests for environmental and cultural reasons and the need to use trees and other forest life for economic development. The Forest Principles also state that forests, with their complex ecological processes, are essential to economic development and the maintenance of all forms of life.

The Rio statement says that forests, with their complex ecological processes, are essential to economic development and the maintenance of all forms of life. They are the source of wood, food and medicine, and are rich storehouses of many biological products yet to be discovered. They act as reservoirs for water and for carbon, that would otherwise get into the atmosphere and act as a greenhouse gas. Forests are home to many species of wildlife and, with their peaceful greenery and sense of history, fulfill human cultural and spiritual needs. (IISD 1998)

Among the forestry principles:
- All countries should take part in "the greening of the world" through forest planting and conservation.
- Countries have the right to use forests for their social and economic development needs. Such use should be based on national policies consistent with sustainable development.
- The sustainable use of forests will require sustainable patterns of production and consumption at a global level.
- Forests should be managed to meet the social, economic, ecological, cultural and spiritual needs of present and future generations.
- The profits from biotechnology products and genetic materials taken from forests should be shared, on mutually agreed terms, with countries where the forests are located.
- Planted forests are environmentally sound sources of renewable energy and industrial raw materials. The use of wood for fuel is particularly important in developing countries. Such needs should be met through sustainable use of forests and replanting. The plantations will provide employment and reduce the pressure to cut old-growth forests.
- National plans should protect unique examples of forests, including old forests and forests with cultural, spiritual, historical, religious and other values.
- International financial support, including some from the private sector, should be provided to developing nations to help protect their forests.
- Countries need sustainable forestry plans based on environmentally sound guidelines. This includes managing the areas around forests in an ecologically sound manner.
- Forestry plans should count both the economic and non-economic values of forests, and the environmental costs and benefits of harvesting or protecting forests. Policies that encourage forest degradation should be avoided.

- The planning and implementation of national forest policies should involve a wide variety of people, including women, forest dwellers, indigenous people, industries, workers and non-government organizations.
- Forest policies should support the identity, culture and rights of indigenous people and forest dwellers. Their knowledge of conservation and sustainable forest use should be respected and used in developing forestry programs. They should be offered forms of economic activity and land tenure that encourage sustainable forest use and provide them with an adequate livelihood and level of well-being.
- Trade in forest products should be based on non-discriminatory, rules, agreed on by nations. Unilateral measures should not be used to restrict or ban international trade in timber and other forest products.
- Trade measures should encourage local processing and higher prices for processed products. Tariffs and other barriers to markets for such goods should be reduced or removed.
- There should be controls on pollutants, such as acidic fallout, that harm forests. (United Nations 1992)

Criteria and indicators for Europe. Since UNCED, criteria and indicators for sustainable forest management have been formulated through several international and national, governmental and nongovernmental processes. The Helsinki Process (officially, the European Process on Criteria and Indicators for Sustainable Forest Management) focuses on the development of criteria and indicators for European forests, which include boreal, temperate, and Mediterranean types. The European countries have agreed upon six criteria, 27 quantitative indicators, and descriptive indicators for sustainable forest management. The criteria for sustainable forest management in the Helsinki Process are as follows:

1. maintenance and appropriate enhancement of forest resources and their contribution to global carbon cycles;

2. maintenance of forest ecosystem health and vitality;

3. maintenance and encouragement of productive functions of forests (wood and nonwood);

4. maintenance, conservation, and appropriate enhancement of biological diversity in forest ecosystems;

5. maintenance and appropriate enhancement of protective functions in forest management (notably soil and water); and
6. maintenance of other socioeconomic functions and conditions.

Those overarching principles for the sustainable management of forests in Europe have been further refined into a set of operational guidelines to be used in forest planning and forest management. The Pan European Operational Level Guidelines provide a common framework that can be used at the field level on a voluntary basis (PEFC 2001). They focus on the basic ecological, economic, and social requirements for sustainable forest management within each of the six criteria, but they are not intended to be used in isolation to determine sustainability in forest management. They are intended to be used in conjunction with individual countries' forest practices codes and standards, environmental legislation, decisions on protected areas, and local ecological, economic, social, and cultural characteristics.

Criteria and indicators outside Europe. The separate but related Montréal Process was initiated as followup to a conference on Sustainable Development of Temperate and Boreal Forests, held in Montréal in 1993 within the framework of the Conference on Security and Cooperation in Europe. The Montréal Process applies to temperate and boreal forests for 12 countries outside Europe—Argentina, Australia, Canada, Chile, China, Japan, the Republic of Korea, Mexico, New Zealand, the Russian Federation, Uruguay, and the United States—that have agreed on seven legally nonbinding criteria and 67 indicators for sustainable forest management, identified for national implementation. The seven criteria are as follows:
1. conservation of biological diversity;
2. maintenance of productive capacity of forest ecosystems;
3. maintenance of forest ecosystem health and vitality;
4. conservation and maintenance of soil and water resources;
5. maintenance of forest contribution to global carbon cycles;
6. maintenance and enhancement of long-term multiple socioeconomic benefits to meet the needs of societies; and
7. legal, institutional, and economic framework for forest conservation and sustainable management.

The United States has made significant progress toward a nationwide assessment of forest conditions and trends relative to an agreed-upon set of criteria and indicators of sustainable forest management based on the Montréal process. In 2004, the U.S. Forest Service published its *National Report on Sustainable Forests*, which incorporated information from the technical and scientific community, academics, and representatives of the forest industry and environmental NGOs to evaluate the status of U.S. forests and forest management (U.S. Forest Service 2004). The report noted, however, that of the 67 indicators, the United States is currently measuring only nine on a systematic basis. For another 20 to 25 indicators, no data are currently available, but data could be collected if sufficient financial resources were made available. For the remaining indicators, technical means for systematically collecting the data do not yet exist. Making the Montréal Process criteria and indicators fully operational in the United States will take some time, and this is likely the case in other signatory countries as well.

Nongovernmental Certification of Sustainability

The intergovernmental processes for developing criteria and indicators of sustainable forestry are designed for the national level. They have been complemented by activities carried out by national and international NGOs to develop and test criteria and indicators for sustainable forestry at the forest management unit level. Many of these are market-based incentive programs aimed at encouraging more environmentally sensitive forest management, referred to generally as forest certification.

Forest certification is a voluntary, nongovernmental process by which forest products are physically identified as having come from forests that are responsibly managed. This determination is made by an independent third party (i.e., not the forest management firm itself, or its direct customer or client) based upon a field evaluation of forest management practices relative to specific standards.

Forest certification differs from government regulation of forest practices (e.g., state forest practices laws) in that it is "market based." It allows consumers of paper and other wood products to play a part in encouraging responsible forest management by providing them a way to distinguish certified wood products from other brands in their purchase decisions. If there is strong enough consumer preference for certified wood products, the companies managing their forests responsibly will outcompete those that do not, forcing them to improve the quality of their forest management so that their products too can be marketed as certified.

U.S. policy is to encourage nongovernmental organizations in the development of certification programs but generally keep government out of the processes for setting forest practices standards or evaluating forest management in the field. Several other countries, particularly in Europe, have established policies to give preference to imports of certified wood and wood products. Concerned that this could be interpreted as a nontariff trade barrier under the rules of the World Trade Organization, the United States has avoided establishing a policy that could be seen as limiting the importation of uncertified wood products. The de facto policy is thus to leave this to individual decisions by the companies that import wood or wood products, based on the purchasing preferences of their own customers. Nevertheless, several public forest management agencies in the United States have sought certification of their forest practices to ensure access to markets for certified wood, to provide independent assurance to citizens that public forest lands are being responsibly managed, and to educate and encourage private forest owners to voluntarily improve forest management practices on the basis of market incentives rather than government regulation.

The initial impetus for "green certification" in forestry arose from the boycott of imported tropical timber by European consumers in the 1980s. The boycott was intended to decrease demand for this timber and reduce tropical deforestation, but this "blunt instrument" made no distinction between tropical timber obtained through exploitive means and that harvested through responsible forest management. It was equally punishing to those companies attempting to practice renewable resource management as to those who had never given renewability a second thought, with unnecessary and unwarranted economic impacts on struggling enterprises in some of the world's poorest developing countries. Recognizing this, representatives of European environmental organizations met with forest industry representatives and tropical timber exporters to examine conditions under which tropical timber could be deemed acceptable for import in Europe.

In the 1990s, two other international events contributed directly to the development of forest management certification as we know it today, the 1992 UNCED and the "Uruguay Round" of the General Agreement on Trade and Tariffs (GATT). At UNCED, the world's governments came to consensus on the goal of sustainable development, acknowledging that sound environments and economies are inextricably linked. This included international acceptance of the Forest Principles as well as a chapter of the

UNCED document, Agenda 21, entitled "Combating Deforestation." Both embraced the concept of sustainable management of the world's forests to meet current needs without compromising the ability of present and future generations to meet their own needs (Society of American Foresters 1995). For a time, Germany, the Netherlands, and the United Kingdom considered trade restrictions in the form of requiring certification of timber originating from tropical countries. This was opposed by tropical timber–exporting nations, who asserted that any such rules should apply to all internationally traded timber, regardless of source. In addition, GATT prohibits the use of trade restrictions based on methods of production to discriminate between "like products," thus preventing the use of any government-imposed tariffs, bans, or quotas in favor of "sustainably" produced wood.

Forest Stewardship Council (FSC). In the absence of governmental action, the World Wide Fund for Nature (WWF) brought together representatives of forest industry and environmental organizations in 1993 to form the Forest Stewardship Council (FSC), whose purpose is to "support environmentally appropriate, socially beneficial, and economically viable management of the world's forests." FSC adopted principles and criteria to apply to the management of tropical, temperate, and boreal forests worldwide and established a process for on-the-ground, performance-based, third-party verification of a forestry operation's adherence to these principles and criteria. Following a successful third-party assessment, the producer is entitled to affix the FSC label directly to the product to inform consumers that it was produced from a forest managed in accordance with the FSC standards.

FSC has developed standards for forest management that are applicable to all FSC-certified forests throughout the world. There are ten principles and 57 criteria that address legal issues, indigenous rights, labor rights, multiple benefits, and environmental impacts surrounding forest management. FSC's ten principles for sustainable forest management are as follows:

1. Forest management shall respect all applicable laws of the country in which they occur, and international treaties and agreements to which the country is a signatory, and comply with all FSC Principles and Criteria.

2. Long-term tenure and use rights to the land and forest resources shall be clearly defined, documented, and legally established.

3. The legal and customary rights of indigenous peoples to own, use, and manage their lands, territories, and resources shall be recognized and respected.

4. Forest management operations shall maintain or enhance the long-term social and economic well-being of forest workers and local communities.

5. Forest management operations shall encourage the efficient use of the forest's multiple products and services to ensure economic viability and a wide range of environmental and social benefits.

6. Forest management shall conserve biological diversity and its associated values, water resources, soils, and unique and fragile ecosystems and landscapes, and, by so doing, maintain the ecological functions and the integrity of the forest.

7. A management plan—appropriate to the scale and intensity of the operations—shall be written, implemented, and kept up to date. The long-term objectives of management, and the means of achieving them, shall be clearly stated.

8. Monitoring shall be conducted—appropriate to the scale and intensity of forest management—to assess the condition of the forest, yields of forest products, chain of custody, management activities, and their social and environmental impacts.

9. Management activities in high conservation value forests shall maintain or enhance the attributes which define such forests. Decisions regarding high conservation value forests shall always be considered in the context of a precautionary approach.

10. Plantations shall be planned and managed in accordance with Principles and Criteria 1–9, and Principle 10 and its Criteria. While plantations can provide an array of social and economic benefits and can contribute to satisfying the world's needs for forest products, they should complement the management of, reduce pressures on, and promote the restoration and conservation of natural forests.

International Standards Organization (ISO). Also in 1993, the International Standards Organization (ISO) created a technical committee to develop standards and guidelines for sustainable forest management, which was carried out by the Canadian Standards Association (Hoberg 1999). ISO standards specify what processes and procedures a company must have in place to produce a quality product but does not certify actual performance under these procedures or the quality of the product itself. "Rather, the ISO certification demonstrates that the company has adopted quality management processes that are consistent and repeatable systems certification" (Berg and Olszewski 1995). The ISO 14001 standards are intended to document that a process or system ensuring continuous improvement in forest management exists, and that management is committed to environmental performance and the achievement of sustainable forestry over time.

Sustainable Forestry Initiative (SFI). Meanwhile, the leading forest industry trade association in the United States, the American Forest & Paper Association (AF&PA) learned, through a nationwide survey of public perceptions of the forest products industry, that Americans were looking for tangible evidence of substantive improvement in the way the industry managed its forests. AF&PA responded with the Sustainable Forestry Initiative (SFI), a set of basic principles of responsible forest management that, by 1996, all member companies were committed to adopt as a condition for continued membership in and representation by AF&PA. Verification of actual performance was first-party, meaning it was based on the assurances of a company itself that it was in fact managing in accordance with SFI principles.

A diverse expert review panel was established to advise the AF&PA board of directors on further improvements to SFI and by 1999 had persuaded the association to make important changes. First, the SFI principles were converted to a formal standard, consistent with ISO 14001. Second, AF&PA developed a licensing program whereby nonmember companies could participate in the SFI standard. Third, SFI program participants could verify their compliance with the SFI standard according to generally accepted auditing and verification procedures, similar to the ISO 14000 auditing standards (Berg and Cantrell 1999). In 2001, the Sustainable Forestry Board was established to oversee standards development, providing a greater degree of independence in improving the features of the SFI program. The current principles for sustainable forestry in the SFI program are as follows:

1. Practice sustainable forestry to meet the needs of the present without compromising the ability of future generations to meet their own needs by practicing a land stewardship ethic that integrates reforestation and the managing, growing, nurturing, and harvesting of trees for useful products with the conservation of soil, air and water quality, biological diversity, wildlife and aquatic habitat, recreation, and aesthetics.

2. Use and promote among other forest landowners sustainable forestry practices that are both scientifically credible and economically, environmentally, and socially responsible.

3. Provide for regeneration after harvest and maintain the productive capacity of the forestland base.

4. Protect forests from uncharacteristic and economically or environmentally undesirable wildfire, pests, diseases, and other damaging agents and thus maintain and improve long-term forest health and productivity.

5. Protect and maintain long-term forest and soil productivity.

6. Protect water bodies and riparian zones.

7. Manage forests and lands of special significance (biologically, geologically, historically, or culturally important) in a manner that takes into account their unique qualities and promote a diversity of wildlife habitats, forest types, and ecological or natural community types.

8. Comply with applicable federal, provincial, state, and local forestry and related environmental laws, statutes, and regulations.

9. Continually improve the practice of forest management and also monitor, measure, and report performance in achieving the commitment to sustainable forestry.

The Future of Certification

Forest management certification in the United States is in a period of sorting out among several systems, each with somewhat different objectives, costs, and benefits. U.S. forest landowners—corporations, private forest landowners, federal and state agencies, and Native tribes—are evaluating the various systems to determine which are most consistent with their own needs and objectives.

Comparisons of FSC and SFI, the two major independent, third-party certification systems in the United States, indicate that there continue to be important differences between them, some of which are intentional. For example, FSC focuses on rewarding exemplary forest management, places greater emphasis on ecological and social components, and is generally more credible with environmental organizations and the general public. On the other hand, SFI seeks to bring along virtually every forest products enterprise in the United States, steadily improving forest management practices by all AF&PA member companies and their suppliers (Mater et al. 2002; Meridian Institute 2001).

Competitive as they are, the two systems are also gradually converging on certain important features. SFI has quickly evolved from being simply a "code of conduct" that AF&PA members pledged to follow to requiring independent, third-party verification of a company's adherence to specific standards before it can use the SFI label on its products, much like FSC. And FSC is gradually moving toward the "continuous improvement" approach to the standards by which it evaluates forest enterprises, long a characteristic of SFI, and is striving toward a more stable, inclusive approach to modifying its standards over time.

Both systems still face similar and substantial challenges to achieving their long-term objectives. Both systems were aimed initially at relatively large forest enterprises, and neither has been able to successfully adapt itself to the needs of private forest landowners, who generally cannot afford the costs of certification but who collectively own 60 percent of U.S. forest land. Primary and secondary wood products manufacturers will continue to have a difficult time becoming certified if a significant fraction of their raw material comes from noncertified forest enterprises. Neither system has created the kind of awareness among American consumers that has long been evident among many European consumers. Although several major wood products retailers have endorsed certification, particularly those that market their products internationally, it is ultimately the preference of consumers for certified wood products that will determine the success of this market-based mechanism.

The incentives for public sector forestry organizations are somewhat different. A number of tribal governments in the United States have sought to have their forest management activities certified as a way of gaining greater independence from federal authorities. Many tribes own substantial areas of forest land, but these are often managed on their behalf by the Bureau of Indian Affairs, an agency in the U.S. Department of the Interior. In 1990, Congress passed the National Indian Forest Resource Management Act to give the tribes greater autonomy in the management of their forests, but it also required an independent evaluation every ten years to assess forest conditions (25 U.S.C. 3101–3120). The second Indian forest management assessment, conducted in 2002, utilized a comprehensive set of field data gathered by teams of forestry experts whose primary purpose was to evaluate the tribes for possible certification. For some, the process led to successful certification. For others, the process provided valuable guidance on how best to improve their forest practices and continue along the path toward greater autonomy and self-reliance.

State forestry agencies that have sought certification have often realized unexpected benefits, in addition to broadening their markets to include those for certified wood. State agencies, like all public forestry agencies in recent years, have come under increased scrutiny and criticism for their forest practices and skepticism about their claims of sound, responsible forest management. Independent, third-party review and evaluation of their forestry activities has helped the agencies better address the concerns of even the most skeptical critics and provided an extra measure of assurance to all citizens. State forest lands, once certified, also become extensive "demonstration forests" with which to educate private forest owners on responsible forest practices. In states with forest practices laws administered by these agencies, certification of private lands often can relieve the state forestry agencies of a significant burden for monitoring forestry operations on these lands. Certification requirements often meet or exceed those of the state forest practices law, meaning that certification provides *prima facie* proof that the private forest owner is also complying with the law.

Federal lands remain, without question, the most controversial when it comes to forest management. Federal land management agencies like the Forest Service and Bureau of Land Management have been hesitant to seek independent, third-party certification. They have been discouraged from doing so both by environmental organizations, which see certification as placing too much emphasis on timber production, and by the forest products industry, which sees certification as placing too little emphasis on timber production. A land and resource management plan must be developed for each national forest every ten to 15 years, under the requirements of the National Forest Management Act (16 U.S.C. 1600, note). Forest management standards and guidelines are established or revised at the forest management unit level during this process. The Forest Service is currently experimenting with developing these standards and guidelines as part of a broader environmental management system, which could serve as the basis for ISO certification. At the same time, several national forests are participating in an experimental certification assessment under both SFI and FSC to determine whether certification is possible, and if so, what changes in forest management practices would be necessary.

In Europe, the Programme for the Endorsement of Forest Certification Schemes (PEFC) has emerged as a means of providing some level of coordination among independent, third-party certification programs being developed in individual countries. As an independent NGO, PEFC uses a rigorous assessment process involving public consultation and examination by independent consultants to determine whether a given certification program will receive PEFC endorsement, which implies mutual recognition of certified wood products traded internationally. Individual certification programs are then licensed to use the PEFC logo. Originally focused on certification programs in European countries, PEFC opened membership to non-European programs in 2001, and membership now includes countries in North America, South America, Asia, Australia, and Africa. Not all members of PEFC have certification schemes actually approved as meeting the PEFC Council's requirements. For example Ireland, Brazil, Canada, Luxembourg, Russia, Slovakia, and the United States are all members of PEFC but do not yet have PEFC endorsement of their certification programs (PEFC 2005).

PEFC has evolved into a truly global organization for the assessment and mutual recognition of independent, third-party certification programs developed in a multistakeholder process. Currently, PEFC has endorsed forest certification programs in 17 countries, accounting for more than 57 million hectares of certified forest land. In the United States, the AF&PA Sustainable Forestry Initiative and the American Tree Farm System (oriented to family forests and other nonindustrial private forests) are awaiting endorsement. Individual states, most notably Oregon, are also considering seeking PEFC endorsement for certification programs that would recognize and differentiate wood products from individual states.

Conclusion

Sustainability in forest management is a dynamic, evolving concept, reflecting changing social values and scientific understanding of the effects of human activities on the functioning of forest ecosystems. As an increasingly broad cross-section of forestry interests comes to accept that truly sustainable forestry must encompass ecological, economic, and social objectives, the most challenging tradeoff for policy makers may be between short-term needs and long-term assurances.

The central idea behind sustainable development—that is, meeting the needs of present human society without unduly compromising the capacity of future human societies to meet their needs (World Commission on Environment and Development 1987)—is not materially different from the basic motivating concept behind sustained-yield forestry in 18th-century Europe or sustainable forestry in 21st-century America.

From a policy-making and operational management perspective, it seems the sustainability challenge will always be to protect the long-term productivity of forest ecosystems—to the best of our biological, social, and economic understanding—without unduly limiting the utilization of forests to meet current needs. Or from an analytical perspective, to operate as close as is socially and politically acceptable to where we think the production possibility curve may lie, neither exceeding ecological capacities nor leaving significant ecological capacity unutilized. How conservative a margin for error is incorporated is as much a political decision as a scientific one.

There has been widespread public debate during the past two decades, in both the United States and Europe, as to what constitutes sustainable forestry. This debate has included scientists, who are coming to a new understanding and appreciation of the complexities of forest ecosystems and their natural dynamics. It has included segments of society that care passionately for certain aspects of forests and advocate strongly for the protection of these important values. It has included policy makers, who must be sensitive to a broad diversity of public concerns about forests and are increasingly called upon to formulate the rules by which forests are managed. And of course, it has included forest managers themselves, who must address the multitude of demands on a finite resource and make the daily decisions that they hope will add up to sustainability in the long run.

Out of this creative ferment of debates and discussions at many places around the world, what has emerged is a basic framework of concepts that is common to the many independent efforts to define sustainable forestry. This framework commonly includes the following components:

- maintenance of productive capacity of forest ecosystems;
- maintenance of forest ecosystem health and vitality;
- conservation and maintenance of soil and water resources;
- conservation of biological diversity;
- maintenance of forest contribution to global carbon cycles;
- maintenance and enhancement of long-term multiple socioeconomic benefits to meet the needs of societies; and
- development of an institutional, legal, and policy framework to support forest conservation and sustainable management.

What can thus be viewed as generally accepted principles of sustainable forest management increasingly forms the core of both government regulatory programs and voluntary, market-based programs. As the dust settles on these definitional debates and the goals of sustainable forestry become clearer, policy makers and forest managers are finding it easier to design forest practices and forest management regimes that will meet those goals. This framework of generally accepted principles of sustainable forest management is also more enduring over time, meaning that forest managers are not struggling to hit a moving target.

Nevertheless, the details will continue to change. The continual evolution of the science of forest ecology, as well as the continual evolution of social values and perceptions relating to forests, means that what constitutes sustainable forest management cannot be defined once and for all time. The concept of sustainability itself will continue to evolve, and so must the laws and policies designed to promote sustainable forest management. Ensuring that forest policy is flexible and dynamic enough to accommodate evolving science and changing societal expectations is critically important for guiding the development of future forest policies that will be both durable and effective.

References

16 U.S.C. 528. Multiple-Use Sustained-Yield Act of June 12, 1960. P.L. 86-517, 74 Stat. 215.

16 U.S.C. 1600, note. National Forest Management Act of October 22, 1976. P.L. 94–588. 90 Stat. 2949, as amended.

16 U.S.C. 1600. Forest and Rangeland Renewable Resources Management Act of August 17, 1974. P.L. 93–378. 88 Stat. 476, as amended.

25 U.S.C. 3101–3120. National Indian Forest Resources Management Act of November 28, 1990. P.L. 101–630. 104 Stat. 4532.

Aplet, G., N. Johnson, J. Olson, and V.A. Sample. 1993. *Defining sustainable forestry*. Washington, DC: Island Press.

Berg, S., and T. Cantrell. 1999. Sustainable Forestry Initiative: Toward a Higher standard. *Journal of Forestry* 97(11): 33–36.

Berg, S., and R. Olszewski. 1995. Certification and labeling: A forest industry perspective. *Journal of Forestry* 93(4): 30–32.

Binkley, C.S., C.F. Raper, and C.L. Washburn. 1996. Institutional ownership of U.S. timberland. *Journal of Forestry* 94(9): 21–28.

Clawson, M. 1983. *The federal lands revisited*. Baltimore: Johns Hopkins University Press for Resources for the Future.

Faustmann, M. 1849 (1995). The determination of value which forest land and immature stands possess for forestry. Originally published in German in *Allegemeine Forest und*

Jagd Zeitung 25. English translation published in *Journal of Forest Economics* 1(1) (Umea, Sweden).

Fearnside, P. 2000. Global warming and tropical land-use change: Greenhouse gas emissions from biomass burning, decomposition and soils in forest conversion, shifting cultivation and secondary vegetation. *Climatic Change* 46(1/2):115–58.

Heske, F. 1938. *German Forestry*. New Haven. CT: Yale University Press.

Hirt, P. 1994. *A conspiracy of optimism: Management of the national forests since World War Two*. Lincoln: University of Nebraska Press.

Hoberg, G. 1999. The coming revolution in regulating our forests. Institute for Research in Public Policy. *Policy Options* (December): 53–56.

Houghton, R. 1999. The annual net flux of carbon to the atmosphere from changes in land use, 1850–1990. *Chemical and Physical Meteorology* 51(2): 298–313.

IISD. 1998. The Earth Summit's agenda for change: Statement of principles on forests. International Institute for Sustainable Development. http://www.iisd.org/rio+5/agenda/principles.htm (accessed March 3, 2005).

Mater, C.M., W.C. Price, and V.A. Sample. 2002. Certification assessments on public and university lands: A field-based comparative evaluation of the Forest Stewardship Council and Sustainable Forestry Initiative programs. Washington, DC: Pinchot Institute for Conservation.

Meridian Institute. 2001. Comparative analysis of the Forest Stewardship Council and Sustainable Forestry Initiative certification programs. Washington, DC: Meridian Institute.

PEFC. 2001. Technical documentation: Pan European operational level guidelines for sustainable forest management. Annex 2 of the Resolution L2. Programme for the Endorsement of Certification Schemes. http://www.pefc.org/internet/html/documentation/4_1311_400.htm (accessed March 3, 2005).

———. 2005. PEFC Members and Schemes. http://www.pefc.org/internet/html/members_schemes/4_1120_59.htm (accessed March 3, 2005).

Pinchot, G. 1947. Breaking new ground. New York: Harcourt, Brace and Company.

Sample, V.A. 1990. *The impact of the federal budget process on national forest planning*. Westport, CT: Greenwood Press.

Schmidheiny, S. 1992. *Changing course: A global business perspective on development and the environment*. Cambridge, MA: MIT Press.

Society of American Foresters. 1995. Forest certification. Summary of report of SAF study group on forest certification. *Journal of Forestry* 93(4): 6–10.

United Nations. 1992. "Non-Legally Binding Authoritative Statement Of Principles For A Global Consensus on the Management, Conservation And Sustainable Development of All Types Of Forests." Report of the United Nations Conference on Environment and Development (Rio de Janeiro, 3–14 June 1992) A/CONF. 151/26 (Volume III). New York: United Nations.

U.S. Forest Service. 2004. *National report on sustainable forests 2003*. Washington, DC: U.S. Department of Agriculture.

Waggener, T.R. 1977. Community stability as a forest management objective. *Journal of Forestry* 75(11): 710–14.

Westoby, J. 1989. *Introduction to world forestry: People and their trees.* Oxford, UK: Blackwell.

Wilson, E.O. 1992. *The diversity of life.* New York: W.W. Norton.

World Commission on Environment and Development. 1987. *Our common future.* New York: Oxford University Press.

ABSTRACT

The present distribution of forests and the degree of their transformation by man are the results of natural factors and cultural development. The limit between forested areas and open spaces, as well as differences between intensively used forests and those showing little or no traces of human intervention, is determined by social needs and values, economic opportunities, and political regulations. This paper considers how forests are currently perceived by the population as physical and social spaces profoundly influenced by timber use and forest management, and it shows that forests' social meaning and political signification is in full evolution.

CHAPTER 12

European Forests: Heritage of the Past and Options for the Future

Franz Schmithüsen

Natural environmental conditions and cultural development processes determine the spatial distribution of forests and the intensity with which forest vegetation has been influenced by human activity. This applies to forests that have been exploited for hundreds of years as well as to wooded areas that, for all appearances, have been barely touched by man. The reasons behind the actual delimitation of the forest and of open spaces are manifold: for instance, a particularly high value given to forests for economic, social, and cultural reasons or, conversely, the lack of economic interest that was attributed to their use in the past. Differences between intensively exploited areas and those showing few apparent human interventions depend on social values and needs, economic potential, and political regulations. In a general way all forests, including those considered to be forests close to the natural state, have been and still are spaces manipulated by man.

The following analysis refers to the social significance of forests as a result of successive and superimposed cultural processes. It brings to the fore the importance of forests as a local environment, a renewable resource, a free space with which one can personally identify, and a representation of a space perceived as natural or at least close to nature. These observations deal principally with forests in western Europe, which are physical and social areas shaped by man over a very long period. The observations are based on historical sources showing the evolution of forestry in Germany and Switzerland as well as on recent empirical studies of people's attitudes and perceptions regarding forests. Among the reference texts and collections of articles giving information on the condition of forests and on their use and management in various countries and regions, one may cite the following: for Europe, Arnould et al. (1997) and Cavaciocchi (1996); for Germany, Hasel (1985), von Hornstein (1951), Küster (1998), Mantel (1990), and Semmler (1991); for Switzerland, Hauser (1972); for Austria, Hillgarter and Johann (1994); and for France, Badré (1983), Bechmann (1984), Centre Historique (1997), Corvol (1987), and Devèze (1965).

Forest Vegetation and Distribution of Forest

The European wood flora consists of more than 300 species and subspecies belonging to almost 100 genera and 45 families. Five large European vegetation zones can be distinguished: the arctic and alpine zone, with dwarf shrub, grass, lichen, and moss vegetation; the boreal zone, with evergreen coniferous forests; the temperate zone, with summer-green deciduous forests; the Mediterranean region, with sclerophyllous evergreen forests and shrubs; and the Pannonic-Pontic-Anatolian zone, with dry forest, steppe, and semidesert vegetation. For the development of vegetation in Europe since the ice ages, one may refer to Lang (1994).

The zone of summer-green deciduous forests covers western, central, and eastern Europe and also parts of the southern European mountain areas. Mixed oak and beech forests are prevalent at the lower altitudes of the Atlantic and sub-Atlantic British Isles, on the continent from northwestern Spain to Denmark and southern Scandinavia, and at the subcontinental and continental parts of central and eastern Europe. Mixed forests of beech and oak, together with spruce and pine forests, are common in western and northern central Europe. In south-central Europe and in the bordering areas to the south, spruce, fir, and pine grow in the mountainous regions along with beech and oak. For the present vegetation pattern and the distribution of forest ecosystems in western and central Europe, see Ellenberg (1996).

The forested areas within central Europe form a mosaic of frequently changing landscape units within which natural forest communities of broad-leaved and coniferous species in various mixtures alternate with largely manmade forests (Pott 1993). Among the species forming extended forest stands are beech (*Fagus sylvatica*), Norway spruce (*Picea abies*), silver fir (*Abies alba*), Scots pine (*Pinus sylvestris*), common oak (*Quercus robur*) and Durmast oak (*Quercus petraea*), hornbeam (*Carpinus betulus*), common birch (*Betula pendula*), and black alder (*Alnus glutinosa*). Frequent but occurring more locally are common ash (*Fraxinus excelsior*), sycamore (*Acer pseudoplatanus*), hedge maple (*Acer campestre*), mountain ash (*Sorbus aucuparia*), aspen (*Populus tremula*), wild cherry (*Prunus avium*), downy birch (*Betula pubescens*), crack willow (*Salix fragilis*), and—mainly in mountainous areas—European larch (*Larix decidua*) as well as stone pine (*Pinus cembra*). More rarely or locally occurring species include wych elm (*Ulmus glabra*), European white elm (*Ulmus laevis*), smooth-leaved elm (*Ulmus minor*), Norway maple (*Acer platanoides*), broad-leaved lime (*Tilia platyphyllos*), small-leaved lime (*Tilia cordata*), white willow (*Salix alba*), black poplar (*Populus nigra*), European wild apple (*Malus silvestris*), wild pear (*Pyrus pyraster*), wild service-tree (*Sorbus torminalis*), whitebeam (*Sorbus aria*), yew (*Taxus baccata*), dwarf mountain pine (*Pinus mugo*), and mountain pine (*Pinus uncinata*).

Climatic zones and altitude levels of vegetation, soil formation, and topography, together with specific ecological demands of plant species, are decisive factors determining forest vegetation development (Ellenberg 1996). Important factors influencing the distribution of forest ecosystems and tree species diversity are the average annual temperature during the growth period, extremes of cold and drought, water regime and soil moisture, and exposure of the site. The natural conditions for the development of vegetation make it clear that large areas in western and central Europe would be covered by forests of deciduous species.

It is not the combination of natural processes alone, however, that explains present-day vegetation. To a large extent, human intervention has determined the actual distribution and floral composition of forests. For thousands of years, human beings have been influencing the spread of tree species and the floristic composition of forests. During the long history of land use, forest areas have been transformed into fields, grassland, and wooded pastures; settled areas, once abandoned, became fallow land and subsequently returned to forest. Varied landscapes have been formed by successions of vegetation that often are still clearly visible today. Consequently, the present distribution of forest areas and species only partially reflects what the vegetation would be without man's intervention. An important indicator of the degree of "naturalness" of forest vegetation is furnished by the soils and the herbaceous flora. As a consequence, the wooded areas of central Europe show a mosaic of varied landscapes, often subject to rapid changes

(Konold 1996; Küster 1995). In particular, one sees associations of forests close to the natural vegetation, formed of deciduous and coniferous species; forests and clearings largely modified by human activities; wooded pastures; and various forms of vegetation succession.

The stages of clearing, as well as of natural succession brought about by the abandonment of previously colonized land, were fundamental in the distribution of fields and forests (Mantel 1990). The greatest period of forest clearing goes back to the seventh to ninth centuries, culminating in the Middle Ages (11th and 12th centuries). The reasons lie in the rapid population increase; for example, in certain regions of Germany, populations doubled between A.D. 900 and 1100 and almost quadrupled by the 13th century. Initially, land had been cleared principally by family groups and villagers and at a later stage by large-scale colonization, which was organized systematically by monasteries and feudal landowners. In the late Middle Ages forest clearing for agricultural land and pastures ceased. Devastating plagues, such as typhus between 1309 and 1317 and bubonic plague (the so-called Black Death) around 1350, caused a large decrease in the population. Whole villages were abandoned, with the population moving toward better living areas and more permanent settlements. The influx of people into the expanding towns depleted rural populations. The abandoned dwellings and fields were almost always won back by the forest. The settlements of the late colonization periods, where land was less suitable for agriculture, were particularly affected. In mountain regions such as the Hohe Rhön and the Solling in Germany, up to 70 percent of the cleared land returned to forest.

After the cessation of medieval land clearing and the subsequent retreat of population, the distribution of forests and fields sustained fewer changes until the beginning of modern times. Great forest clearances still occurred at the initiative of sovereigns and feudal landlords, who were promoting preindustrial enterprises, such as glassworks and ironworks in the Spessart and the Black Forest, in order to use the wood resources that were still available. Often these undertakings necessitated that laborers and their families be brought in, leading to a secondary land colonization in the mountain regions. On the other hand, the decrease in population occasioned by wars, particularly the Thirty Years' War, limited the need for agricultural land. Toward the end of the 18th century, another wave of forest clearing was favored by the liberalization of land use and by sales of property from the state. Almost simultaneously, reforestation on a large scale began, following the decrease in profitability of certain agricultural lands and the improvement of economic conditions in wood production. At the beginning of the 19th century, the introduction of stabling and improvements in agriculture led to a focus on the more fertile soils. A little later the decrease in sheep rearing, due to wool imports, favored the abandonment of surfaces that had previously been used as pasture. During the second half of the century the profitability of agricultural exploitations in certain regions decreased even more following the importation of cereals. This favored reforestation, or the return of the forest, as a natural succession.

Those tendencies have persisted until the present day, with some interruptions—above all, the two World Wars. The concentration of agriculture in suitable, cost-effective locations and the increase of forests on hillsides and in mountain areas have accelerated in recent decades. The increase in agricultural productivity, more open agricultural products markets, the restructuring and concentration of farming units, and the use of comparative advantages in a large single market have brought about a long-term increase of forest land in most countries of the European Union. Once again, these tendencies can mostly be seen in the mountains, where there is a rapid change between the repartition of forests, agricultural land, pasture, and fallow land. A number of studies show the dynamic increase in wooded areas and the consequences for the landscape in regions such as the Black Forest in Germany or in the Swiss Alps (Bund 1997; Fischer 1985; Gerber 1989; Kempf 1985; Ludemann 1990; Schmidt 1989).

The increase in surface area of forests—above all, in mountain regions—contrasts with considerable losses in regions of intensive cultivation or with a high population density. Quite a different development occurs in the neighborhood of towns and larger built-up areas where forests were and still are cleared and reconverted into building land. The wooded surfaces near large towns and industrial zones or situated close to town perimeters have thus considerably diminished in size; yet the intensive building activities increase their value as an indispensable part of cityscapes. It is precisely in these areas that the multifunctional services of trees and forests have progressively gained in importance and, therefore, that the forests have come to need preservation.

The dynamic processes leading to the formation of a mosaic of landscapes demonstrate alternations between great reductions in the surface area of forests during certain periods and large-scale reforestations, as well as an expansion of the wooded areas by natural succession in other times. When farmers had to decide which surfaces to cultivate and which to abandon, climate, soil, and topography played an important role. Forests were cleared, above all, on terrain offering good conditions for settlement and high productivity, whereas land of poor agricultural productivity was destined for reforestation or left to regenerate naturally. In both cases the effects of economic and social changes on the vegetation are multiple and often contradictory.

Local Resources Complementary to Agricultural Production

Throughout history the forests of western and central Europe have served as local resources available to the whole population, meeting many needs of everyday life and providing essential components of nutrition (Hasel 1985; Hauser 1972; Irniger 1991; Mantel 1990; Selter 1995). Local forest uses were an important basis for the rural economy. Forests also offered a means of expanding and protecting agricultural production, which resulted in sizable incomes. In bad years with low average crop yields, collecting forest products helped many families survive.

The supply of firewood and building materials came from the immediate neighborhood of the villages, but the more distant zones gained in importance with the growing demands of an increasing population. Stands of deciduous trees, particularly species whose sprout shoots from stumps could be regularly exploited, were favored. Sayings such as "wood and weeds grow for everybody" or "wood and suffering grow every day" express the evident reality of collecting firewood but also the elementary necessity of wood for daily uses. Providing towns with wood grew increasingly difficult. The towns acquired logging rights and property titles to forest in the surrounding countryside, or they drew up supply contracts to ensure delivery of the enormous quantities of firewood and construction timber that were necessary every year.

Other forest products were collected on a large scale. Until the 19th century, hazelnuts, wild fruits, berries, mushrooms, beech mast, and acorns were widely harvested. Roots, leaves, bark, and branches were used in the pharmacopoeia or as dyes, as well as to manufacture or clean household utensils. Wood, hard or soft and from many species, was used by households and local artisans. Many people, particularly artisans, knew precisely the utility of diverse products from many different tree, shrub, and herbaceous plant species. Collecting honey from wild bees and beekeeping in the forest took place to an extent that is difficult to imagine now. Used to prepare food and drinks, honey was the only sweet substance available before the introduction of cane sugar and, later, sugar beets. Beeswax was necessary to fabricate candles and wax tapers and as raw material for manufacturing writing tablets, seals, and other products. Beekeepers favored species such as the lime-tree and the willow as well as mixed stands of pine, oak, beech, aspen, and hazel. Reports detailing bee breeding in the Nuernberger Reichswald, upon which the production of vast amounts of gingerbread was based, give clues about forest vegetation during the 14th and 15th centuries, and they allow us to assess the changes that have taken place since then.

The practice of fattening pigs in the forest on acorns and beech mast was important in many regions for food supply and for generating cash income. Its economic importance is demonstrated by, among other things, the formerly widespread distinction of trees as fructivorous (*ligna fructifera*) or barren (*ligna infructifera*). The first group included, above all, oak species and beech, whereas the second grouped together the softwoods and conifers. In oak and beech stands, trees with large crowns promising rich fruit were systematically protected. The estimated annual fruit supply determined the number of pigs allowed to be driven into the forest in autumn. Years with rich seed harvest were described as full mast years and offered favorable predictions for the winter. Years with only average or low seed production meant that farmers had to worry about how to survive the winter. In some regions, pig fattening in forests gained so much importance that the earnings resulting from it became a more important measure for the value of forests than the earnings from wood use. Sometimes the size of a forest was assessed by the number of pigs that could be driven into it during full mast rather than by its area. For some communities the significance of years with full fruit from forest trees was comparable to good wine years. It has been reported that the citizens of Hagenau in the Alsace celebrated the eagerly awaited full mast in 1649 with a high mass (Hasel 1985).

Contrary to today's structured landscape, with its perceived spatial division of land, the borders between forests, fields, meadows, and pastures were in the past often vague and shifting. Presumably, it would have been difficult for villagers and city dwellers to tell where the forest ended and where the open land began. Did the wooded vegetation, used for gathering fruits and berries, represent forest areas or fallow fields? Was the land to pasture cattle an open forest or a wooded pasture? Could one separate the forest from the fields when agriculture and forestry alternated regularly? The answer might have been that it was more important to separate the intensively used areas, such as home gardens and plowed fields, from the common land, which was accessible to the community and where the villagers had a say in use and management.

In vast regions forests were used intensively for pasture. Pasturing led to soil compaction in many places and caused long-term changes in forest sites. With the extent of great forest domains reduced for more profitable sheep rearing and with mountain forests transformed into cattle pastures, the foundation was laid for economic development and widespread foreign trade. Large-scale pasturage is the origin of scattered forests, of open pastures with copses or isolated trees, or even the total disappearance of all forest vegetation over entire regions. Because the age-old practice of pasturage, the actual upper limits of alpine forests are in many places well below the timberline, which would correspond to natural conditions. One use of trees already widespread in the Middle Ages, particularly in the mountains, which persisted until modern times, was the harvesting of fodder by plucking leaves by hand or lopping with a billhook. Repeated over several years, this led to the formation of pollards. Certain impressive ash trees with stumpy branches, which can be seen between fields and along streams, bear witness to their earlier use as sources of fodder.

The exploitation of animal litter from the forest intensified in the 18th century. The introduction of stabling necessitated large quantities of litter that could hardly be garnered from agriculture at that time. With the increase in marketable fodder crops, the proportion of cereals diminished, and thus the available quantities of straw did as well. Without the contribution of leaves and needles gathered in the forests surrounding the villages as a substitute for fertilizer and litter, many small farms would not have been able to survive. Where there were deciduous trees, the litter was gathered using rakes. In coniferous forests raking the forest floor and cutting branches was common. From an agricultural point of view, the gathering of litter in the forest was recommended as a welcomed contribution to the increase in yield. It was only later that the damage caused to forest floors and the negative effects on stand development were recognized.

The multiple uses of the forest as a local resource complementary to agricultural production have marked the landscape in numerous ways. They have favored the preservation of deciduous forests, particularly stands of beech and oak, and the mixed forests in the neighborhood of villages and towns. Forests were less dense than they are now on account of their intensive use. Traces of agroforestry and silvopastoral systems are still visible in numerous forests. The vegetation that developed under the influence of diverse forms of historical exploitations—pollarding and lopping, gathering of fodder, stripping the bark from oaks, the use of forest litter, silvopasture—is often perceived by today's population as something attractive, representing a state close to nature. It is worthwhile remembering that such forests have for a very long time been influenced by man, who has considerably modified the selection of species, the structure of the stands, and the edaphic conditions. These forests, just like the stands that succeeded them, reflect the social and economic needs of the past.

The separation between the systems of agricultural production and forestry, clearly seen today, has developed gradually since the beginning of modern times. Its beginnings, in the 18th century, corresponded to the demands of agrarian reformers for an increase in agricultural yields by an intensive use of arable land and pastures. Forest management followed this process. Ways were sought to limit factors harmful to the development of forests and to create more favorable conditions for an increased timber production. In both cases this led to serious consequences for the structure of the landscape and the diversity of species. Biotopes that had developed within the framework of a mixed exploitation have disappeared or have at least lost ground, in both agricultural zones and the forest. Without doubt, the separation among arable land, pastures, and forests has been a major factor in landscape change.

Wood as an Energy and Raw Material Resource

From the 15th to the 18th century, large-scale wood exploitation combined with a systematic exploration for more accessible areas to satisfy the growing preindustrial demands for energy and raw materials. Radkau and Schäfer (1987) demonstrate the importance of wood for the development of numerous technologies. Sieferle (1982) indicates the major role it played as an energy source until the Industrial Revolution at the beginning of the 19th century. Mantel (1990) and Hasel (1985) describe its multiple uses in construction, in the mining industry, and in salt works. For regional studies showing various ways of using wood as a raw material indispensable to social and economic development, one can cite the following: for Germany, the works of Schenk (1996), Schoch (1994) and Textor (1991); for Austria, those of Hafner (1979), Johann (1968), Koller (1970, 1975a, 1975b), and Oberrauch (1952); and for Switzerland, the studies of Bill (1992) and Parolini (1995).

The economic potential of forests has been claimed by various interest groups, often in contradictory ways. The greatest divergence was between the demands of the local population to take advantage of the forest for their own needs and the efforts of sovereigns and local landlords to lay their hands on new exploitable wood resources to supply factories and long-distance trade. Until the 19th century, this opposition was apparent in the long and serious conflicts that took place over user and property rights. In addition, there was an increasing competition between the use of wood for energy and the demands for timber by craftsmen and preindustrial enterprises; there was also a strong rivalry between towns that wanted to ensure their annual supply and private entrepreneurs prospecting for new accessible resources.

Many forest regions were profoundly influenced by the rapid development of the economy and new technologies (Hasel 1985; Mantel 1990). The competition between alternative uses of forests as a resource for energy and raw material supply was conditioned by production and transport costs as well as by the added value of manufactured goods. The production of charcoal and potash and resin tapping could take place in distant and relatively inaccessible regions; however, the supply of large quantities of firewood and timber for the cities as well as for large glass factories, salt factories, and the metallurgical industry required the construction of complex transport systems. Floating logs down rivers and building channels and sophisticated installations to regulate water flow were essential for transporting wood over large distances.

The production of charcoal was probably the most widely used way of tapping wood as an energy source. It reduced the weight of wood to a quarter or a fifth and facilitated transport by carts on tracks at the edges of still-inaccessible forests to cities, factories, or the nearest floatable watercourse. Charcoal was used for cooking and heating in the cities and was indispensable to the mining industry, forges, metallurgy, salt works, brickyards, and lime kilns and for potassium and glass production. Production took place in kilns, in which up to 300 m^3 stacked volume of wood could be charred. During the 19th century, the use of mineral coal as well as the increased value of wood for new products made charcoal production uneconomical. It survived in some remote areas up to the 20th century.

Tapping, another essential forest use, was particularly common in the spruce forests of mountainous regions and in the eastern pine forests. Conifers could be tapped up to three times per year to extract one to two pounds of resin, or half that amount of pitch, per tree. Resin and pitch were important goods for distant and overseas trading, primarily to be used in shipbuilding. For the inhabitants of remote regions, tapping was one of the few opportunities to earn money; however, it had to compete with other forest uses, and it had to contend with the rapidly increasing demand for timber. The damage to forests caused by tapping was enormous. The removal of strips of bark caused the trees to decay and die prematurely. The increase in wood demand brought about by the growing population and town developments led to restrictions in the more accessible forest areas. In remote areas tree tapping continued up to World War I.

Large amounts of wood were burned to produce potash, which was used for textile bleaching, soap production, and above all, glassmaking. Potash production was common until the 18th century and did not lose its significance until the 19th century, when the potash deposits in central and northern Germany were developed. For the production of 1 kg of potash, about 1,000 kg or 1.5 m^3 stacked volume of wood was required. The ratio between raw material weight and product weight made potash production, like resin tapping, one of the first uses in not-yet-expoited forests. In more densely populated areas where wood could be used more profitably, potash production was forbidden or restricted.

Extensive forest was the prerequisite for setting up glassworks. The required amount of wood was enormous relative to the end product. The production of 1 kg of glass required between 1 and 2 m^3 of wood, of which more than 90 percent was used for potash production and less than 10 percent for melting over a log fire. The glassmakers competed with the charcoal burners and the potash producers and migrated to other places when the forests became worn out or when firewood and timber were made accessible by rafting logs downriver. After the Thirty Years' War landlords encouraged the establishment of large-scale commercial glassworks to develop new sources of capital from their forests. Because the concentrated demand of such facilities caused the clearcutting of thousands of hectares within just a few decades, glassworks are known in the history of forest uses as *xylophage*—"wood-eating industries." In the second half of the 18th century, they were gradually given up for economic reasons. A decisive factor was recognition of the detrimental effects on forest stands and the disadvantages of this kind of resource exploitation.

The saying "no wood, no boiling," or its French version, *point de bois, point de sel*, indicates that salt production required a continuous supply of large amounts of wood: for the fire to heat the boiling pans, for barrels to pack and handle the salt, for the construction of drift constructions and pipelines, for brick and lime kilns, and for installations related to salt mining. Salt factories were located in the Alps along the rock salt deposits (e.g., Hall in the Tyrol, Aussee and Hallstatt in the Salzkammergut, Hallein close to Salzburg, and Reichenhall in Bavaria). In northern and central Germany they were established in a number of locations (e.g., Lüneburg, Soest, Oldesloe, Halle, Sooden). In southern Germany, Schwäbisch Hall in Württemberg was also an important salt producer. The rights for salt production were bestowed upon the sovereigns as regalia and were exploited by them or conceded to entrepreneurs. The sovereigns, in turn, ensured the supply of wood by expanding forest property and usage rights and by granting subsequent wood supply contracts to private entrepreneurs.

To understand the dimensions, consider the annual firewood requirements of the Reichenhall salt factory around 1600: more than 200,000 m^3 stacked volume. After a massive decline during the Thirty Years' War, consumption rose again to 140,000 piled meters per year and then declined to 50,000 piled meters at the beginning of the 19th century. Comparable amounts were used in other places, such as the salt factory of Lüneburg, for which the annual wood intake is stated to have been between 100,000 and 300,000 piled meters of firewood. When the wood supplies of the surroundings were exhausted, waterways were extended to make more distant areas accessible. To ensure the supply of wood of the Lüneburg salt factory, channels were built to extend as far as to Mecklenburg. Another method was to divert the salt spring via big wooden pipelines to newly built boiling installations in forested areas that had yet to be exploited. At the beginning of the 17th century, brine pipelines were laid from Reichenhall to Traunstein, and in 1818 they were continued as far as Rosenheim.

The amounts of wood required by mining and metallurgy were similarly enormous. Wood was used to shore up pits and galleries, to activate the machines, and for numerous processing installations. Prior to the invention of gunpowder, wood was even needed for heating the rock and then for blasting it by suddenly cooling it by throwing water on it. Wood and charcoal were used as reducing agents for smelting and for energy production in smelters, furnaces, and hammer works. Areas where large-scale mining and ore processing were carried out include upper Palatina, upper Franconia, the Thuringian Forest, the Harz, the Nordeifel and the Siegerland, and the upper part of the Rhine valley. In the Alpine region Tyrol, Carinthia, and Styria were particularly important regions for mining and iron production.

These industries' demand for wood was enormous. Toward the end of the 18th century, the annual wood consumption for mining and smelting in the Harz is estimated to have totaled more than 20,000 trunks for mine shafts; 9,000 trunks for building; more than 30,000 m^3 for reduction charcoal; and another 30,000 m^3 for firewood as an energy resource (Mantel 1990). Sovereigns and landlords ensured the necessary wood supply by granting felling rights to miners and entrepreneurs and by establishing special-use regimes in forests destined to supply mines and metalworking industries. As a result, forest exploitation in such areas was particularly severe; but on the other hand, it was in the mining regions that the first efforts were made to replant and manage forests. At the end of the 18th century, with the arrival of coal, metallurgical techniques changed radically. This implied a great reduction in the demand for wood and created new conditions for forestry development.

Naval dockyards and the extension of ports and structures in contact with water created a more selective demand for wood, but in great quantities. Boat building required oak timber above all, but wood from coniferous trees was increasingly being used for the construction and upkeep of ships. The tallest trunks were particularly sought after as masts; straight trunks with strong appearance and dense annual rings fetched high prices. Particularly in the 18th century, trade with Holland and the expansion of rafting of logs downriver in the Rhine basin illustrate the possibility of transporting wood over long distances to meet the demand of political and economic centers. Commercial networks extended to the tributaries of the Rhine, leading to regular timber exports on a large scale. The enormous quantities of deciduous and coniferous trees, of well-specified dimensions and quality, served to build the towns and cover industrial requirements. Several recent studies go deeply into the subject of log floating and the timber trade with Holland (Ebeling 1992; Keweloh 1988; Rommel 1990; Scheifele 1988, 1993, 1996).

The centuries-long demand for wood—as a source for energy and raw material for villagers and citizens, preindustrial mining and metallurgy, salt and glass production, and shipbuilding and port installations—led to a systematic exploitation of forests in many European regions. Such exploitation caused landscape changes that are still visible today. Information about exploited forests and clearcut areas can be obtained through the descriptions from forest users, the reactions of uninvolved observers, and statements and comments of the public. Large-scale and concentrated felling not only changed the forests exploited at a given time, it also had severe consequences for the structure and composition of forest stands that developed naturally after clearcutting or that were later reforested. Deciduous forests and mixed forests retreated in regions like the Harz mountains, the Black Forest, and the Alps, where there was a particularly high, concentrated, and long-term demand for wood for preindustrial processing; coniferous species, mainly spruce, superseded beech. To a lesser degree the spread of other tree species, such as fir, was also influenced. Where logging concentrated on selective cutting of large dimensions, such as for shipbuilding and for long-distance floating trade, differentiated tree stand structures and species compositions could be at least partially preserved.

The Move to Sustainable Forestry

The evolution of forest cultivation and sustainable wood production is dealt with by Brandl (1970 and 1987), Hasel (1985), Hausrath (1982), Mantel and Pacher (1976), Mantel (1980), Mantel (1990), Rubner (1967), Seeland (1993), and Seling (1997). Concerning regional aspects, one may refer to the works of Allmann (1989), Baum (1995), Ernst (2000), Fenkner-Voigtländer (1992), Hachenberg (1988, 1992), Kremser (1990), Kunz (1995), and Loderer (1987) for Germany; for Switzerland, Bürgi (1997), Kasper (1989), Müller (1990), Schuler (1977), and Stuber (1996). Whether the shortage of usable wood was real or assumed as a decisive factor in the transition to a sustainable wood production is controversial (Grewe 2002; Schäfer 1992; Schmidt 1997). Among the studies that show similarities but also differences compared with the evolution of forestry in France, one may refer to the works of Bonhôte (1997), Pagenstert (1961), Rubner (1965), and Woronoff (1990).

Establishing a comprehensive political and legal framework to regulate forest uses and forestry development has been a tedious, difficult, and often conflicting process—particularly because sovereigns and nobles had continuously extended their authority and jurisdiction by claiming unexploited wood resources for operating mining industries, commercial salt production, glassmaking factories, and shipbuilding. They secured juridical control over vast areas, created their own forest administrations, and imposed close supervision on communal and, to a lesser degree, private lands. The growing influence of the state created tensions among forest services, entrepreneurs, peasants, and villagers.

Public provisions referring to forest uses over more than one generation are probably among the oldest forms of long-term environmental and natural resource policy. Customary law, already codified in the first half of the 14th century, regulated forest uses in accordance with the demands and options of their times (Mantel 1990). As early as 1295, a local rule of Landau (Palatina) provided that wood cut in the area should be for the use of the local inhabitants. The *Frankenspiegel*, which chronicled the laws that were customary around 1330, stipulated that fellings should be done moderately and without devastation. Similar principles were expressed later on in many other local laws of villagers' associations, convents, municipalities, and towns. Use regulations explicitly prohibited the felling of fructiferous trees and species that were important for local wood supply. Forests surrounding settlements were intended for local users only and were subdivided into annual felling units. After logging, such a unit was protected against grazing until regrowth was ensured.

During the 15th and 16th centuries, the cities as well as preindustrial entrepreneurs had to face the fact that wood supplies from as-yet unexploited forests could not sufficiently meet the growing needs for firewood and construction timber, salt production, and metallurgy. An unprecedented increase in demand led to high prices in regional and international trade in logs and sawn timber; this fact progressively had repercussions in many parts of central Europe. As a consequence, the essential conditions for a more stable forest regime were established in the 17th and 18th centuries. By 1850, one could say that most forest areas had come under some form of long-term forest production system. This evolution was initiated and supported by an increasing number of forest and timber ordinances issued by the sovereigns. The issues at stake were meeting local needs, long-term availability of raw materials and energy, and increased outputs through better forestry practices. Public policies and law established the requirements of sustainable wood production, which meant stopping mere exploitation of what was available, recognizing the long-term nature of forests, and promoting the involvement of several generations in forestry production. Increasingly, these policies provided for planning, management, and measures of regeneration and reforestation.

Step by step, policy and law introduced principles of renewable natural resources use as we understand them today. During the course of three centuries, starting at the beginning of the 18th century, forestry and wood processing have become productive sectors of the economy, using a renewable resource in a sustainable manner as the basis of business management. As observed by Zürcher (1965), the term *sustainable* was used as early as 1713 by von Carlowitz, who worried about maintaining mining activities and wrote (translation by the author), "The greatest art, science, diligence and institution of these countries will rely on the manner in which such conservation and growing of wood is to be undertaken in order to have a continuing, stable and sustained utilization, as this is an indispensable cause without which the country in its essence cannot remain."

In 1804, Georg-Ludwig Hartig had already formulated the principle of sustainable forestry in its classic intergenerational perspective, remarking in his textbook *Taxation of Forests* (translation by author), "It is not possible to think and expect sustained forestry if the wood allocation from the forests is not calculated according to sustainability.... Any wise forest direction consequently needs to tax (assess) the woods as high as possible, but aiming at using them in a way that the descendants can draw at least as many advantages as the now-living generation appropriates." In 1841, Carl Heyer referred to the technicality of sustainability of wood production in saying that a forest is "managed in a sustainable manner if one takes care of the regeneration of all logged stands in order to maintain the soil that is destined to forest production."

During the 20th century, the meaning of the principle of sustainable forest management expanded from wood production to include all aspects of forest uses and values. In a modern business management-oriented definition, as formulated by Speidel (1984), sustainable forestry means the ability of forest enterprises to produce wood, infrastructural services, and other goods for the benefit of present and future generations. It means maintaining and creating the entrepreneurial conditions necessary for a permanent and continually optimal fulfillment of economic and extraeconomic needs and goals. Sustainability addresses the time perspective (permanent and continuing), the kinds of activities (maintaining and creating), the objectives (needs and goals), and the qualifying criteria (optimal fulfillment).

Continuity and increase of wood supply required considerable private and public efforts and investment, but that long-term investment could not be obtained without security of forest tenures. Having the formal aspects of forest ownership rights clarified and unified is probably the most significant contribution of forest laws adopted during the 19th century. Generally, the laws often had a strong tendency to restrict or abolish usufruct rights and to transform collective tenure into clearly defined private and public landownership. Private property rights were legally registered, and forests still under collective tenure were divided among the users. In other cases communal and state forests were maintained or newly created. Quite often a combination of private and public tenures developed, characteristic of the prevailing ownership of forests in most European countries. The laws defined the landowner's wood production and management rights in using the forest as a productive asset for generating profit and income. They also determined responsibility for maintaining collective uses in the public interest, such as access to forests and protective values in the mountains, which were important to a large part of the population.

The long move to sustainable forest uses has led to a system of natural resources management that has kept its exemplary value. It is based on scientific models that adjust harvesting intensities to the long-term potential of forest sites, species composition, age classes, and forest stand structures. The principle of sustainable wood production was implemented by applying these models over large areas and gradually in all forests. In regions where oak and beech forests dominated, the coppice-with-standards system was a typical example of systematic management on a large scale. This approach combined production of firewood from new sprout shoots with production of construction timber from trees retained over several cycles of firewood harvests. The coppice-with-standards silvicultural system, developed since the 16th century, still constitutes an important method of management and is used, for instance, in France. On the other hand, numerous forests in Germany and Switzerland where the system was once practiced were converted into high forest from the middle of the 19th century onward.

Most important, however, were the regeneration of forests over large areas and the management of uniform stands. In the plains and lower mountains the introduction of sustainable wood production during the 19th century quite often favored an organization of stands by predetermined periods of rotation, allowing regeneration of clearcut areas. Seeding of conifers and large plantations of spruce or pine permitted afforestation of exploited and devastated surfaces where natural regeneration was difficult or even impossible at the time. In general, conifers were systematically favored because they corresponded to the economic aims, according to which the thinning and final felling of even-aged stands allowed a rapid increase in wood production. In the Alps and, to a lesser extent, in other mountains of central Europe, the practice of selective logging was combined with natural regeneration. By now these practices have evolved toward various forms of silvicultural practices close to nature, such as selection forests.

The rapid expansion of mineral coal use by the mid-19th century and of fossil fuel use during the 20th century had major consequences. The diminishing pressure on wood as an energy source radically modified the conditions under which forests would be used for the industrial and economic expansion of a country. This has been a decisive element in the passage from a locally governed exploitation to a modern sector of the economy, functioning according to the principle of sustainable management of a renewable resource. The solutions for putting the principle into practice, developed from scientific models, allowed the intensity of felling to be adjusted to the long-term production potential of forest stands and sites. Silvicultural techniques to ensure regeneration by plantation or sowing, natural regeneration, tending and thinning of young stands, and species selection according to prevailing site conditions have advanced progressively. A modern forest economy, which has developed by successive stages, thus provides an increase in wood production, maintains or increases soil fertility and species diversity, and provides protection and recreation services in various combinations.

Modern silviculture implements a range of exploitation and regeneration methods, with the aim of bringing the forest to a stable and well-balanced condition. For several decades the use of natural regeneration has intensified along with efforts to increase the proportion of deciduous trees in the coniferous forests created in the past. Conservation of the genetic pool, with the aim not only of protecting biodiversity and the particularities of the landscape but also of maintaining the capacity of the forest to adapt to changing environmental conditions, has become a major factor in management. The planning system aimed at sustainable forestry, which has gradually become widespread, has been and remains a model for the valorization of natural renewable resources. In the same way, close-to-nature forestry practices make an important contribution to the process of sustainable development. They maintain the diversity of stands and at the same time combine flexibility in production with a long-term outlook; they also offer the population attractive and varied forests and landscapes. Because of the social utility of forests, their uses and management have become regulated to an extent that is uncommon in other economic activities. Legal requirements focus primarily on protecting the forest cover, on minimum standards for sustainable management, and on measures contributing to increased productivity. These provisions have proved their usefulness and will remain valid.

Multifunctional Forest Management

The traces of earlier colonization and abandoned farms reveal the dynamics of needs and values that determined the actual distribution of forests. The alternating processes of the forest cover's reduction and expansion modified the limits between forests and open space and led to the formation of varied landscapes. In regions under intensive cultivation, as well as around large towns and in the periurban space, forests now occupy only a small part of their initial range. On the other hand, in mountain regions and in the Alps the forest has remained or has again become a primordial element. In these regions it determines to a considerable extent the economic and social potential as well as the specificity of the landscape.

An ambivalent connection exists between the impacts of past uses and the perceptions and attitudes of people regarding the present state of forests and forest management. Knowledge of how the modern forest economy evolved to meet changing needs quite often contrasts with the significance our largely urban population places on the forest of today. To understand present and future options in managing forests, one has to be aware of the historical context. On the other hand, it is essential to understand today's needs and values and to grasp the economic utility and social significance of forests in modern societies. What forests mean at the present time to the population, landowners, and specific user groups has become an interesting and topical subject of research (Braun 2000; Corvol et al. 1997; Kalaora and Poupardin 1979). Empirical studies of the perceptions and attitudes of people regarding forests and forest management give information about the evolution of their social significance. Obviously, the individual and collective preferences of citizens living in largely urbanized societies are different from those of people who used the forests in the past.

Researchers have studied the perceptions and attitudes of the population at the scale of a country, of certain regions within a country, or of specific localities (Dufour and Loisel 1996; Elsasser 1996; Jensen and Koch 2000; Oesten and Roeder 1995; OFEFP 2000; Rocek 1999; Schelbert-Syfrig and Maggi 1988; Schmithüsen and Kazemi 1995; Schmithüsen et al. 2000; Silva 1997; Suter Thalmann 2000; Wild-Eck 2002; Wild-Eck and Zimmermann 2000; Zimmermann et al. 1996; Zimmermann et al. 1998). Other studies deal more specifically with the expectations and comportments of visitors in forests considered to be places reserved for leisure and recreation (Elsasser 1996; Kalaora 1981; Laffite 1993; Loesch 1980; Nielsen 1992; Schmithüsen and Wild-Eck 2000). On the whole, the studies analyzing the reasoning and purposes of citizens, owners, and users of the forest area have gained in importance in the domains of management and politics (Jacobsen and Koch 1995; Jensen 1993; Rocek 1998; Schmithüsen et al. 1997; Terrasson 1998; Wiersum 1998).

The findings confirm, first of all, that the forest remains for most people a usable and productive part of man's environment and that its management is notably conditioned by economic preferences. If wood formerly constituted an indispensable source of energy and was a major construction material, it is now replaceable, from a technical point of view, with the use of fossil fuels and new materials. Its use depends on its ability to win a place despite national and international competition. Coming from a renewable resource with a largely neutral carbon dioxide production cycle, wood production is today an essential political option in the context of protecting the environment and fostering sustainable development.

The results of the empirical studies show as well that forests have acquired a new and more global meaning in modern society, going beyond their role as a productive and usable resource. For a growing part of the population the forests represent a free space for recreation that is different from other transformed areas. At the same time, forests are more and more identified as a natural environment, perceived by many people to have little or no human influence. They represent the free interplay of natural forces, in contrast with inhabited areas and land intensively exploited by agriculture. This new development reflects the needs and preferences of contemporary society and the desire of an increasingly urban population for relaxation in natural surroundings. It expresses the preoccupations provoked by the impending threats to the environment and to biodiversity, resulting from personal experience and sensitivity toward global-scale phenomena in our world. Also, it is founded on the individual values of a large number of people for whom the forest represents a place for meditation, reflection, and freedom.

The wish to preserve the forest, a symbol of nature, is expressed by demands for limiting forest exploitation and protecting areas near to the natural state. For many people the protection of environment and landscape has become a major criterion in judging overall performance in forest management. The surveys confirm as well the importance of the social facilities provided by urban forests and the two important perspectives from which the green spaces within and around the urban space are seen by citizens. Forests suffer less from outside influences and can counterbalance and compensate for effects to which other intensively used areas are subject. Forests constitute a space permitting a greater liberty of movement and more spontaneous activities than other parts of an urban landscape.

Motives of people interviewed vary according to individual preferences and their social and economic conditions. The emphasis is usually on the forest as a place where one may walk, practice various sports, study nature, or breathe and relax; it is also a place where one feels happy and can rest from daily stress. The answers to the inquiries underline the importance of the forest as a place where one can withdraw and express one's love of nature, as a quiet place for personal reflection, and as a realm of physical and emotional sensations. Although visitors to the forest come for many reasons, the significance of emotional, spiritual, and mystical values is growing.

Opinions on the current role played by the forests of the interviewees' home regions show, for example, that in Switzerland the mountain forest is considered by almost everyone as a natural area and as an element in environmental protection (Schmithüsen et al. 2000). To the same extent, it is considered a place for recreation, an element of the landscape, and a renewable resource for wood production. The respondents' answers show as well that the importance of forests as a natural environment and a local area of liberty determines the priorities they give to management and forestry activities. Silvicultural care and regeneration, as well as repairing damage caused by natural disasters, are considered by more than 90 percent of respondents as important or very important. Activities aiming to protect or restore flora or fauna receive the same priority.

The available information brings into evidence the often contradictory expectations and demands surrounding forests and forestry management. For town dwellers the forest represents, above all, a favored area for leisure and relaxation. Inhabitants of mountain regions see it as protection against natural dangers and as a tourist attraction, and forest owners, farmers, and industry see it primarily as a source of revenue from harvesting wood. For one part of the population the forests are unique, and the necessity of conserving them predominates. Another part considers the economic aspects of wood production's providing employment and a source of revenue most important.

Sustainable management furnishes a concrete example of how the forest economy can react to diverse social interests and adapt to local conditions. The goal of sustainability is a fundamental condition for a forest economy that can keep open multiple options with respect to market trends and the changing needs of the population. If the conflicts generated by land use were previously at the fore, the very purpose of the forest and how it is managed currently make up the essential part of debates about man's relationship with his environment. The various fundamental concepts and management systems can now be found at the center of political debate. In the face of more and more pressing demands for environmental protection and conservation of biodiversity on a large scale, it is not the principle of sustainable wood production that is in question but certain forestry practices that are judged to be incompatible with sustainable development. From this point of view, a forest economy capable of taking into account profound currents of opinion in our society will benefit from the approval and acceptance of the population.

With regard to multifunctional uses, one may state that public policy objectives and modern forest laws have become more diversified and comprehensive during the past 20 years. Moving from a focus on wood as a sustainable resource, forest laws are now addressing a wider range of private and public goods and values. They acknowledge the equal importance of production and conservation, and their goals refer to the multifunctional resources of the forest, its economic potential, and to its importance to the environment. Increasingly, these policy objectives address the variety of ecosystems, the need to maintain biodiversity, and the development potential of forests in rural and urban areas. There is a trend to shift or delegate constitutional competencies on forestry matters to regional governments. Where the national level remains responsible, regional entities are becoming more involved in policy implementation. A similar process is occurring between the state level and communities or associations. The transfer of responsibilities favors multilevel political decisions and the negotiation of locally adapted solutions. It acknowledges that forests are of national concern but at the same time are local resources in which rural and urban people have immediate interests.

Sustainability provides an imperative for the use of natural resources. It starts from the principle that the present level of consumption and its effects on the environment must respect an equilibrium that makes the necessary room for maneuvering for future options. From this point of view, the forest economy does not represent a gratuitous mobilization of natural resources and production means. A sustainable management of forests requires investments that permit the maintenance of productivity and the adaptation of wood production to a long-term potential. It necessitates a framework of conditions that allow the harmonization of present interests with future potential. A sustainable use of natural resources is thus linked to concrete economic and technical conditions and, therefore, depends in the same way on fundamental human perspectives and social norms. Sustainability does not of itself express an intention for the use of resources; rather, it represents that which people and social and political communities recognize as worth saving and managing responsibly.

Conclusions

The varied landscapes found in Europe and the successive forms of forest uses observed during different historical periods indicate the diversity and intensity of multiple needs; they also demonstrate the importance of spiritual values and of social and political realities. Certain changes resulting from past human interventions appeared over a relatively short period of time, and their consequences on the extent and composition of forest stands rapidly became clear. Other changes, often those with the heaviest consequences, came to light indirectly, and their effects could be appreciated only after long periods.

Manifold uses of the forest have followed and often superseded each other over the course of centuries. Forests have been and still are a local resource, complementary to agricultural and pastoral production, energy and raw material resources, and the foundation of modern forestry and wood-processing industry developments. Use and management of the natural potential of forests have made possible many economic and social activities, which in return have shaped the forest to a large degree. Thus, European forests bear witness to cultural processes and developments, and they show evidence of the impact of numerous and constantly changing human needs. Forests have quite a different meaning for today's population from that which prevailed in past centuries. Shaped by the past, the actual forest stands offer multiple alternatives for satisfying today's economic and social demands, and present-day multifunctional management will leave further options and a different development potential for future generations.

The following aspects are of particular relevance in assessing current trends in European forestry development:

- In the course of the past two centuries sustainable forest management has made great progress, thanks to the efforts of forest owners, professionals, and scientists. Step by step, it has integrated incremental societal demands into current management practices. Multifunctional objectives and close-to-nature practices are pursued in many European forests. Altogether, the patient work of foresters has led to productive forests and diversified landscapes.
- Wood production remains the center of forestry. It provides economic opportunities, maintains a valuable labor force in rural areas, and contributes to a regular regeneration of protection forests in mountainous regions. Rational and economically feasible wood production is the prerequisite for an expanding European wood-processing sector. Thanks to considerable investments in new production technologies, the sector's competitiveness in world markets increases steadily.
- Wood production and the use of wood products imply a largely neutral production and consumption cycle with regard to emissions of carbon dioxide. Expanding the forest and wood products sector is an essential option in the context of environmental protection, climate change, and maintenance of the renewable resource base. Accumulating additional

biomass under good forestry practices and by afforestation is an important political requirement for implementing the Kyoto Protocol.
- A new challenge is to develop forestry practices and wood production within the larger context of rural development and landscape management. Doing so will make an indispensable contribution to maintaining the renewable natural resource base for our own use as well as for future options in using forests. Private enterprise and public policies, as much as private and public investment, need to be coordinated to use natural resources more efficiently on a landscape scale.
- New forms of combined land uses at local and regional levels are needed. Conservation of the genetic pool is henceforth a major factor in management. It aims to protect biodiversity and maintain the capacity of forest ecosystems to adapt to changing environmental conditions. Consensual and nonconflicting strategies for providing accessible space that satisfies the demand for leisure and tourism should also be developed and strengthened.
- The level of integration between environmental requirements and efficient economic productions processes is the benchmark for modern forestry. Legal and economic instruments balancing rights and responsibilities in private and public land management are indispensable for generating an optimal combination of benefits from sustainable forest management. Marketable products and services can be financed from market proceeds. Public goods and services for which no markets exist or for which none can be developed, for whatever reasons, need public investment or must be financed by those benefiting directly.
- To solidify the content of multifunctional forestry development, we must agree on the foundation for individual and collective decisions. Equal consideration of economic, social, and environmental goals is essential in determining the framework for forest protection and forestry development. This requires democratic decision making and private and public arbitration processes among a large group of stakeholders. It also needs an explicit understanding of what is to be considered the goal of individual as well as societal progress.

References

Allmann, J. 1989. *Der Wald in der frühen Neuzeit. Eine mentalitäts- und sozialgeschichtliche Untersuchung am Beispiel des Pfälzer Raumes 1500–1800.* Schr. z. Wirtschafts- und Sozialgesch. 36. Berlin, Duncker und Humblot. (416 S.)

Arnould, P., M. Hotyat, and L. Simon. 1997. *Les forêts d'Europe.* Collection fac. géographie. Paris, Edition Nathan. (413 p.)

Badré, L. 1983. *Histoire de la forêt française.* Paris, Arthaud. (186 p.).

Bechmann, R. 1984. *Des arbres et des hommes—La forêt au Moyen-Age.* Paris, Flammarion. (385 p.)

Baum, C. 1995. *Der Klosterwald von St. Blasien. Eine forstgeschichtliche Untersuchung über die Waldverhältnisse im Stiftsbann der ehemaligen Benediktinerabtei St. Blasien im südlichen Schwarzwald bis zu Beginn des 19. Jahrhunderts.* Freiburg Brsg., Hochschul-Verlag. (314 S.)

Bill, R. 1992. *Die Entwicklung der Wald- und Holznutzung in den Waldungen der Burgergemeinde Bern vom Mittelalter bis 1798.* Dissertation ETH Zürich.

Bonhôte, J. 1997. *Forges et Forêts dans les Pyrénées Ariègeoises—Pour une histoire de l'environnement.* Collection Universatim, F-31160 Aspet, PyréGraph éditions. (320 p.)

Brandl, H. 1970. *Der Stadtwald von Freiburg. Eine forst- und wirtschaftsgeschichtliche Untersuchung über die Beziehungen zwischen Waldnutzung und wirtschaftlicher Entwicklung der Stadt Freiburg vom Mittelalter bis zur Gegenwart.* Freiburg Brsg., Poppen & Ortmann. (258 S.)

———. 1987. Zur Geschichte der Wirtschaftlichkeit in der Forstwirtschaft. AFZ 42: 1019–1023

Braun, A. 2000. *Wahrnehmung von Wald und Natur.* Opladen, Leske und Budrich.

Bund, B. 1997. Der Wandel der Kulturlandschaft Nordschwarzwald seit der 2. Hälfte des 19. Jahrhunderts. Eine historische Raum-Zeit-Analyse mit Hilfe eines geographischen Informationssystems (GIS). Dissertation Universität Freiburg Brsg. (180 S. und Kartenband)

Bürgi, M. 1997. Waldentwicklung im 19. und 20. Jahrhundert—Veränderungen in der Nutzung und Bewirtschaftung des Waldes und seiner Eigenschaften als Habitat am Beispiel der öffentlichen Waldungen im Zürcher Unter- und Weinland. Diss. ETH Nr. 12 152, Zürich. (226 S.)

Carlowitz, H.C. von. 1713. *Sylvicultura Oeconomica oder Hausswirthliche Nachricht und Naturgemässe Anweisung zur Wilden Baum-Zucht.* Leipzig.

Cavaciocchi, S. (ed.). 1996. *L'uome et la foresta, Secc. XIII–XVIII.* Atti delle Settimane di Studi 27, Instituti Internationale di Storia Economica F. Datini, Firenze, Le Monnier. (1234 p.) – with numerous contributions in English, French and German.

Centre Historique des Archives Nationales, 1997. *Histoire de Forêts—La Forêt Française du XIII° au XX° Siècle.* Paris, Société Nouvelle Adam Biro. (158 p.)

Corvol, A. 1987. *L'homme aux bois—Histoire des relations de l'homme et de la forêt, XVII°–XX° siècle.* Paris, Fayard. (585 p.)

Corvol, A., P. Arnould, and M. Hotyat. 1997. *La Forêt: perceptions et représentations.* Paris, Editions l'Harmattan. (401 p.)

Devèze, M. 1965. *Histoire des forêts.* Paris, Presses Universitaires de France. (128 p.)
Dufour, A., and J.-P. Loisel. 1996. Les opinions des Français sur l'environnement et sur la forêt. Enquête « Conditions de vie et aspirations des Français » réalisée à la demande de l'Institut Français de l' Environnment (IFEN), Paris, CREDOC.
Ebeling, D. 1992. *Der Holländerholzhandel in den Rheinlanden.* Stuttgart, Steiner. (241 S.)
Ellenberg, H. 1996. Vegetation Mitteleuropas mit den Alpen. 5. Aufl. Stuttgart, Ulmer. (1095 S.)
Elsasser, P. 1996. *Struktur, Besuchsmotive und Erwartungen von Waldbesuchern—Eine empirische Studie in der Region Hamburg.* (Vol. 1). Hamburg: Institut für Ökonomie der Bundesforschungsanstalt für Forst- und Holzwirtschaft.
Ernst, C. 2000. *Den Wald entwickeln. Ein Politik- und Konfliktfeld in Hunsrück und Eifel im 18. Jahrhundert.* Reihe: Ancient Régime, Aufklärung und Revolution 32. München. (408 S.)
Fenkner-Voigtländer, U. 1992. *Forsteinrichtung und Waldbau im Elmsteiner Wald unter deutschen und französischen Einflüssen 1780–1860.* Ein Beitrag zur Forstgeschichte des Pfälzerwaldes. Dissertation Universität Freiburg Brsg. Mitteilungen der Landesforstverwaltung Rheinland-Pfalz 10. (341 S.)
Fischer, A. 1985. *Waldveränderungen als Kulturlandschaftswandel. Fallstudien zur Persistenz und Dynamik des Waldes in der Kulturlandschaft des Kantons Luzern seit dem Forstgesetz von 1875.* Basel, Wepf und Co. (214 S.)
Gerber, B. 1989. *Waldflächenveränderungen und Hochwasserbedrohung im Einzugsgebiet der Emme.* Geographica Bernensia G 33, Geographisches Institut der Universität, Bern. (99 S.)
Grewe, B.-S. 2002. Der versperrte Wald—Vorindustrieller Waldressourcen-mangel am Beispiel der bayerischen Rheinpfalz 1814–1870. Dissertation Universität Trier.
Hachenberg, F. 1988. *Waldwirtschaft und Forstliche Landschaftsgestaltung im vorderen Hunsrück in zwei Jahrhunderten. Zur Forstgeschichte des Forstamtes Kastellaun in den Jahren 1815–1985.* Schutzgemeinschaft Deutscher Wald Obermoschel/Pfalz Nr. 6. (425 S.)
———. 1992. 2000 Jahre Waldwirtschaft am Mittelrhein. Selbstverlag des Landesmuseums Koblenz, (214 S.)
Hafner, F. 1979. *Steiermarks Wald in Geschichte und Gegenwart. Eine forstliche Monographie.* Wien, Österr. Agrarverlag . (396 S.)
Hasel, K. 1985. *Forstgeschichte—Ein Grundriss für Studium und Praxis.* Hamburg und Berlin, Parey. (258 S.)
Hauser, A. 1972. *Wald und Feld in der alten Schweiz.* Zürich und München, Artemis. (422 S. und Bildanhang)
Hausrath, H. 1982. *Geschichte des deutschen Waldbaus—Von seinen Anfängen bis 1850.* SchrR. Inst. für Forstpolitik und Raumordnung, Uni. Freiburg, Freiburg Brsg. (416 S.)
Heyer, C. *Die Waldertrags-Regelung.* Giessen.
Hillgarter, F.-W., and E. Johann. 1994. *Österreichs Wald—Vom Urwald zur Waldwirtschaft.* 2. völlig überarb. und erweiterte Auflage. Wien, Eigenverlag Autorengemeinschaft. (544 S.)

v. Hornstein, F. 1951. *Wald und Mensch—Waldgeschichte des Alpenvorlandes Deutschlands, Österreichs und der Schweiz.* Ravensburg, Maier. (283 S.); Reprint 1984.

Irniger, M. 1991. *Der Sihlwald und sein Umland—Waldnutzung, Viehzucht und Ackerbau im Albisgebiet von 1400–1600.* Mitt. Antiquarische Ges. in Zürich 58, Zürich.

Jacobsen, C. H., and N. E. Koch. 1995. Summary Report on Ongoing Research on Public Perceptions and Attitudes on Forestry in Europe (Summary Report). Horsholm: Danish Forest and Landscape Research Institute.

Jensen, F. S. 1993. Landscape Managers' and Politicians' Perception of the Forest and Landscape Preferences of the Population. *Forest and Landscape Research* 1993: 79–93.

Jensen, F. S., and N.E. Koch. 2000. Measuring Forest Preferences of the Population – A Danish Approach. *Swiss Forestry Journal* 151, 11–16.

Johann, E, 1968. *Geschichte der Waldnutzung in Kärnten unter dem Einfluss der Berg-, Hütten- und Hammerwerke.* Klagenfurt. Verlag des Geschichtsvereines für Kärnten. (248 S.und Bildanhang)

Kalaora, B. 1981. *Le musée vert ou le tourisme en forêt – Naissance et développement d'un loisir urbain, le cas de la forêt de Fontainebleau.* Editions Anthropos. (304 p.)

Kalaora, B., and E. Poupardin. 1979. *La forêt et la ville—Essai sur la forêt dans l'environnement urbain et industriel.* Paris: INRA, station de recherche sur la forêt et l'environnement.

Kasper, H. 1989. *Der Einfluss der eidgenössischen Forstpolitik auf die forstliche Entwicklung im Kanton Nidwalden in der Zeit von 1876 bis 1980.* Dissertation ETH Zürich; Mitt. EAFV (65) 1: 3–180, Birmensdorf.

Kempf, A. 1985. *Waldveränderungen als Kulturlandschaftswandel—Walliser Rhonetal. Fallstudien zur Persistenz und Dynamik des Waldes zwischen Brig und Martigny seit 1873.* Basler Beiträge zur Geographie 31. Basel, Wepf und Co. (229 S. und 33 S. Anhang)

Keweloh, H.-W. (Hg.). 1988. *Auf den Spuren der Flösser—Wirtschafts- und Sozialgeschichte eines Gewerbes.* Stuttgart, Theiss. (286 S.)

Kremser, W. 1990. *Niedersächsische Forstgeschichte—Eine integrierte Kulturgeschichte des nordwestdeutschen Forstwesens.* Rotenburger Schriften Sonderband 32. Rotenburg/Wümme, Selbstverl. Heimatbund. (965 S.)

Koller, E. 1970. *Forstgeschichte des Salzkammergutes. Eine forstliche Monographie.* Wien, Österr. Agrarverlag. (558 S. und Bildanhang)

———. 1975a. *Forstgeschichte Oberösterreichs.* Linz, Oberösterr. Landesverlag. (269 S. und Bildteil)

———. 1975b. *Forstgeschichte des Landes Salzburg.* Salzburg, Verlag der Salzburger Druckerei. (347 S. und Bildanhang)

Konold, W. (Hg.). 1996. *Naturlandschaft—Kulturlandschaft: Die Veränderung der Landschaften nach der Nutzbarmachung durch den Menschen.* Landsberg, ecomed. (322 S.)

Kunz, J. 1995. *Der Gemeindewald von Hassloch. Ein Beitrag zur Geschichte des Kommunalwaldes in Rheinland-Pfalz mit wirtschafilichem Schwerpunkt.* Veröffentlichungen des Arbeitskreises Forstgeschichte in Rheinland-Pfalz Nr.2/1995, Hassloch. (291 S.)

Küster, H. 1995. *Geschichte der Landschaft in Mitteleuropa.* München, Beck. (424 S.)

———. 1998. *Geschichte des Waldes—Von der Urzeit bis zur Gegenwart.* München, Beck. (267 S.)

Lafitte, J.-J. 1993. Sondage d'opinion sur les forêts périurbaines. *Revue Forestière Française,* 35(4), 483–492.

Lang, G. 1994. *Quartäre Vegetationsgeschichte Europas.* Jena, Fischer. (462 S.)

Loesch, G. 1980. Typologie der Waldbesucher—Betrachtung eines Bevölkerungsquerschnitts nach dem Besuchsverhalten, der Besuchsmotivation und der Einstellung gegenüber Wald. Dissertation, Universität Göttingen.

Loderer, A.A. 1987. Besitzgeschichte und Besitzverwaltung der Augsburger Stadtwaldungen—Ein Beitrag zur Augsburger Stadtgeschichte. Dissertation Universität München. (381 S. mit Anlagen)

Ludemann, T. 1990. Im Zweribach—Vom nacheiszeitlichen Urwald zum "Urwald von morgen". Dissertation Universität Freiburg Brsg. (268 S.)

Mantel, K. 1980. *Forstgeschichte des 16.Jahrhunderts unter dem Einfluss der Forstordnungen und Noe Meurers.* Berlin, Parey. (1071 S., 32 Abb.)

———. 1990. *Wald und Forst in der Geschichte—Ein Lehr- und Handbuch.* Mit einem Vorwort von Helmut Brandl. Nach dem Tode des Verfassers für den Druck bearbeitet von Dorothea Hauff. Alfeld-Hannover, Schaper. (518 S.)

Mantel, K., and J. Pacher. 1976. *Forstliche Biographien vom 14. Jahrhundert bis zur Gegenwart—Zugleich eine Einführung in die forstliche Literaturgeschichte.* Hannover, Schaper. (441 S.)

Müller, U. 1990. Schutzwaldaufforstungen des Staates Freiburg im Senseoberland—Forstpolitische Massnahmen des Staates Freiburg seit 1850 am Beispiel der Schutzwaldaufforstungen im Flyschgebiet des Senseoberlandes. Diss. ETH Nr.9001, Freiburg i.U., Kantonsforstamt. (258 S.)

Nielsen, C. 1992. *Der Wert stadtnaher Wälder als Erholungsraum—Eine ökono-mische Analyse am Beispiel von Lugano.* Chur: Rüegger.

Oberrauch, H. 1952. *Tirols Wald und Waidwerk. Ein Beitrag zur Forst- und Jagdgeschichte.* Innsbruck, Universitätsverlag Wagner. (328 S.)

Oesten, G., and A. Roeder. 1995. Wertschätzung des Pfälzerwaldes. *Allgemeine Forstzeitschrift,* 50(2), 105–107.

OFEFP. 2000. Les attentes de la société envers la forêt suisse—Enquête d'opinion. *Cahier de L'Environnement* Nr. 309; Office fédéral de l'environnement, des forêts et du paysage, OFEFP, Berne.

Pagenstert, G. 1961. Forstliche Beziehungen zwischen Deutschland und Frankreich im 19. Jahrhundert. Dissertation Universität Freiburg Brsg. (236 S.)

Parolini, J.D. 1995. Zur Geschichte der Waldnutzung im Gebiet des heutigen Schweizerischen Nationalparks. Diss. ETH Nr. 11 187, Zürich. (227 S.)

Pott, R. 1993. *Farbatlas Waldlandschaften—Ausgewählte Waldtypen und Waldgesellschaften unter dem Einfluss des Menschen.* Stuttgart, Ulmer. (224 S.)

Radkau, J., and I. Schäfer. 1987. *Holz—Ein Naturstoff in der Technikgeschichte.* Reinbeck bei Hamburg, Rowohlt. (313 S.)

Rocek, I. 1998. Les attitudes des habitants de la République Tchèque envers la forêt et la gestion forestière. Document de travail, Série internationale 98/3 . Chaire de politique et économie forestière, École polytechnique fédérale de Zurich.

———. 1999. Les opinions des propriétaires forestiers – Résultats d'une enquête en République Tchèque. Document de travail, Série internationale. Chair de politique et économie forestière, École polytechnique fédérale de Zurich.

Rommel, W.-D. 1990. *Die Flösserei auf dem Kocher, insbesondere die Versorgung der Salinen Hall und Friedrichshall und ihre forstwirtschaftliche Auswirkung auf das Limpurger Land vom Ausgang des Mittelalters bis zur Industrialisierung.* Diss. Universität Freiburg Brsg., Schwäbisch-Hall- Gelbingen. (218 S.)

Rubner, H. 1965. *Untersuchungen zur Forstverfassung des mittelalterlichen Frankreichs.* Berlin, Schrift. Wirtsch.- u. Soz.gesch. 8.

———. 1967. *Forstgeschichte im Zeitalter der industriellen Revolution.* Berlin, Duncker und Humblot. (235 S.)

Schäfer, I. 1992. *"Ein Gespenst geht um"—Politik mit der Holznot in Lippe.* Detmold, Selbstverlag des Naturwissenschaftlichen und Historischen Vereins für das Land Lippe. (328 S.)

Scheifele, M. 1988. *Die Murgschifferschaft—Geschichte des Flosshandels, des Waldes und der Holzindustrie im Murgtal.* Gernsbach, Katz. (521 S.)

———. 1993. *Die Flösserei auf der Ettlinger Alb.—Aus der Geschichte des Albtales.* Gernsbach, Katz. (148 S.)

———. 1996. *Als die Wälder auf Reisen gingen—Wald-Holz-Flösserei in der Wirtschaftsgeschichte des Enz-Nagold-Gebietes.* Karlsruhe, Braun. (368 S.)

Schelbert-Syfrig, H., and R. Maggi. 1988. Wertvolle Umwelt—Ein wirtschaftswissenschaftlicher Beitrag zur Umwelteinschätzung in Stadt und Agglomeration Zürich (Wirtschaft und Gesellschaft 3). Zürich: Zürcher Kantonalbank.

Schenk, W. 1996. Waldnutzung, Waldzustand und regionale Entwicklung in vorindustrieller Zeit im mittleren Deutschland. Historisch-geographische Beiträge zur Erforschung von Kulturlandschaften in Mainfranken und Nordhessen. *Erdkundliches Wissen* Heft 117, Stuttgart, Steiner. (325 S.)

Schmidt, U.E. 1989. Entwicklung in der Bodennutzung im mittleren und südlichen Schwarzwald seit 1780. *Mitteilungen der Forstlichen Versuchs- und Forschungsanstalt Baden-Württemberg*, Heft 146 Band 1 und 2. (206 S. Text und 109 S. Anhang)

———. 1997. Das Problem der Ressourcenknappheit—dargestellt am Beispiel der Waldressourcenknappheit in Deutschland im 18. und 19. Jahrhundert—eine historisch-politische Analyse. Habil.-Schrift Forstw. Fak. Universität München. (434 S.)

Schmithüsen, F., and Y. Kazemi. 1995. Analyse des rapports entre les attitudes des gens envers la forêt et leurs attitudes envers la gestion forestière. *Schweizerische Zeitschrift für Forstwesen*, 146(4), 247–264.

Schmithüsen, F., and S. Wild-Eck. 2000. Uses and Perceptions of Forests by People Living in Urban Areas—Findings from selected Empirical Studies. *Forstw. Cbl.* 119 (2000), 395–408.

Schmithüsen, F., Y. Kazemi, and K. Seeland. 1997. *Enquêtes sur les Attitudes de la Population envers la Forêt et ses Prestations Sociales—Origines sociales et thèmes de recherche des enquêtes sociologiques réalisées en Allemagne, Autriche et Suisse de 1960–1995. Schweizerische Zeitschrift für Forstwesen* 148(1), 1–43.

Schmithüsen, F., S. Wild-Eck, and W. Zimmermann. 2000. Einstellungen und Zukunftsperspektiven der Bevölkerung des Berggebietes zum Wald und zur Forstwirtschaft. Beiheft 89; *Schweiz. Zeitschrift für Forstwesen.*

Schoch, O. 1994. *Von verschwundenen Waldgewerben im Nordschwarzwald—Beispiele aus dem oberen Fnztal.* Neuenbürg, Müller. (163 S.)

Schuler, A. 1977. *Forstgeschichte des Höhronen.* Gut & Co. Verlag, Stäfa. (180 S.)

Seeland, K. 1993. Der Wald als Kulturphänomen—Von der Mythodogie zum Wirtschaftsobjekt. Arbeitsberichte, Allgemeine Reihe 93/3, Professur Forstpolitik und Forstökonomie ETH Zürich.

Seling, I. 1997. *Die Dauerwaldbewegung in den Jahren zwischen 1880 und 1930. Eine sozialhistorische Analyse.* Schriften aus dem Institut für Forstökonomie der Universität Freiburg Bd. 8, Freiburg Brsg. (128 S.)

Selter, B. 1995. *Waldnutzung und ländliche Gesellschaft—Landwirtschaftlicher "Nährwald" und neue Holzökonomie im Sauerland des 18. und 19. Jahrhunderts.* Paderborn, Schöningh. (482 S.)

Semmler, J. (Hg.), 1991. *Der Wald in Mittelalter und Renaissance.* Düsseldorf, Droste. (239 S.)

Sieferle, R.P. 1982. *Der unterirdische Wald—Energiekrise und Industrielle Revolution.* München, Beck. (283 S.)

Silva, M.-A. 1997. La signification de l'arbre pour la ville et les habitants de Genève Arbeitsbericht: Allgemeine Reihe 97/3. Zürich: Professur Forstpolitik und Forstökonomie.

Speidel, G. 1984. *Forstliche Betriebswirtschaftslehre.* 2. Auflage, Paul Parey, Hamburg und Berlin.

Stuber, M. 1996. "Wir halten eine fetter Mahlzeit, denn mit dem Ei verzehren wir die Henne"—Konzepte nachhaltiger Waldnutzung im Kanton Bern 1750–1880. Diss. phil.-hist., Uni Bern, Zürich, Beih. *Schweiz. Zeitschrift für Forstwesen.* 82. (275 S.)

Suter Thalmann, C.-L. 2000. Erkennen der gesellschaftlichen Ansprüche an den Schweizer Wald im Wandel der Zeit – eine Buwal Studie. *Schweiz. Zeitschrift für Forstwesen* 151, 17–20.

Terrasson, D. (ed.). 1998. *Public Perception and Attitudes of Forest Owners towards Forests in Europe. Perception publique et attitudes des propriétaires envers la forêt en Europe.* Commentaires et synthèse du groupe de travail COST E 3 – WGI – 1994–1998. CEMAGREF Editions, Antony. 243 p.

Textor, H. 1991. *Die Amorbacher Zent. Eine wald-, forst- und wirtschaftsgeschichtliche Untersuchung des Klosterwaldes, des Mitmärkerwaldes, des herrschafts- bzw. landesherrlichen Waldes vom frühen Mittelalter bis zur Säkularisation 1802/03.* Mitteilungen des Naturwissenschaftlichen Museums der Stadt Aschaffenburg. (358 S. und 138 S. Anhang)

Wiersum, F. (ed.). 1998. Public Perception and Attitudes of Forest Owners toward Forest and Forestry inEurope. *Proceedings of the COST Conference—Working Group 1*; Hinkeloord Report 24; Sub-Department of Forestry, Agricultural University Wageningen, The Netherlands.

Wild-Eck, S. 2002. Statt Wald—Lebensqualität in der Stadt. Bedeutung naturräumlicher Elemente am Beispiel der Stadt Zürich. Seismo Verlag, Zürich. (454 S.)

Wild-Eck, S., and W. Zimmermann. 2000. COST- und Monitoring-Projekt: Zwei neue forstliche Meinungsumfragen im Vergleich. *Schweiz. Zeitschrift für Forstwesen*, 151 (1), 1–10.

Woronoff, D. (ed.). 1990. Forges et forêt—Recherches sur la consommation protoindustrielle de bois. Paris, *Recherches d'histoire et de sciences sociales* 43. (261 p)

Zimmermann, W., St. Wild-Eck, and F. Schmithüsen. 1996. Einstellung der Bergbevölkerung zu Wald, Forstwirtschaft und Forstpolitik. *Schweizerische Zeitschrift für Forstwesen*, 147(9). 727–747.

Zimmermann, W., F. Schmithüsen, and St. Wild-Eck. 1998. Main Findings and Policy Implications from the Research Project Pubic Perceptions of Mountain Forests in Switzerland. In: Wiersum, F. (ed.), 1998. *Public Perception and Attitudes of Forest Owners towards Forest and Forestry in Europe*; Hinkeloord Report 24: 47–59; Agricultural University Wageningen, The Netherlands.

Zürcher, H.U. 1965. Die Idee der Nachhaltigkeit unter spezieller Berücksichtigung der Gesichtspunkte der Forsteinrichtung. *Mitt. der Schweizerischen Anstalt für das forstliche Versuchswesen. Band.* 41, Heft 4. Zürich.

ABSTRACT

Ever-increasing demand for wood and wood products will characterize the next century, as world population grows and developing countries gain economic strength. Foresters must meet new needs, primarily biomass for energy production. The optimal use of wood requires concerted efforts to improve utilization and recycling of forest products. Exciting innovations include new types of laminated products that turn small-diameter and other formerly worthless trees into added-value structural materials of great durability and strong mechanical performance. New preservative treatments that are safe for the environment are being developed. All this bodes well for the wood products sector, but ultimately, the industry will have to compete with people's equally insistent demands for forest preservation, too. Thus environmental policies will constrain forest operations.

CHAPTER 13

The Future of Wood in Our Evolving Societies

Michel Vernois

Despite all the high-tech monitoring tools we have to assist us in tracking societal trends and making forecasts, predicting even the immediate future can prove risky. As we attempt to foresee the future over a time scale covering the next century, forecasting will probably become one of the major challenges of our society. This is partly because our needs and aspirations, whether explicit or implicit, are constantly evolving, and partly because of the pace of technological advancements.

Without being completely wrong, we are nevertheless able to anticipate the major trends and challenges that are likely to manifest themselves in the short term and become the main issues of the 21st century: the increasing number of people who will soon have access to wealth, education, and new technologies; the harmonious management of resources, some of which are in inevitable decline; the implications of the greenhouse effect and global climate change for ecosystems; and our desire for security and reassurance in a world that has so few bearings.

The love that man has for the forest and wood is intensifying in the context of continuing urbanization. Recent storms that hit Europe, Slovakia, and southern Sweden, destroying forest landscapes, have reminded us of the fragility of our ecosystems.

Wood is precious, and not wasting it is of capital importance. It should be used advisedly and be the subject of effective management schemes developed to perpetuate the resource and the diversity of ecosystems. To this end, we must take the time and make the effort to share information with a large audience—to educate and communicate with them, using various means—and dispel misunderstandings or fears pertaining to industrial or recreational activities within the forestry sector.

Current data concerning the consumption rates of wood as a raw material show that in 1995, for the entire planet, demand was estimated at 4 billion m^3 for round wood, of which the 25 nations of Europe accounted for 300 million m^3. We are forecasting a global consumption rate of 5 billion m^3 in 2010, and the forecast for Europe in 2025 is in the region of 500 million. Certain changes in the wood products industry, namely the increasing needs of Asia, will probably alter the regions from which demand for wood comes. In 1995 only 3 percent of the world's forested area generated 25 percent of all forest products. Fast-growing plantations will probably account for 5 percent of this forested area by 2010 and will probably generate new products, such as biofuel and other extractions from biomass, while significantly contributing to a reduction in the greenhouse effect.

What possible long-term difficulties can we anticipate in the forest products sector? First of all, we have reason to believe that competition between potential wood users may pose a serious risk of price destabilization and difficulties in mobilizing the raw materials in demand. We predict a high increase, in Europe and elsewhere, in the demand for fuelwood for generating units of several dozen megawatts, some of which would operate with a combined power generation system. The demand for softwood should thus increase. At the same time, demand for pulpwood will remain stable within the paper and panel industries, with plants in eastern Europe and Russia. Reconstituted wood, which is being sold in the countries of southern Europe, is likely to have greater presence in the market once there is a demand for constant characteristics and once techniques using small-diameter wood begin to mature.

The increase in demand for energy and the growing shortage of petroleum products in the future will pave the way for the mass production of biofuels, much of it from forest products rather than from agricultural crops. For the same reasons, we anticipate the advent of a "green chemistry" era based on the use of renewable carbon as a substitute for fossil carbon. The same is true of wood-plastic composite materials; the polymer matrix could even be of plant origin, making the products 100 percent natural.

The generally favorable impact of environmental protection policies will inevitably put a sharp curb on forest operations. It goes without saying that certain control procedures are required for felling and harvesting, particularly in some parts of the world, since the values associated with the primary forest and the biodiversity of species are of vital importance. Because of increasing social demand for natural products and its ability to store carbon, wood has unquestionable advantages. We should expect to have to provide tangible proof of this last point and better communicate the ecological benefits of using wood; life-cycle analyses will have to become systematized to support the argument.

Another point that merits our attention is that forest products can help reduce the greenhouse effect from emissions of carbon dioxide. In certain countries, such as France, there are incentives, if not regulatory measures, to increase the quantities of wood allocated to the construction industry. Hence, in France an act that seeks to limit air pollution in line with the Kyoto Protocol agreements stipulates that public buildings consist of at least 12.5 percent wood.

Although it is a sensitive subject in France, and probably also in many other countries, the trend toward "cocooning" and the preference for living in single-family houses are yet another positive factor for the forest products industry.

Current technological advancements in the utilization of wood and its by-products focus on obtaining the maximum yield of the raw material in all its forms, including recycled wood products. Added to this is the need for reliable products, particularly in the construction industry, that pose no hazards either to human health or to the environment, and that have well-defined and controlled characteristics. I feel that this trend will inevitably continue over the next few decades.

Naturally, we cannot ignore the environmental or human health and safety issues associated with forest products, and consequently, we will have to prove their harmlessness, particularly with regard to preservation agents used to enhance the durability of wood. In some sectors, we will need to train professionals in working with certain materials. This may become an issue that will need to be dealt with on a long-term basis.

Although they have been developed only over the past two decades, wood-based reconstituted materials have considerable potential for technological advancement and already have an established position in the market; I believe they should be encouraged. Examples that illustrate the rational and optimal use of wood—sawmill industry by-products, small-diameter trees, and fast-growing species—include reconstituted solid wood in the form of panels; I-beams, of which the members are generally made of solid wood and the web of plywood; highly compressed fibreboard; and oriented-strand board (OSB) panels, which have excellent structural characteristics.

New types of panels have been available in North American markets for a few years now, though few have reached European markets because of a lack of industrial investment on the continent. I think that we will eventually see products that particularly lend themselves to architectural developments, such as "cathedral" ceilings, as people seek homes with large interior spaces. Among the products satisfying the demand, we could mention Lamibois, Parallam, and Intrallam, which consist of very long strips of wood measuring approximately 30 cm and have excellent mechanical properties One of the difficulties encountered with the introduction of these types of panels and beams on the European market appears to be cultural in nature. In particular, carpenters are not used to working with material with cross sections four times smaller than solid wood beams of equivalent resistance. Although these products are more expensive per cubic meter, because of their mechanical characteristics, they will remain competitive nonetheless.

Such materials produce very slender structures. This is true of Intralam, almost unknown in Europe, and Parallam, which in addition to its excellent physical properties has very desirable aesthetic characteristics. We are hoping that a factory able to manufacture this type of structural panel will soon be built in Europe. Only laminated-veneer lumber (LVL, in which the grain of each layer of veneer runs in the same direction, rather than the cross-lamination typical of plywood) is currently produced in northern Europe, and the assembly production capacity is likely to be doubled shortly.

High-load structures can be made using Lamibois or LVL. These materials allow for new architectural designs that are not conceivable with solid wood. To a lesser extent, the same is true of OSB core I-beams, of which the studs consist of Lamibois.

Medium-density fiberboard (MDF) has really taken off on the world market over the past few years. It is made of uniform wood fibers, to which is added a small amount of synthetic resin; mats of this fibrous material are pressed under high pressure into panels, which can then be cut and shaped like solid wood. Aware of the problems associated with the use of adhesives incorporating formaldehyde, which may be released into the atmosphere, the adhesive and panel industries are researching the use of adhesives that are formaldehyde free or natural. One possibility is to pretreat the wood chips with enzymes, making the reactivated lignin "self-adhesive" and eliminating the need for synthetic resins altogether. Recent industrial tests were performed in Sweden for this, with a view to developing adhesive-free MDF.

The transfer of technologies has enormous potential for the wood sector. The scope for investigation in this field is considerable, since wood specialists very rarely resort to technologies from other industrial sectors that are so fundamentally different. Take, for instance, wood molded trims for interior car parts—a perfect example of a use for wood that nobody would imagine even exists. Few people are aware that the door panels, rear window shelf, and rear floor pan of their cars may be made of wood fiber mats pressed to give them a three-dimensional shape. Some of these mats are made with the carding-batting and needling devices used in nonwoven textile technologies, and the Madison Forest Products Laboratory has been making every effort to promote these techniques. In this way, wood products are being designed for the automotive industry to substitute for synthetic materials, such as plastic.

We cannot omit mention of wood-plastic composite materials, especially since demand is increasing at a rate that exceeds 20 percent per year in North America. There are now 52 factories in the United States and Canada that produce wood plastic materials. Although until recently these products consisted of 40 to 50 percent polymers, current technologies enable manufacturers to extrude products containing 80 percent wood.

A new product manufactured in Austria is among the more innovative: it is a wood composite of which the polymer matrix is made up of natural products (80 percent wood flour and 20 percent a modified starch associated with a natural resin). I think this product, which is not thermoplastic but is thermosetting, has a bright future. Such products can be extruded or injected into complex forms, and new wood molding and malleability techniques allow for ingenious and novel designs.

But we cannot ignore this imperative: We need to ensure the durability of the material used, whether by chemical treatment with low impact on the environment, or by physical treatments that enhance durability. Over the past ten years, France and Finland in particular have been producing wood treated at high temperatures without chemical products. Some high-temperature heat treatment furnaces reach 230 degrees C. Depending on the treatment temperatures, different degrees of durability are achieved.

Research is being conducted in France on what I would call an old yet innovative wood treatment process, "bi-oleothermy." The wood undergoes a procedure similar to deep-frying: it is treated in a bath of hot oil (110 degrees C or more), then plunged directly into a quench bath that may contain a different kind of oil; this treatment improves wood durability.

Innovative uses for wood products forecast for 2010 include the production of ethanol as a biofuel. A refinery in Skelefton, Sweden, now produces ethanol in relatively large quantities. Certain programs seek to optimize wood yield: in addition to the production of ethanol for vehicles, the surplus energy recovered from the process would be used in cogeneration facilities to produce electricity and supply urban heating networks, the overall process having an energy coefficient of 75 percent.

There are many ways to produce ethanol using ligno-cellulosic biomass. Analytical modeling in Sweden has shown that we could produce enough ethanol from wood resources in this country (around 2.5 million m^3 per year) to supply 4 million vehicles. France has undertaken to substitute 5.75 percent of its petrol consumption with biofuels by 2010.

In the long term, our society will probably turn to hydrogen for energy. This may happen sooner than anticipated if market prices for gasoline remain as high as they are. Given the environmental issues at stake, considerable effort has been devoted to fuel cells, which, for the time being at least, are easier to supply with ethanol by means of catalytic reforming than with liquefied or pressurized hydrogen. In this sense, wood is being offered an important role for the future.

Chemical processes using natural products are being discussed. The pretreatment of cellulose with enzymes, for example, could significantly reduce the amount of energy required for wood refining processes carried out to obtain fiber for the paper industry or for MDF panels.

In summary, there are strong arguments in favor of an increase in the use of wood and a real future for the industry, with all consideration for demographic, environmental, and health aspects. We can predict a highly competitive market for softwoods from fast-growing plantations. In silviculture (apart from niche or local markets) there will probably be a higher demand for softwood, but remaining competitive means growing homogeneous stands and regular growth.

In this context, wood quality is less important. Hardwoods will be mainly used for decorative objects and furniture, which require timber of very high quality.

Wood packaging faces a different sort of future. Materials used for packaging are facing strong competition as they are becoming increasingly multifunctional.

We are also going to have to address questions about wood that comes into contact with foodstuffs. Likewise, the need to heat the wood to 56 degrees C in accordance with the new health standard (NIM P 15) pertaining to pallets may hinder the use of wood for this application.

The wood products sector confronts a number of handicaps in view of a rapidly evolving society. The sector will probably be slow to adapt to current challenges. Technologies from other sectors of industry will provide many opportunities for using industrial wood, but at the moment, the wood sector does not seem to be showing much interest in wood-plastic composites, even though demand is increasing. Compared with other industrial sectors, forestry devotes less funding to research and development.

There are many actors along the chain from forest to wood product, and their interests are sometimes conflicting. Certainly a major challenge for the sector will probably be reconciling the several competing demands: wood for manufacturing products, forests to combat the greenhouse effect, and forest preservation to satisfy the desires of an environmentally conscious society.

ABSTRACT

In France, two major policy-making efforts are underway in spring 2005, the National Forest Program and the National Strategy for Biodiversity. Both are part of the ongoing commitments of Europoean countries to United Nations and pan-European environmental initiatives. Both are built on the presumption that sustainable forestry must address economic, social, and ecological issues. In fact, the concepts of sustainable forest management, multifunctional purpose of the forest, and ecocertification are now so accepted that they are no longer topics for debate.

CHAPTER 14

The National Forest Program and the National Strategy for Biodiversity in France

Cyrille Van Effenterre and Jean-Jacques Bénézit

*I*n talking about the political aspects of the debate about forestry in France, I wish to highlight several themes involving the government and the different stakeholders in forestry and preservation of the nature. There are two policy documents under discussion: the National Forest Program, under the supervision of the Ministry of Agriculture and Fisheries; and the National Strategy for Biodiversity, implemented by the Ministry of Ecology and Sustainable Development.

National Forest Program

The National Forest Program is above all a European commitment. It was decided by the ministerial conference on the protection of forests in Europe that every country in Europe should elaborate a national program. It is also a way to contributing to other international instruments, such as the United National Forum for Forests, the Kyoto Protocol, and the pan-European ecological networks in forests. It is a followup to the forest law that was passed in 2001 by the French government.

Naturaly, the working out of this statment is considered an opportunity and a tool for common discussion among stakeholders. The elaboration of the program began in early 2005 and is continuing. It has involved several participants: the state, the federation of local authorities that own forests, private owners, l'Office National des Forêts (the national forest agency), organizations of wood industries, the organization that represents forest cooperatives (through which multiple private owners come together for management, harvesting, or business operations), forestry research centers, and other stakeholders, including environmental NGOs.

The participants organized five workshops in their meetings:
- wood production and economical services;
- value of forest amenities and nonwood products;
- biodiversity;
- economic and industrial wood activities; and
- forests of the overseas territories.

The workshops related to wood production and economic issues deal with the typical issues, such as management, harvesting, industrial transformation, markets, and business. The workshops on amenities and biodiversity are more unconventional: the concept of forest amenities is quite new and is used by private owners and the organization of local forest authorities to promote the recognition of the nonmarket services coming from the forests—those not currently funded by society, the taxpayers, or the users.

The workshop on tropical forests was a special one, considering the uniqueness of these territories, the forests, and their stakeholders.

The National Forest Program is still underway and has not yet been approved by the French government. The draft is nevertheless well along, and the main topics are the following:
- *Developing timber procurements contracts between forest managers and wood industries.* The best way to encourage sustainable harvesting is to bring together sellers and buyers for the long term.
- *Improving management planning.* We must simplify our land planning regulations.
- *Supporting forest investments.* This requires maintaining public funding coming from the regions, the government, and the European Union.
- *Implementing land planning charters for foresty.* The charters are contracts between local stakeholders and local authorities that prescribe coordinated use and planning of the aera. They are comparable to stewardship contracts.

- *Developing contractual tools about amenities.* This topic involves ways to develop financial treatments of the supply and demand for recreational and environmental services.
- *Promoting new outlets and uses of wood.* Renewable energy and greater use of wood in construction (which can be sharply increased in France) are two possibilities.
- *Reinforcing industrial investments.*
- *Monitoring activities and results.*
- *Boosting research and development.*

What is remarkable is that the notions of sustainable forest management, multifunctional purpose of the forests, and ecocertification are now so natural that they are no longer a subject for debate.

National Strategy for Biodiversity

The other new policy document in France is the *National Strategy for Biodiversity*. It, too, is an international commitment, following the international Convention on Biodiversity (signed at the 1992 "Earth Summit" in Rio de Janeiro), and the pan-European biological and landscape diversity strategy. And it is also a governmental statement, the French government having decided in February 2004 to engage this plan and to implement it during 2005–2010. At the same time, this strategy is meant as a tool for communication and consciousness raising, and a schedule for specific actions and government funding in different areas.

The national strategy was approved by the French government last year, and several plans—dealing with natural heritage, agriculture, the oceans, urbanism, transportation, and agriculture—are already written and approved. The need for a specific section on forestry was agreed upon, and a group for forestry and biodiversity was therefore set up, comprising representatives of the Ministry of Agriculture and Fisheries, the Ministry of Ecology and Sustainable Development, nongovernmental organizations for protection of the nature, forest owners, l'Office National des Forêts, forestry research centers, and the national museum of natural history.

I was appointed as chairman of this group, which is also handling the biodiversity section of the National Forest Program. The method for the carrying out the work was to break down the issue of biodiversity in forest in different parts, as illustrated by the following matrix:

Issues	Stakes		
	Species	Spaces	Landscape & Territories
Conservation, sustainable use management			
Institutional organization and regulations			
Knowledge evaluation and monitoring			

For each issue, we identified several items: reports of the situation, stakes, goals, objectives, actions, measures, schedule, and indicators.

This exercice is not yet complete, but tentative results of this workshop—the main outcomes of our discussions—include the following ideas:

- The highest value of biodiversity in French forests is overseas—in Guyana, the Caribbean, New Caledonia, and Réunion.
- Natura 2000 Network provides the tools to protect European biodiversity.
- Biodiversity in cultivated forests could be enhanced by improving forest practices (such as increasing the percentage of old trees).
- The enhancement of biodiversity in cultivated forests can give forest owners amenities and indirect values.
- The maintenance of ordinary biodiversity is also important and has to be part of multipurpose management.

Eventually, the main areas of the strategy should probably be the following:

- improving knowledge of ecology (species, spaces, and ecosystems);
- achieving a national network of protected areas;
- implementing new plans for restoration of endangered species;
- taking biodiversity into account in large-scale planning;
- improving management practices;
- monitoring impacts with indicators and evaluations; and
- developing vocational training, educational programs, and information for forest owners and the public.

The two main issues still under discussion seem to be the percentage of totally or partially protected areas (about 15 percent of the land) and the effects of reducing the lifespan of trees through forestry management at the expense of biodiversity associated with pioneer stages and old-growth trees, and the need to extend the age of harvesting or, alternatively, to maintain stands of old trees throughout the forests.

Outlook

If we want to be pessimistic, we will say that these two programs are just wishful thinking that will not have any concrete consequences. In fact, talking about money was forbidden by the Ministry of Finances, and so far there are no figures in those papers!

But if we want to be optimistic, we can say that in France, we are not too far from a general agreement, involving all stakeholders—from NGO representatives and government officials to industry leaders—about what we shall do, what we ought to do, what we are allowed to do, and what we will do in the field of forestry in the next decade.

ABSTRACT

For many years, forestry in Germany experienced stable development without any abrupt changes. During the past five years, however, this has changed. Virtually every state forest service has found itself confronted with serious cuts in resources and the need to restructure its organization. In this article, the current state of the forests in Germany is analyzed, first from the viewpoint of natural resources, and second through the framework of economic conditions. Many forests are in good condition, displaying high degrees of biodiversity and tree species mixture, with a high total amount of growing stock and an annual increment that exceeds annual utilization. Economically, however, Germany's forests are less viable, and the willingness of politicians to cover the economic losses of the forest sector is decreasing.

CHAPTER 15

New Challenges for Forestry in Germany

Konstantin von Teuffel

After several centuries of exploitation and devastation in central Europe, it became the major task of the emerging profession of forestry to build up and restock forests throughout Germany. Generations of foresters strove to reestablish and increase the growing stock in our forests. The first Federal Forest Inventory using systematic sampling was conducted in 1987. In 2002, 15 years later, the second inventory was taken, thus representing a repeated inventory, though only for the old states of the Federal Republic of Germany, since reunification had taken place in 1989.

Germany's forests today constitute a total area of 11.1 million hectares, or almost one-third of the total land area. This has grown since 1987 by 0.7 percent, or 54,000 hectares. Naturally, some forest land is lost to development in Germany, but a far larger area has become forest through afforestation, as well as natural succession on former agricultural lands. Higher percentages of forested area are found in the southern states of Germany (42.1 percent in Rhineland-Palatinate, 41.7 percent in Hessia, 38.8 percent in Baden-Wuerttemberg, and 36.3 percent in Bavaria).

For historical reasons only 44 percent of the forests in Germany are in private ownership. More than half of them (57 percent) belong to "enterprises" of less than 20 hectares. Thirty percent of the forested area is the property of the federal states, 20 percent belongs to local communities, and finally, 6 percent belongs to the federal government (including *Treuhandwald*—that is, the forests to be reprivatized in eastern Germany, and military training grounds).

Forests in Germany are dominated by softwoods, even though most sites would naturally be stocked with broad-leaved forests. This is because almost all forests had to be replanted after dramatic overuse from the Middle Ages well into the 18th century. For this purpose, coniferous tree species were preferred because they are easier to cultivate, grow faster, and are less susceptible to browsing by game. Today, 41 percent of Germany's forests are broad-leaved, and 59 percent are coniferous forests. Norway spruce (*Picea abies*) covers 29 percent, followed by Scots pine (*Pinus sylvestris*), 24 percent; beech (*Fagus sylvatica*), 15 percent; and oak, 10 percent. Other broad-leaves cover 16 percent, and other conifers, 6 percent. We can identify larger forest areas that typify Germany: the North and East are dominated by pine forests, whereas the South is rich in spruce. The mountainous center of Germany is characterized by beech and other broad-leaved forests. Seventy-three percent of all of the forests within Germany are mixed-species structures.

During the past 15 years, we have seen a significant shift toward more hardwoods. For example, in Baden-Wuerttemberg, the area of broad-leaves has increased from 37 to 42 percent. A substantial part of this significant tree species change is attributable to several severe storms during the inventory period, which gave the opportunity to change tree species composition. Importantly, the second major element is active forest management. This applies to all ownership categories, even though, generally speaking, for economic reasons the share of conifers in private forests is significantly higher. This holds especially true for public owners who have decided to put a constant and ongoing effort into the conversion of their forests, with the goal of enriching the forest tree species mixture, as well as vertical structure (Spiecker et al. 2004; von Teuffel et al. 2005).

The volume of growing stock has increased over the past 15 years. The present average in Germany is 320 m^3 per hectare, with some states, such as Baden-Württemberg, reaching 350 m^3 per hectare. Together with Austria and Switzerland, Germany has the highest level of growing stock per hectare in Europe. This is a result of more than 200 years of sustainable forestry and can be considered a major success. However, there are some intrinsic risks and challenges associated with these achievements. The forests in Germany and vast other parts of Europe have been severely affected by storms in recent years, and it is clear that the level of growing stock increases this type of risk significantly. The ecological diversity of these forests, however, is high, and they have the potential of growing larger volumes of valuable timber.

During the past 15 years, we have observed a clear shift toward large-dimension timber, especially in southern Germany. For decades, German forestry had been striving to have forests with large-dimension timber, since these size classes were generally more valuable and generated higher prices on the timber market. For the first time in forest history, there now is this high volume in growing stock of large-dimension timber (dbh > 50 cm). The increase in large-dimension timber was especially significant in southern Germany. In Baden-Wuerttemberg the growing stock of spruce with a dbh of more than 50 cm almost doubled in only 15 years, from 1987 to 2002. During the past decade, however, the demands of the timber-processing industries have increasingly shifted toward small and medium-size softwood timber, which has gradually depressed prices for large timber. An analysis of the situation shows that the ongoing rise in average dimension of growing stock is in response to the demands of the wood-processing industry. The attractive size classes are harvested, but trees of larger dimension, which command a lower price per cubic meter, are left standing in the forest as foresters wait for rising prices. The shift in growing stock structure does not apply to coniferous species alone; the same trend appears in hardwoods, too. Moreover, the situation applies not just in public forests but in all ownership categories, including large privately owned forests.

A description of the state of forests in Germany would be incomplete without looking at the level of increment, in comparison with the amount of utilization. This important aspect of the performance of our forests allows an analysis of sustainability in forest management.

The total volume of increment in German forests is 12.1 m^3 per hectare per year, a level that has been rising constantly during the past decades for a variety of reasons (Spiecker et al. 1996). Yet only 8.3 m^3 per hectare was harvested annually (Figure 1). Thus it can be broadly stated that sustainability has been attained during the past 15 years. Looking more deeply into the matter, we find a very heterogeneous picture. Since there are no repeated inventory data for the eastern federal states, our information is restricted to the Federal Republic of Germany as defined by its borders before 1989, so we examine *pars pro toto* data from Baden-Wuerttemberg, in the southwestern part of Germany. With regard to market demands, the professionally managed forest owners overutilized spruce, cutting significantly more small and medium timber and leaving the large dimensions in the forests. Beech and most other tree species were underutilized; that is, cutting

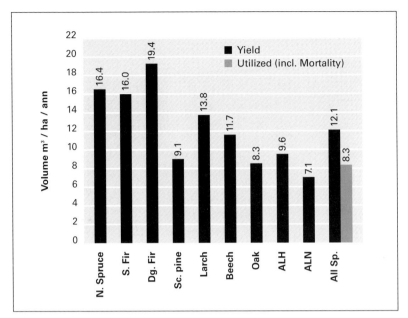

Figure 1. Total forests in Germany.

rates were lower than increment during the observed period. All other species, with the exception of pine, which does not play an important role in tree species composition in the South, show a higher rate of increment than utilization. In the ownership categories of community and small private holdings, there was a marked degree of underutilization, where only two-thirds of the annual increment was harvested and brought to the market.

In summary, the results of the Federal Forest Inventory show that after more than 200 years of reestablishing forests in Germany, the outcome is generally satisfying. There is a high level of growing stock with many forests in good vital condition, displaying high degrees of biodiversity and tree species mixture. The total amount of growing stock in Germany (3.4 billion m^3) exceeds that in Sweden (2.9 billion m^3) and Finland (1.9 billion m^3). Annual increment is also at a high level and one-third higher than annual utilization. Sustainability is secured, though some tasks in the conversion of forests for tree species mixture and vertical structure remain.

There has been an intense stock reduction in the middle diameters of spruce and, at the same time, a marked stock accumulation of large-dimension spruce and beech timber. This can also partly be interpreted as a result of concerns about nature conservation and ecology.

Appraisal of Future Utilization

Again because of the availability of data, we look more closely into the situation in Baden-Wuerttemberg, though the general trend holds true for the vast part of Germany, and especially for southern Germany.

By request of the German Federal Ministry for Agriculture, the Forest Research Institute of Baden-Württemberg developed a model designed to assess future harvesting potentials in German forests. The model allows for different utilization scenarios.

In one scenario, "business as usual," we extrapolate the utilization pattern applied during the past 15 years for a further 15 years for spruce and beech, the most important species in southern Germany. Because of the shift in age class distribution, the amount of harvestable spruce timber is reduced, and more importantly, the level of growing stock of medium-size timber is reduced dramatically, jeopardizing sustainable spruce production.

The alternative scenario, targeting a more or less sustainable supply of smaller-dimension spruce as well as a significant increase in the harvest of large timber, results in an even more dramatic reduction of small and medium spruce and some more harvesting opportunities of large spruce. Applying the two scenarios to beech brings much the same effect with respect to harvesting possibilities of large beech, without the breakdown of growing stock in smaller dimensions.

The appraisal of future utilization opportunities suggests that in the state-owned forests in Baden-Wuerttemberg, there will be a decrease in the total allowable cut in spruce of approximately 15 percent. This is also the case for large privately owned forests. There will be a dramatic alteration of harvesting potential with respect to species as well as size classes toward more hardwoods in general, and additionally toward more large timber in both spruce and beech. However, there will be a marked drop in small and medium timber. Community-owned and small private holdings have a higher harvesting potential. Interesting questions are whether timber markets in central Europe will be able to use this supply, and what the economic consequences for the affected forest owners will be.

Recent Economic Situation

In economics, it is useful to examine the relationship between the value of the main product, timber, and labor costs. In 1955, one cubic meter of timber could pay for 43 person-hours worked in the forest (Figure 2). Today it pays

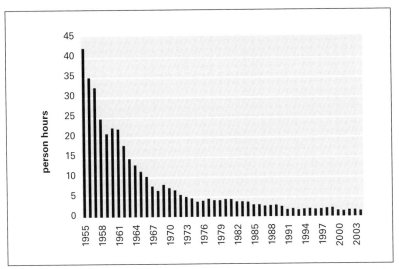

Figure 2. The number of person-hours to be paid from the price of one cubic meter of timber.

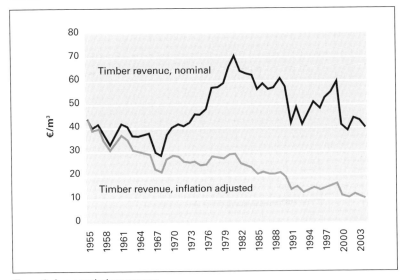

Figure 3. Average timber revenue.

for only 1.5 hours. Even in the past five years, the relationship has undergone another significant reduction. Timber prices today are almost exactly at the same level as in 1955 (Figure 3), and with inflation adjustments, we receive only one-quarter of the 1955 value for one cubic metre of timber. Just to show that the sector has not been dormant, productivity has made substantial

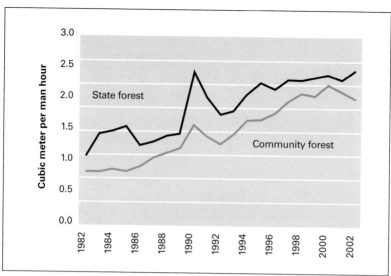

Figure 4. Productivity.

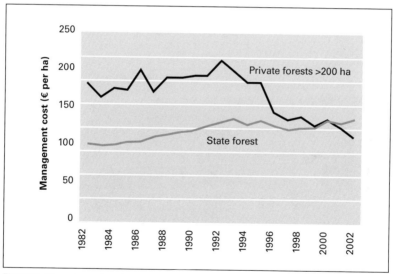

Figure 5. Management cost.

progress over the past 20 or so years (Figure 4). Probably the most important problem can be defined in the level of management (administrative) costs, which currently total 130–140 € per hectare per year in both state and private forest holdings in Baden-Wuerttemberg (Figure 5). We can see a large difference (Figure 6) between direct labor expenses in state and private forests.

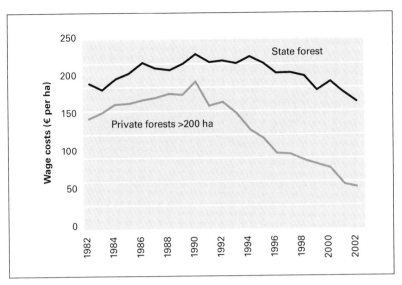

Figure 6. Wage costs.

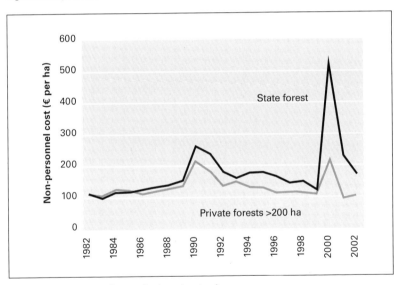

Figure 7. Non personnel costs (incl. contractors).

The state forest spends some 200 € per hectare each year on wages, whereas private forests are run with almost one-quarter of these costs. One should expect that in return, private forests spend more money in nonpersonnel costs, including contractors. But this is not the case (Figure 7). The peaks in this chart relate to the two severe storm events in 1990 and 1999. The solution

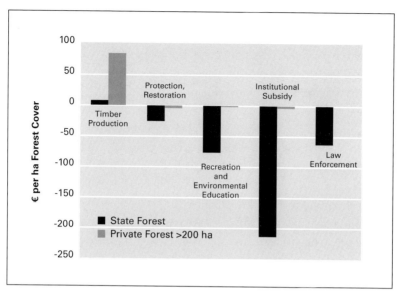

Figure 8. Net income according to product areas.

to this seeming inconsistency is that whereas in state-owned forests the work is principally done with own personnel, private forests use either the services of contractors or, more importantly, sell their timber standing.

Because of this, the economic success of state-owned forests is smaller compared with private holdings. Moreover, the state offers its services to other owner categories and has to put resources into law enforcement activities as well as recreation, environmental education, protection, and restoration of forests. However, we can also see (Figure 8) that even well-managed private holdings make only 70 € per hectare a year, a sum that does not really allow for big reinvestments.

Summarizing the economic situation, four aspects can be stated:
- Forest ownership in Germany is in an economically critical situation.
- Public as well as private forests are affected by this scenario.
- The situation will become even more critical regarding the shift to less valuable timber classes.
- Conservation and recreation functions are currently not self-funding but are included in the costs and benefits of forest enterprises.

Consequences

Thus we see that forestry in Germany in many enterprises, especially forests in public ownership, is currently economically unattractive. A general trend is evident (and here lies probably the most important difference between Germany and the United States): the willingness of politicians to cover the economic losses of the forest sector is constantly decreasing. Forests and their various functions are highly respected, but since the general state of forests seems to be satisfactory, the willingness to put financial resources into them is reduced, especially when compared with the situation 25 years ago, when forest decline (*Waldsterben*) initiated huge investments in forestry. The principal reason for this development is the financial crisis of government (federal and state) budgets, which has worsened during the past two decades, together with the fact that forestry is not a high-priority political issue. Consequently, we can observe a constant decrease in institutional as well as direct financial subsidies for the forest sector. However, neither private nor public forest enterprises are likely to survive with ongoing deficits. Therefore, a clear separation between enterprise required expenses/earnings and public spending (for other non economic forest functions) is needed, especially for all forests in public ownership. The model of the past 50-plus years of running state-owned forests and assisting community and privately owned forest holdings while conducting law enforcement activities in one single administration, the state forest service, retains less and less support. During the past five years, major organizational restructuring has passed through most state forest services in Germany, all of which targeted a significant reduction of deficits. There is a general trend toward separating the management of the forests from other tasks, like outreach and law enforcement activities. A partial privatization of public forests would no longer appear unrealistic.

One of the unsolved challenges is how to make nature conservation, ecological requirements, and recreation self-funding. Part of a solution may come with the further development of the "eco-account system", under which deforestation for development or infrastructure must be compensated by payments into an eco-account, which is then used to improve forests for ecological or recreational purposes.

Since timber receipts are strongly dependent on the development of international markets, regional submarkets (as they existed until 20 years ago) can no longer be relied upon. Unless abrupt major changes in the underlying conditions occur, no marked increase of timber prices can be expected in the near future. Consequently, the rationalization efforts of all forest owners will have to be increased. The remaining potential in shifting to a more "close-to-nature" management, by using natural processes and avoiding high inputs of labor and machinery, is minute. There is certainly some potential in expanding highly mechanized forest operations as well as returning to a certain extent to more conifer cultivation, thus reducing the tremendous input into the cultivation of broad-leaves. The biggest potential doubtless lies in the reduction of administrative as well as direct labor costs, which will lead to substantial and ongoing personnel cuts. Moreover, the vertical integration with the timber-processing industries will be deepened. Since we are rather unlikely to see bigger forest areas being sold to timber companies, there seems to be some potential for public-private partnerships of various kinds.

In summarizing the state of forestry in Germany, it is clear that the state of Germany's forests is satisfactory when seen from the perspective of growing stock, yield, and sustainability of the forest resource as a whole. There has been tremendous progress over the past few decades in improving their ecological state. The biggest problem in German forestry today, however, is the economic viability of forest enterprises. With the declining willingness of politicians to put more resources into the sector, major efforts are needed to secure economic viability.

Today's Challenges and Opportunities

The economic situation summarized above is subject to a very dynamic development. The German wood processing industry has undergone a severe concentration process that has, partly induced by foreign investments, led to the construction of huge processing plants with a processing capacity of several hundred thousand m^3 each. The traditional small sawmills in rural areas in some regions of Germany have been almost completely erased from the market. Overall processing capacity has risen to a degree that even recent storms like "Kyrill," in January 2007, hardly disturbed the market, since the plants were easily able to absorb the 25 million m^3 of timber that the storm brought. In the midterm, one might even expect a shortage of the resource in Germany, especially for the most desirable timber assortments, such as medium-sized softwood logs. What sounds at first glance contradictory to the results of the national forest inventory, still indicating a higher increment compared with the exploitation in Germany, leads in fact to one of the major challenges for German forestry in the coming years: the mobilization of the resource in small to medium-sized private forest ownerships. Although the harvesting potential in the public forests and especially the large private forests is more or less exhausted, an increasing number of urban, absentee forest owners have lost interest in managing their forests, many of which have excellent site conditions but have been divided into nonmanageable units. The result is a high volume of standing timber in private forests that is virtually out of reach of the industry, even though this timber is close to the mills.

Finding a solution to that problem is of high importance: the opening of the European Union to the east places Germany in the center of a still-expanding market, with more than 500 million consumers and changing market conditions due to the influence of aggressive actors in Russia and China. The prospects for exports appear excellent.

The battle for the resource has also started in another area: the increase in the price for oil and other fossil energy sources and the new energy policy of the German government (abandoning nuclear energy) and the European Union have dramatically increased interest in timber as a renewable energy source. The price of wood pellets and chips has steadily risen over the past few years, and large investments have been made in production plants. The capacity for producing wood pellets doubled in 2006 in Germany. Now, the market for wood for bioenergy competes with the use of timber as a source of fiber. After some trials in the 1980s, Germany is seeing a revival of the idea of establishing fast-growing species in wood plantations for both energy and fiber, mainly on abandoned agricultural land. All this and the economic pickup after a long recession led to an increase in timber prices in 2006, suggesting that forestry in Germany could well be a profitable business again, provided that labor and administrative costs can be controlled.

ABSTRACT

More than a century ago, Europe and the United States approached forest management quite differently. Europe's degraded forests were being replanted and managed for sustained yields of timber, through strong forest protection laws and conservation programs imposed by authoritarian governments. North America still had vast tracts of untouched timber, and in its democratic, capitalist culture, a conservation movement was only beginning to have an effect. By the end of the 20th century, both continents had embraced multiple-use forestry to meet the complex demands of their respective societies. Europeans and North Americans are now revisiting forest management because the political and social context has changed. Ecosystem management and sustainable forestry seek to maintain the composition, structure, and functions of forest ecosystems in the long term, guided by explicit goals, executed by well-researched practices, and made adaptable to human environmental interactions and processes. Differences remain. But full convergence is neither necessary nor desirable. We can better face the future if we learn from each other's successes and inevitable failures in forest management. The knowledge gained from our different approaches to forest management will help us contribute to forestry beyond our borders, especially in the development of the world's boreal forests and the forests of the Southern Hemisphere.

CHAPTER 16

The Continuing Evolution in Social, Economic, and Political Values Related to Forestry in the U.S. and Europe

Dennis C. Le Master and Franz Schmithüsen

European influence on forestry and forestry education in the United States during the late 19th and early 20th centuries was very strong (Dana and Fairfax 1980). Europe was the center of higher education in the world at the time, and European universities alone offered formal curricula in forestry. Prospective U.S. forestry students, such as Gifford Pinchot, had little choice but to go to Europe to study if they wanted a diploma in forestry.

Sustained-yield forest management was at the core of European forestry (defined here as managing a forest to achieve and maintain a balance between timber growth increment and cutting). It had made sense since the high Middle Ages (13th and 14th centuries) because the difficulty of transportation made local consumption virtually dependent upon local production. Hence, communities had to be self-sufficient in their timber and fuelwood requirements. During the 15th and 16th centuries, cities as well as preindustrial entrepreneurs found that the current wood supply from forests that were already under exploitation could no longer cover the growing need for firewood and construction timber for domestic use, salt production, and metallurgical factories. As a consequence, the essential conditions for a more stable forest regime were established in the 17th and 18th centuries. Public policies and law determined the framework conditions for sustainable wood production, which meant stopping mere exploitation of what was available in the accessible forests.

The concept of sustained-yield forest management was transferred to the United States from Europe at a time of national concern over the possibility of a timber famine. Nevertheless, the underlying assumptions of sustained-yield management did not fit American circumstances at the time. Neither land nor timber was scarce, and whether the demand for wood was stable was at least questionable. Accordingly, appropriate standards for forest management were uncertain. Still, in the late 19th century, wood was the principal construction material and energy source in the United States, and large forest areas had been cleared in a kind of forest mining, with no thought of reforestation. Local timber supply shortages were becoming increasingly frequent. A real concern was whether a regional or national timber shortage would limit economic growth. Sustained-yield forest management addressed that concern. If it were applied, timber supply would not be a problem.

Forest Clearing in the United States and Europe

Forest clearing for agriculture by European settlers began almost as soon as they arrived in North America and continued throughout the 17th and 18th centuries as more and more settlers arrived. The 19th century was a period of rapid population growth. The U.S. population increased from 5.3 million in 1800 to 76.2 million people in 1900, a more than 14-fold increase (World Almanac 2004). Roughly 230 million acres (93 million hectares) of forestland were converted to cropland and pasture during the same time span (Powell et al. 1993). The demand for timber must have seemed almost insatiable after 1850 because of rapid industrialization, urbanization, and the growth of railroads, whose tracks were spreading across the country.

Deforestation continued until about 1920, when the acreage devoted to cropland stabilized. In turn, the extent of forestland stabilized at about 732 million acres (296 million hectares), about a third of the landscape (Powell et al. 1993). The area covered by forest grew during the later part of the 20th century (Floyd 2002). Additions to forest land base since the 1930s have been smaller but significant. Converted cropland and pasture tend to be offset by losses due to development, both commercial and residential.

In 1992, forests covered 737 million acres (298 million hectares), 33 percent of the total land area (Powell et al. 1993). If Alaska is subtracted from the total, the forest area of the lower 48 states is 608 million acres (246 million hectares), or 32 percent of the landscape. Total forest area in the United States expanded by 0.2 percent during the period 1990–2000 (FAO 2001). The extent of forest land today is about 70 percent of what it was in 1600, just prior to European settlement (Powell et al. 1993). [Discrepancies are common in forest databases. For example, *Global Forest Resources Assessment 2000* shows the land area for the United States as 915,895,000 hectares (2,273,177,000 acres) with 225,993,000 hectares (558,429,000 acres) forested, 24.7 percent (FAO 2001, 236). Differences are typically due to alternative definitions of variables.]

In contrast, the great period of deforestation in Europe began in the eighth and ninth centuries, reached its culmination in the 12th and 13th centuries, and then effectively stopped in the 14th century (Mantel 1990). The reason for clearing the land was rapid population growth in Europe, and the reasons for its cessation were the typhus epidemics between 1309 and 1317 and the bubonic plague, or Black Death, in the middle of the century (1347–1350), which resulted in dramatic population declines. About a third of the population of Europe—an estimated 25 million people—died from the plague in the mid-1300s (http://black-death.biography.ms/). Whole villages were abandoned in Europe in areas that had been only recently colonized. Dwellings and fields were left to be reclaimed by forest.

The large population loss brought about economic changes due to increased social mobility. It eroded the peasants' already weakened obligations under the feudal system in western Europe. Further, the sudden scarcity of cheap labor provided incentives for innovation and the substitution of capital for labor, contributing to the Renaissance, which began in the early 15th century. Forest clearing was less important during the 15th and 16th centuries. During the 17th century, population increases were held in check by war, in particular the Thirty Years' War (1618–1648), and hence there was a corresponding restraint in the need for additional cropland and pasture. In the 18th and 19th centuries forest clearing for agricultural purposes occurred in some areas but was offset by reforestation activities and natural succession as a result of decreased demand for farmland due to imports of wool and, later, cereals.

During the 20th century, with interruptions during the two world wars, forest land was cleared for agricultural purposes in areas suited for agricultural production, but forests expanded on less productive land, particularly in mountainous regions. In recent decades, the increase in agricultural productivity and the expansion of international markets, together with restructuring and concentration of management units, including collective farms in eastern Europe, have led to abandonment of agricultural production on substantial areas that are suitable for reforestation or favor a return to forest through natural succession. These changes occur mostly in the mountains, where rapid change is affecting the balance among cropland, pasture, and forest.

The net result is that after the huge changes in the Middle Ages, the European forest land base has stabilized at a little over 1 billion hectares, which accounts for about 46 percent of the landscape (FAO 2001). If Belarus, the Republic of Moldova, the Russian Federation, and Ukraine are subtracted from the total, Europe has 168,548,000 hectares of forestland, which equals 34 percent of the landscape (FAO 2001). However, in regions under intensive cultivation, as well as in the areas around large towns and in the surrounding open area, the forest now occupies only a small part of its initial expanse. On the other hand, in hillier areas and in the mountains, such as the Alps and the Carpathians, the forest has remained or has become a primordial element of the area, largely determining its economic and social potential and the specificity of the landscape. Total forest area in Europe expanded by 0.1 percent during the period 1990–2000 (FAO 2001).

Sustainable Wood Production in Europe

European forests have historically represented a combination of resources available to the local people that went well beyond domestic fuelwood and construction material (Schmithüsen 2006). These resources were essential to meeting daily needs. Forests were a complement to agricultural production, and as such they provided a direct food source for humans in the form of nuts, berries, and mushrooms, a source of medicinal plants, pasturage for domestic ungulates and forage for swine, and a source of animal bedding material. Forests were used in industry in Europe as a source of energy in the form of both charcoal and fuelwood in glass works, salt works, and the production of ferrous and nonferrous metals. Timber was used in mining to shore up pits and galleries. Wood was reduced to ash to produce potassium for bleaching textiles and making soap. Forests were a source of dyes for textile manufacturing. Importantly, for both commerce and national defense, forests supplied ship masts and naval stores (pitch, tar, spirits of turpentine, and rosin).

Conflicts over the use of forest resources were common and centered on the demands of local populations to take advantage of the forest for their own needs, as a complement to agricultural production, and the efforts of landlords and local overlords to develop the forest as a commercial or industrial enterprise. This struggle continued up to the 19th century and was apparent in legal struggles concerning user and property rights. By 1850 at the latest, most European forests were under a system of sustainable wood production.

Since about 1700, forestry and wood processing have become productive sectors of the economy, using a renewable resource in a sustainable manner as the basis of business management (Schmithüsen 2006). Step by step, policy and law introduced the principles of renewable natural resources use as we understand them today. The term *sustained utilization* was used as early as 1713 by von Carlowitz, who worried about maintaining wood supply for large-scale mining activities. In 1804, Georg-Ludwig Hartig had already formulated the principle of sustainable forestry with its intergenerational perspective when he remarked that descendants should be able to draw at least as many advantages as the then-living generation appropriated.

In the 20th century, the meaning of sustainable forest management expanded from wood production to include all aspects of forest uses and values. In a modern business management-oriented definition—as formulated, for instance, by Speidel (1984)—sustainable forestry means the ability of forest enterprises to produce wood, infrastructural services, and other goods for the benefit of present and future generations. It means maintaining and creating the entrepreneurial conditions necessary for the permanent and continually optimal fulfillment of economic and noneconomic needs and goals.

U.S. Westward Expansion and the National Forests

The 19th century in the United States featured the westward expansion made possible in large part by purchase of the Louisiana Territory in 1803 (for approximately $15 million), rapid population growth (primarily by immigration), and the clearing of approximately 120 million hectares of forest land for agricultural production. Forests were abundant at the outset of the century, stretching unbroken for hundreds of miles. They were recognized as a source of valuable resources, but they were also seen as a barrier to development. As in Europe, forests were a source of timber, fuelwood, wildlife, pasturage for domestic ungulates, and forage for swine, food for humans in the form of nuts, berries, and mushrooms, and some medicinal plants. They were also important on the East Coast as a source of naval masts and stores. But with the possible exception of ship masts, the relative value of forest resources in the United States was very low in 1800.

By the late 1800s, concern—but only among a minority—developed about destructive logging and the loss of forest lands. Since wood was still the major building material and the predominant source of fuel in the United States, a timber shortage would have a major negative effect on the economy. There were also less market-oriented reasons. Aesthetics was one. The economic exploitation of timber was basically an ugly process, especially at the scale at which it was done. The impacts of destructive logging and loss of forests on water quality and soil productivity were also reasons.

George Perkins Marsh published *Man and Nature* in 1864, and it was republished in 1874 as *The Earth as Modified by Human Action*. The book detailed the impacts of forest destruction on climate, water quality and timing, and loss of soil fertility, using the Mediterranean region as an example. Marsh concluded that ancient Mediterranean civilizations had sown the seeds of their own demise by the wanton clearing of their forests. Careful American readers did not miss the parallel with what was occurring on U.S. forest lands. James Pinchot found the book fascinating, and it confirmed his belief in the direct relationship between a nation's forests and its general welfare (Miller 2001). He arranged for his son Gifford to receive a copy of *The Earth as Modified by Human Action* as a gift on his 21st birthday.

On November 13, 1889, after graduating from Yale University earlier in the year, Gifford Pinchot was accepted at L'École Nationale Forestière. His experience there was mixed, but he became certain of one thing: the circumstances in which forestry was practiced in Europe was substantially different than in the United States. He returned home late in 1890. The following year, 1891, the Creative Act (26 Stat. 1095) was passed, which authorized the president to set aside forested areas of the public domain as forest reserves. The Organic Administration Act (30 Stat. 11, 34) was passed six years later, in 1897, and it specified the purposes for which forest reserves might be established and provided for their protection and use.

Pinchot succeeded Bernhard E. Fernow as chief of the Division of the Forestry in 1898. The Transfer Act of 1905 (33 Stat. 628) transferred administration of the forest reserves from the secretary of the Interior to the secretary of Agriculture. Later that same year, the Act of March 3 (33 Stat. 861, 872–873) changed the name of the Bureau of Forestry to the Forest Service, and in 1907, the Act of March 4 renamed the forest reserves the national forests. In the space of 17 years—1890 to 1907—through several incremental changes in policy and events—forestry in the United States took a significant turn in its development toward the sustained-yield forestry practiced in Europe.

Management of the national forests was largely custodial until the mid-1940s. Control of wildfire was the major activity on national forests in the West, and it was amazingly successful (MacCleery 1992). In the East, the challenge was restoring cutover forest and abandoned farmland to forest cover in national forests acquired under authority of the Weeks Law of 1911 (36 Stat. 961). This effort, too, was very successful (Shands and Healey 1977).

The concept of sustained-yield management began to be accepted by private timber companies in the 1940s. There was no apparent future in the past practice of buying forests, liquidating the timber, and abandoning the land. If private timber were to be available on any significant scale, it had to be grown and managed, like an agricultural crop. Since private timber inventories had been steadily drawn down throughout the 20th century and at an accelerated rate during World War II, private timber companies looked to the national forests for their timber supply until enough timber had been regrown on private lands to supply raw material needs. The prevailing notion at the time was that by maintaining a continuous supply of timber and protecting the basic productivity of the soil, a broad set of forest values would automatically be made available. Since growing trees and protecting the soil would spontaneously or naturally produce other values, there was little need to focus upon them. Management efforts should be directed at the trees—other values would be addressed incidentally.

U.S. Forest Legislation in the 1960s and 1970s

Incomes rose during the 1950s, and people had more leisure time (Le Master 1984). They wanted places to recreate. Urban populations increased, and natural environments—areas comparatively untransformed by human activity—became increasingly scarce. Wildlife and the opportunities to view them in natural places became important. Millions of Americans went to the national forests to enjoy natural surroundings, to recreate, and to view wildlife.

The Multiple-Use Sustained-Yield Act of 1960 (74 Stat. 215) marked a major change in national forest management in the United States. It authorized and directed that equal and active consideration be given to five renewable surface resources, namely (in the order they appeared in the statute), outdoor recreation, range, timber, watershed, and wildlife and fish, and that they be "utilized in the combination that will best meet the needs of the American people" (sec. 4(b)). Incidental production of one these five would no longer be acceptable. The national forests were to be actively managed for multiple objectives. Further, the act directed that the five renewable surface resources produce high yields that could be sustained in perpetuity without impairment of the productivity of the land (sec. 4(b)).

Sixteen years later, after a major lawsuit (*Izaak Walton League v. Butz*) that effectively suspended timber harvesting in large parts of the National Forest System, Congress passed the National Forest Management Act of 1974 (90 Stat. 2949, as amended). This act established standards and guidelines for national forest planning and management. Among other things, limits were set on timber production, opportunities were provided for public participation in national forest planning, and plans were required to "provide for diversity of plant and animal communities based on the suitability and capability of the specific land area in order to meet overall multiple-use objectives ..." (90 Stat. 2949, as amended, sec. 6).

Implementation of the National Forest Management Act initiated a transformation of the Forest Service:

- from an agency concerned about outputs, economic development, and commodity clients to an agency concerned about healthy ecosystems and diverse, changing market and nonmarket values;
- from an agency that emphasized a line-staff organizational scheme to an agency whose line-staff design is flexible, which shares power and engages in partnerships both inside and outside government;
- from an agency that emphasized science-based management to an agency that recognizes the importance of both biological and social sciences in its management programs and activities;
- from an agency that prided itself as being a "can-do," mission-oriented organization to an agency responsive to changing social values in the

context of sustainable, interrelated ecological, social, and economic systems; and
- from an agency dominated by white male foresters to an organization that is diverse in terms of gender, ethnicity, culture, and the disciplinary training of its employees.

National forest planning also served to sharpen the debate over sustainable forestry. Biological correctness was necessary for sustainable forestry, but it was not sufficient. Also required were economic viability and social responsiveness. Hence, national forest plans also had to take into account the local economic and social systems to gain public support and minimize conflict. In the early 1990s, three characteristics of sustainable forest management were identified. Such management would be characterized by being ecologically sound, economically viable, and socially responsible (Aplet et al. 1993).

Another transition occurred in the early 1990s, which was an evolution from multiple-use sustained-yield management to ecosystem management. Ecosystem management is defined as "management guided by explicit goals, executed by policies, protocols, and practices and made adaptable by monitoring and research based on the best understanding of ecological interactions and processes necessary to sustained ecosystem composition, structure, and function over the long term" (Helms 1998).

One essential difference between the two strategies is that the former is oriented largely toward resources while the latter is based on ecosystem structure and functioning. A second difference is the greater emphasis in ecosystem management on protection of biodiversity. A third is the application of adaptive management in ecosystem management based on monitoring and research.

Changing European Values in the 20th Century

As stated by the Food and Agriculture Organization of the United Nations in *State of the World's Forests, 1995,* European forestry is in a position to provide multiple goods and services from forests that have been used and managed over centuries, and to adjust to changing demands and values in modern societies:

> Most European forests are managed to produce a wide range of goods, notably wood, as well as many locally important non-wood goods, and

services such as recreation, protection (of soils, watersheds and transport infrastructure in mountainous regions), and nature conservation. The role of Europe's forests as an important "carbon sink" is increasingly recognized. As growth exceeds fellings, there is a net uptake/storage of carbon in the biomass. Human management over the centuries has shaped the forests of Europe, creating forests of great beauty and rich biodiversity (such as the selection forests of central Europe and English ancient woodlands), as well as efficient wood production forests which are often also valuable for the non-wood goods and services they provide (FAO 1995).

Empirical studies indicate that forests have, in fact, acquired a new significance in European societies during the 20th century (Schmithüsen 2006). Their significance has gone beyond their role as a source of valuable construction material and energy, which continues to be important for forest owners, the wood products industry and its workers, and farmers. Environmental protection and services such as soil stabilization and watershed protection, already an important aspect of forestry during the 19th century, together with new concerns for maintaining biodiversity and measures for carbon sequestration, are today central themes of forest management. For an ever-larger component of the population, mainly urban residents, forests represent free space for recreation and a place for meditation and reflection in natural surroundings.

Public policies for protecting and managing forest resources as well as the corresponding laws have been revised during recent years in practically all European countries (Le Master and Owubah 2000; Schmithüsen 2000). Major changes have occurred in central and eastern European countries. In transition to an open civil society, democratic institutions, and a market economy, they have had to develop a completely new policy and legal framework for addressing agriculture and forestry, nature conservation, and environmental protection (Mekouar and Castelein 2002). Societal demands on private and public forests, together with responses from within the forestry community and from the public at large, have received considerable attention from politicians and the forest administration.

Multifunctional forest management practices implementing the principles of sustainable development are current in many European countries. They increasingly involve forest owners, forest users, and environmental groups on an equal footing; contribute to balancing private and public interests with workable arrangements for landowners facing public demands; and facilitate a shift from governmental and hierarchical regulatory systems to negotiation, public process steering, and joint management responsibilities. Close-to-nature forestry practices are used as a land management strategy allowing adaptation to changing societal values. Favoring flexible and long-term production cycles and relying to a large extent on natural site factors, they contribute to maintain biodiversity, varied ecosystems, and diversified landscapes, and they leave options for alternative uses and new developments. Acknowledging economic necessities and multiple social and environmental demands, multifunctional and close-to-nature forest management offers a flexible range of land-use options for the future.

For more than 40 years, Europe has been engaged in building or rebuilding a common continental space in which people and nations can live peacefully together. It is for Europeans a revolutionary process with great successes but also drawbacks. Europe's move toward open civil societies, democratic rule, progressive economic development, and common political institutions has many faces. Contributing to cooperation in many domains and to a new European identity is the European Union. It is a driving force toward a more permeable and integrated continent in which people can move according to their personal choice and in which transnational and national political institutions coexist. The expansion of the European Union in 2004 that has resulted from the joining of eight countries from central and eastern Europe has brought new geographical and political horizons. With two other new members, Cyprus and Malta, the European Union now has 25 member countries. Its population amounts to more than 450 million people and its land area covers around 3.8 million km^2, extending from Ireland to the eastern borders of Poland, and from northern Finland to Portugal and Cyprus in the south.

The growing economic, social, and political integration of Europe has far-reaching implications for forestry and the wood-processing sector. For the wood products industry, a continental European sphere offers opportunities and challenges involving new and larger markets, more market competition, gains in efficiency and productivity, and stronger positions in world markets (European Commission 2000). The forest development perspectives are manifold and lead to a new vision for European forests, forest ecosystem networks covering large European regions, progressive adaptation in national policies and laws, common management principles and standards, and research and education networks on a European scale.

Changing U.S. Values in the 20th Century

Substantial changes have also occurred in the ways Americans value forests, especially during the second half of the 20th century (Le Master et al. 1997). More values of various kinds have been articulated, and nonmarket values—those not exchanged in markets—have increased relative to market values. Bengston et al. (2004) examined three forest value orientations—clusters of interrelated values and basic beliefs about forests—that emerged from an analysis of the public discourse about forest planning, management, and policy during the period 1980–2001. The three value orientations are anthropocentric, biocentric, and moral-spiritual-aesthetic. The study found that the share of expressions of anthropocentric forest value orientations (e.g., livestock forage, recreation, and timber) declined over 1980–2001, while the share of biocentric value expressions (e.g., biological diversity and carbon storage) increased. Moral-spiritual-aesthetic value expressions (e.g., getting back to nature) remained constant over time.

Management of public lands has changed as a result of changes in public values, slowly at first, then quite rapidly. Forest management changed on industrial lands as well, following the lead of public forest land management. Nonmarket values such as protection of water quality and wildlife habitat became a formal part of management strategies of large industrial holdings. Management of nonindustrial forest lands was and continues to be very diverse in terms of goals and methods. Yet there is significant evidence that nonindustrial private forest owners are quite responsive to public values toward wildlife, water quality, and biological diversity. These landowners are managing for such values on the basis of altruism, incentive programs, and the fear or the fact of government regulation.

In brief, many of the value and policy changes going on in Europe are also occurring in the United States.

UN Conference on Environment and Development

The 1992 United Nations Conference on Environment and Development (UNCED, the "Earth Summit" in Rio de Janeiro) was unprecedented in both size and the scope of its concerns about economic development and environmental degradation, particularly as they related to forests and forestry

practices. Three major agreements directed at changing the traditional approach to development were adopted:
- The Rio Declaration on Environment and Development, a series of principles that define the rights and responsibilities of states;
- Agenda 21, a comprehensive program of action for sustainable development, and whose Chapter 11 concerns stopping deforestation; and,
- The Statement of Forest Principles, a set of principles for sustainable management of forests worldwide.

In addition, two legally binding conventions aimed at preventing global climate change and maintaining biological diversity were agreed to—namely, the Framework Convention on Climate Change and the Convention on Biological Diversity. The United Nations was also called on to negotiate an international agreement on desertification, which has subsequently occurred.

Significant progress has been made since UNCED. The Intergovernmental Panel on Forests, from 1995 to 1997, and the Intergovernmental Forum on Forests, from 1997 to 2000, were established within the United Nations to implement the Forest Principles and Chapter 11 of Agenda 21. The two processes generated 270 proposals for action for management, conservation, and sustainable development of forests. The UN Forum on Forests was established in October 2000 to provide a coherent, transparent, and participatory global framework for policy implementation, coordination, and development, including carrying out those proposals. It now serves as the main forum for international policy deliberations on forests in the absence of a global forest convention.

The number of legally and nonlegally binding agreements having a bearing on forests and forest management grew significantly with UNCED. Ten legally binding forest-related agreements, including the UN Forum on Forests, now exist, and four of them were initiated either during or after UNCED (Braatz 2002). In addition, more than 20 other forest-related international conventions and agreements exist that have a bearing on forests and forest management, and the majority of these were initiated during the 1990s as a direct result of UNCED.

The list of international agreements is impressive on its face and encourages the view that countries of the world are converging toward a common set of generally accepted principles for sustainable forest management directed toward common goals in forest conservation. However, they do not constitute a coherent and holistic set of principles for forest management worldwide. A comprehensive, international, legally binding instrument is still lacking.

Convergence on Sustainable Forestry: Fact or Wishful Thinking?

The efforts in North America and Europe, as well as in the United Nations, to adapt forestry practices to expanding and new social demands have led to more diversified and comprehensive forest policy goals. Moving from a focus on wood as a sustainable resource, they now address a wide range of private and public goods and values and acknowledge the equal importance of production and conservation. On both continents and in a worldwide perspective, the actual policy goals are incremental and refer to the role of forests as multifunctional resources, for their economic potential and their importance to the environment. Increasingly they address the variety of ecosystems, the need to maintain biodiversity, and the urgency of safeguarding the natural renewable resource base for future generations. New policies and laws favor multifunctional land-use strategies capable of addressing divergent social interests and adapted to local conditions.

However, although steps have been taken to establish a common international base for the protection and sustainable management of forests, many more are necessary before a generalization of this kind can be made comfortably. For example, the fifth session of the UN Forum on Forests, which was anticipated to be a milestone event, ended on a very disappointing note on May 27, 2005. The forum's chairman, Manuel Rodriguez Becerra, said, "At the end, we didn't agree.... It's a collective failure.... We were not able to agree in making decisions that are relevant for addressing the huge deforestation, the huge degradation of forests...." Becerra continued: "Developing countries say we need strong means of implementation and the developed world says we need strong objectives, and there is not a strong point of encounter between means and goals" (http://p128news, scd, jahoo.com/s/afp/20050528/sc_afp/unforestagriculture_05052800383). The gulf between the Northern and Southern Hemispheres, between developed and developing countries, remains large and very real, and in the meantime, the global deforestation rate is currently estimated at 9.4 million hectares a year, overwhelmingly in the Southern Hemisphere.

The main forest certification programs operating in the United States, the Forest Stewardship Council (FSC) and the Sustainable Forestry Initiative (SFI), have also exhibited some convergence, as noted by Sample (2006). Although important differences remain, SFI, like FSC, now requires independent, third-party verification of a company's adherence to the standards required for use of the SFI label. And FSC, like SFI, is moving toward a "continuous improvement" approach to the standards by which it evaluates forest enterprises and a more stable, inclusive approach to modifying its standards over time. Both systems face similar and substantial challenges. Neither has been able to design an effective program for private landowners, who own 60 percent of the forest land in the United States, and neither has created much awareness among American consumers comparable to that of European consumers.

In Europe forests are certified under two competing processes, the Forest Stewardship Council and the Pan European Forest Certification System. The latter is based on the common definition of sustainable forest management, agreed upon at a 1994 followup meeting to the "Helsinki Process," on the protection of forests in Europe, which is the most important pan-European institution in forestry matters and involves more than 40 countries. It uses six specific evaluation indicators combined with quantitative and qualitative criteria that were accepted at the Third Ministerial Conference in Lisbon in 1998.

Despite the many successes of the sustainable forestry movement, certain trends of serious concern remain. Millions of acres of American forestland are being taken out of sustainable forest management, largely through land sales and land-use exchanges that result in parcelization and fragmentation. Large blocks of forest are being sold for business reasons, and many end up being held in smaller, sometimes disjointed parcels that cannot produce the multiple benefits that contemporary society has learned to expect from forests.

Indeed, one of the biggest problems in the United States today is the parcelization and fragmentation of forest land, which is substantially eroding the structure and function of forest ecosystems. Industrial firms are divesting themselves of forest land in a substantial way, and timber investment management organizations (TIMOs), which were expected to help, have proven to be driven by the same economic incentives common to any private enterprise. Agreement may exist among scientists and forestry practitioners on what constitutes sustainable forestry, but no such agreement is apparent among U.S. forest landowners.

Conclusions

The changing conditions for sustainable forest management are to be seen in the overall perspective of maintaining the natural resource base, in a holistic understanding of forests and landscapes, and as part of the overall goal of protecting the environment and improving the quality of life for present and future generations. This is in fact the central theme of wise use of forests and ecosystem management, which builds on the legacy of the past and provides opportunities for the future (Farrel et al. 2000). An integrative approach to gaining more knowledge about the interactions among social systems and human behavior, ecosystem processes, and environmental change is essential to capture more closely the impacts and feedbacks between man and his natural resource base.

One has to understand the locally differentiated interactions between society and forests, the dynamically changing social and cultural meaning of forests, their physical potential for providing different combinations of goods and services, and the political and economic conditions for maintaining their stability and biodiversity under alternative management systems (Schmithüsen 2004). As it is for other land management sectors, sustainable development is the overarching political principle and the benchmark for judging to what extent the forest sector and forest policies contribute to economic and social welfare and to a safe environment that benefits present and future generations. Its essence is that economic growth, social integration, and caring for a livable environment be on equal footing. Economic growth, social integration, and protection of the environment depend on each other, cannot be substituted for, and are fundamental to social progress and common welfare. The specific managerial possibilities and political commitments of individual countries and regions toward sustainability have to be seen in their respective stage of development, their cultural traditions, and the extent of their natural resource base.

Europeans and North Americans are both revisiting sustainable forest management because the political and social context has changed. European multifunctional forestry corresponds to multiple-use, sustained-yield forestry and its extension, ecosystem management—a term coined in North America and now readily accepted in Europe as well. It defines a strategy necessary to sustain the composition, structure, and functions of forest ecosystems in the long term. It is guided by explicit goals, executed by policies, protocols and practices, and made adaptable by monitoring and research based on the best available understanding of human environmental interactions and processes.

In 1900 Europe and the United States were divergent in their collective views of forest management for several reasons:

- The United States was rich in its forest wealth, and Europe was not.
- The United States was fundamentally a democratic society, while the nation-states of Europe were not, at least for the most part, and their forests were managed authoritatively.
- European countries had promoted conservation during the 19th century through strong forest protection legislation, and through land conservation and reclamation programs focusing mostly on replanting and afforestation.
- In the United States, large-scale deforestation and forest overexploitation had prompted a conservation movement, which developed as a countervailing force against 19th-century American capitalism. It had no counterpart in Europe, where the economic system was mixed, combining capitalist elements with strong state interference.

By 2000 Europe and the United States were converging in their collective views of multiple-use, sustained-yield forest management for these reasons:

- Europe and the United States today are both fundamentally democratic in their form of governments.
- Europeans have invested in and expanded their forest wealth, while the United States has drawn down its forest capital so it is now comparable to that of Europe.
- Both Europe and the United States participate in global markets for forest products.
- European and U.S. societies both make complex, multiple demands upon their respective forests.

On the other hand, European and U.S. societies will not fully converge in their conceptual bases of multiple-use, sustained-yield forest management for three reasons:

- Forests play a significantly greater role in European culture—as the setting of many European fairy tales and the basis for institutions such

as the "free man's rights" in Scandinavia—whereas in the United States, forests were historically regarded as a barrier to economic development, something to be cleared.
- Private property rights tend to be overdrawn in the United States and often obscure the public character of many forest resources.
- The scale of forests is much different in Europe, where forests are still developed and managed on the basis of nation-states, a considerable number of them within the European Union, and will likely continue to be in the foreseeable future.

But full convergence is neither necessary nor desirable. We can better face the future if we learn from each other, from both our successes and our inevitable failures in forest management. We can also better contribute to the world through the knowledge, experiences, and convictions gained from our different approaches to multiple-use, sustained-yield management, especially in the development of the world's boreal forest as well as the many forest management challenges of the Southern Hemisphere.

With the forests of Russia extending from eastern Europe to the Pacific, a virtually intact belt of boreal and temperate forests encircles the globe. It is in our common interest, as Europeans and North Americans, to respond to the immense efforts required to achieve sustainable management and preservation of unique ecosystems of the world's boreal and temperate forests. Europeans, like North Americans, are engaged within the UN system, seeking to implement international agreements, legally binding and otherwise, that constitute an emerging international forest regime, and to reach consensus on how forests around the world are to be used and managed, given the growing public perception of forests as a common heritage.

The challenges can be captured by the term *sustainable development*. The venue is global. We need to participate, and collaborate, in addressing these (among many other) critical global issues:
- How is Russia going to develop its great expanse of boreal forest?
- How are the tropical high forests of central Africa and Central and South America going to be developed?
- China and India, which combined have 2.4 billion people, are net importers of timber and huge net contributors to carbon in the earth's atmosphere. Are they going to be so forever? Or will they invest in

forests and forestry to sequester carbon and meet more of their own timber needs as well as receive the environmental services of the forests they desperately need?

References

Aplet, G.H., N. Johnson, J.T. Olson, and V.A. Sample (eds.). 1993. *Defining sustainable forestry*. Washington, DC: Island Press.

Bengston, D. N., T. J. Webb, and D. P. Fan. 2004. Shifting forest value orientations in the United States, 1980–2001: A computer content analysis. In *Environmental Values* 13(3)(August): 373–92.

Braatz, S. 2002. *National reporting to forest-related international instruments: Mandates, mechanisms, overlaps and potential synergies*. http://www.fao.org/DOCREP/005/Y4171E/Y417E53.HTM.

Dana, S.T., and S.K. Fairfax. 1980. *Forest and range policy*. Second edition. New York: McGraw-Hill.

European Commission. 2000. *Competitiveness of the European Union woodworking industries—summary report*. Luxembourg: Office for Official Publications of the European Communities.

Farrel, E.P., E. Führer, D. Ryan, F. Andersson, R. Hüttl, and P. Piussi. 2000. European forest ecosystems: Building the future on the legacy of the past. In *Pathways to the wise management of forests in Europe*. Amsterdam: Elsevier.

Floyd, D.W. 2002. *Forest sustainability*. Durham, NC: Forest History Society.

Food and Agriculture Organization of the United Nations (FAO). 1995. *State of the world's forests, 1995*. Annex 2: European Forests and Forestry. Rome: FAO.

———. 2001. *Global forest resources assessment 2000*. Rome: FAO.

Helms, J.A. (ed.). 1998. *The dictionary of forestry*. Bethesda, MD: Society of American Foresters.

Le Master, D.C. 1984. *Decade of change*. Westport, CT: Greenwood Press.

Le Master, D.C., and C.E. Owubah. 2000. *Nation states and forest tenures—an assessment of forest policy tools in eastern European countries*. IUFRO World Series 10: 28–38, Vienna: IUFRO Secretariat.

Le Master, D.C., J.T. O'Leary, and V.A. Sample. 1997. *Forest Service response to changing public values, policies and legislation during the 20th century in the United States*. IUFRO World Series 7: 43–76. Vienna: IUFRO Secretariat.

http://black-death.biography.ms/.

http://p128news,scd,yahoo.com/s/afp/20050528/sc_afp/unforestagriculture_05052800383.

MacCleery, D.W. 1992. *American forests: A history of resiliency and recovery*. Durham, NC: Forest History Society.

Mantel, K. 1990. *Wald und Forst in der Geschichte*. Alfeld-Hannover: Schaper.

Marsh, G.P. 1882. *The earth as modified by human action, A new edition of "Man and Nature."* New York: Charles Scribner's Sons.

Mekouar, A., and A. Castelein. 2002. Forestry legislation in Central and Eastern Europe—A comparative outlook. In *Forest science contributions of the chair forest policy and forest economics*. Volume 26: 1–26. Zurich: Swiss Federal Institute of Technology.

Miller, C. 2001. *Making of modern environmentalism*. Washington, DC: Island Press.

Powell, D.S., J.L. Faulkner, D.R. Darr, Z. Zhu, and D.W. MacCleery. 1993. *Forest resources of the United States*, 1992. General Technical Report RM-234, Washington, DC: USDA Forest Service.

Sample, V.A. 2006. The emerging consensus on principles of sustainable forest management: Common goals for the next century of conservation. In *Working toward common goals in sustainable forest management—the divergence and reconvergence of European and American forestry*. Washington, DC: Pinchot Institute for Conservation.

Schmithüsen, F. 2000. *The expanding framework of law and public policies governing sustainable uses and management in European forests*. IUFRO World Series 10: 1–27, Vienna: IUFRO Secretariat.

Schmithüsen, F. 2004. Forest policy development in changing societies—political trends and challenges to research. In *Towards the sustainable use of Europe's forests—forest ecosystem and landscape research*. EFI Proceedings (2004) No. 49: 87–99. Joensuu/Finland: European Forest Institute.

———. 2006. European forests—heritage of the past and options for the future. In *Working toward common goals in sustainable forest management—the divergence and reconvergence of European and American forestry*. Washington, DC: Pinchot Institute for Conservation.

Shands, W.E., and R.G. Healey. 1977. *The lands nobody wanted*. Washington, DC: The Conservation Foundation.

Speidel, G. 1984. *Forstliche Betriebswirtschaftslehre*. Hamburg und Berlin: Paul Parey.

World almanac 2004. 2004. New York: World Almanac Education Group, Inc.

Glossary: Definitions of Management Strategies Mentioned in This Chapter

ecosystem forest management a strategy guided by explicit goals, executed by policies, protocols, and practices, and made adaptable by monitoring and research based on the best available understanding of interactions and processes between human activities and forest ecosystems necessary to sustain the composition, structure, and multiple functions of forests over the long run

forest management the practical application of biological, physical, quantitative, and qualitative information required for implementing managerial and political principles related to the use and regeneration of forests to meet specified economic goals and social objectives while maintaining the productivity of the resource

multiple-use, sustained-yield forest management a strategy focusing on sustained production of multiple resource outputs as determined by economic demands and social values that best meets the needs of land-owners, forest users, and of the public

sustainable forest management the practice of meeting forest resource needs and values of the present without compromising the similar capability of future generations

sustained-yield management managing a forest to achieve and maintain a balance between timber growth increment and cutting

ABSTRACT

A few outstanding American foresters and conservationists were instrumental in incorporating forestry into the mandate of the Food and Agriculture Organization, which was created in 1945 as one of the specialized agencies of the United Nations. A brief presentation of the machinery governing FAO helps illuminate how member countries can exert influence on its forestry policies and programs. Factors of influence include the relative size of the statutory contribution of the country, its extrabudgetary contribution in the form of "funds-in-trust," its expertise in the relevant field, the importance it attributes to this part of the mandate of FAO, and its location. Here, the influence of Europe and the United States at FAO is tracked through time: prior to 1945 and the decision to include forestry in FAO mandate; from 1945 to 1951, when the headquarters was moved from Washington to Rome; from the early 1950s to the early 1970s and the Stockholm Conference on the Human Environment; and then up until 2000. This historical review shows that through active participation in its governing and statutory bodies, western European countries have had a more direct involvement in forestry at FAO and a much larger representation in its Secretariat and extrabudgetary contributions, whereas the United States has exerted influence through the work of individuals, universities, and research groups and, more generally, thanks to the importance and diversity of its forestry expertise.

CHAPTER 17

European and U.S. Influence on Forest Policy at the Food and Agriculture Organization of the United Nations

Jean-Paul Lanly

The Food and Agriculture Organization was created in 1945 as one of the specialized agencies of the United Nations. Although FAO's primary objective is to contribute to food security worldwide, it is of interest also to foresters and conservationists because its mandate included forestry from the very beginning—a decision that, as we shall see below, had to be fought through.

The constituency of FAO, consisting in 2005 of 188 member countries and the European Community, directs its policies and activities through a governing structure that is somewhat complex and hierarchical, given the broad and worldwide mandate of the institution. The Conference, in which all members are represented, is the supreme governing body and meets every second year. The FAO Council, composed of 49 member countries, none of them permanent, acts as an executive organ, meeting at least four times between regular Conference sessions. The Council is assisted by its Programme and Finance Committees and "sectoral" committees, such as the Committee on Forestry. Farther down in the organizational structure are subsidiary and advisory bodies (the other "statutory" bodies and the panels of experts), whose discussions can be more fully and efficiently held on matters particular to a geographic region (e.g., the FAO European and North American Forestry Commissions, or the "Silva Mediterranea"

Committee) or theme (e.g., the FAO Advisory Committee on Paper and Wood Products).

Like most organizations, FAO has two main components: its governing "machinery," which we have sketched above, and its Secretariat, or staff. The basic relationship between these two components can be presented schematically as follows: the staff prepares Secretariat papers to assist the governing and statutory bodies in their deliberations, conclusions, and recommendations, and in return takes action on such recommendations and is accountable to the governing bodies. The terms and balances of this interaction depend on the matters at stake and on the persuasiveness of the Secretariat. On most forestry issues, whether of a policy or a technical nature, the proposals of the FAO Secretariat are generally endorsed but always subject to evaluation and review, sometimes with amendments. This outcome is due primarily to the experience, wisdom, and self-restraint exercised by the Secretariat. However, particularly on politically sensitive issues about which consensus is difficult if not impossible to reach, FAO is not allowed to propose a position, let alone serve as a forum for its members. A classic example of such issues in forestry was the international instrument on forests (the so-called Forest Convention)[1] in the early 1990s.

Factors of Influence

The above description of the FAO system helps in understanding how member countries, either individually or collectively, can exert influence on forest policies at FAO. In this regard, the exact wording of the title of this paper is important: if it had read "European and U.S. influence on *FAO* forest policy(ies)," or "on forest policy(ies) *in* FAO," it could have led to two misinterpretations arising from the fact that most people equate FAO solely with its Secretariat: first, that the Secretariat and not the whole "FAO system" has policies of its own; second and correlatively, that to have an influence, a given country or group of countries needs only to exert it on the Secretariat. In fact, the influence of countries on forest policy(ies) adopted by FAO derives mainly from their ability to have their proposals and positions endorsed by the organization's governing bodies.

The say that countries have in the recommendations of these bodies and, hence, in sector policies depends on various factors. First, and most important, is the relative size of their contribution to the budget of the organization (its budgetary funding). Although the "one country, one vote" rule prevails in the UN system, there is in reality, of course, some relationship between, on one hand, the importance of a country or group of countries measured in terms of gross domestic product—and hence its statutory contribution to a UN agency like FAO—and, on the other hand, its influence in that agency and on the policies it adopts. The statutory contribution of the United States to the budgetary funding of FAO (i.e., to its regular, or "normative", program) has remained by far the highest of all countries, though its percentage of the organization's budget has decreased from 33 percent in 1945 to 22 percent at present.[2]

In theory, the size of contribution, especially in the case of large contributors, also has an influence on the nationalities of professional staff at the Secretariat, through the recruitment system by quotas. However, for various reasons, the number of U.S. citizens among FAO professionals at headquarters and in the regional offices, particularly in the forestry field, has always been lower, not to say much lower, than what the contribution of their country would have allowed. Among western European countries, Germany has remained throughout in the same position as the United States, whereas France, the United Kingdom, Belgium, and the Netherlands have always been well represented.

Another direct factor of influence has come with the development of the "multibilateral," extrabudgetary funding—those funds entrusted by donor countries to FAO, in addition to their statutory contribution to its budgetary funding—for the execution of specific normative or field activities, and for the positioning of junior, or less junior, experts to help implement them.[3] Several European countries, particularly the Nordic nations, the Netherlands, and Italy and to a lesser extent Belgium, France, Germany, and Switzerland, have thus contributed from the early 1970s to forestry projects and programs designed jointly with the Secretariat. Much funding has thus been channelled by European countries to important global FAO programs, particularly those on community forestry and on forest resources assessment, whereas the United States has had no trust funds with FAO in forestry.

The relative influence of U.S. and European countries on forest policy at FAO has remained, at least indirectly, a function of their experience and knowledge in the various forestry fields, and the results of the relevant research and study work carried out by their institutions. Both sides of the Atlantic have had their comparative advantages in this respect.

The long and institutionalized experience of France and the United Kingdom, and to a lesser extent of the Netherlands, Belgium, and Germany, in tropical regions gave them a certain degree of superiority over the United States until the mid-1960s, notwithstanding the significant work on tropical forests carried out at the International Institute of Tropical Forestry in Puerto Rico and by American ecologists like Leslie Holdridge and Eugene Odum and foresters like Tom Gill, Frank Wadsworth, and Larry Hamilton. The European countries, which were still colonial powers during the first years of FAO, were the depository of the largest body of forestry knowledge and experience in the humid and dry tropics across Africa, Asia, the Pacific, and the Caribbean and had their own tropical forestry research and education institutions: Imperial Forestry Institute at Oxford, Centre Technique Forestier Tropical at Nogent-sur-Marne near Paris, Gembloux and Louvain Universities in Belgium, Wageningen University in the Netherlands, and the Federal Research Center for Forestry and Forest Products in Hamburg, among others.

The gap in knowledge and experience in tropical forestry disappeared gradually with the increase in work by U.S. research groups and think tanks on the subject, the establishment of international forestry groups at the U.S. Forest Service and the U.S. Agency for International Development, and the growing number of American foresters with tropical experience, particularly in Latin America. The positioning of the United States was reinforced by the advocacy work of large American NGOs in tropical forest conservation. However, its influence in this field was exerted essentially on the relevant policies of the US-based international funding institutions—the World Bank, the Inter-American Development Bank, and the UN Development Programme.

Though the conservation and development of forests and other wooded lands was never a top priority for the governments of the southern and eastern Mediterranean countries, and hence for international funding agencies, FAO has always devoted some attention to North Africa and the Near East, particularly in the early years, from the 1950s to the mid-1970s. The United States, despite its long involvement in the Near East and its experience of similar climatic and ecological conditions at home, never actively supported or tried to influence FAO policy and fieldwork in this part of the world. On the contrary, European countries, especially those more directly concerned and bordering the Mediterranean Sea, like France, and to a lesser extent Italy and Spain, were always the ones to propose and fund initiatives in this region.

On temperate and boreal forestry, the relative influence of the United States (together with Canada) and of European countries on policies at FAO and the UN Economic Commission for Europe (UNECE) has remained equally essential, particularly through the work of the UNECE Timber Committee and the European Forestry Commission (and their secretariat, the Joint FAO-UNECE Division in Geneva) and, to a lesser extent, of the North American Forestry Commission. This was the more so as the Soviet Union, though a member of UNECE, never joined FAO. (As of this writing, Russia had not yet joined.)

If we look at the relative strength and influence by forestry subject rather than on an ecoregional basis, both sides of the Atlantic appear somewhat equal, though there is of course more institutional dispersion in Europe, and some comparative advantage of the United States in some fields like forest economics and, until the 1980s, information technology (particularly remote sensing).

Many other factors determine the influence of the United States and the European countries on forest policies at FAO. Two more are worth mentioning at the end of this brief account. The first one is the different relative importance given by those countries to FAO's forestry mandate. The United States, as well as France, the United Kingdom, and other western European countries—and later on, the European Community as a member organization of FAO from 1991—always gave overwhelming priority to the policies and activities at FAO in the fields of agriculture *stricto sensu* (including trade in agricultural products) and nutrition. However, other European countries—the Nordic group and the western-central European countries (Germany, Austria, and Switzerland)—have always paid special attention to forestry at FAO.

The other factor in the relative influence of the United States and European countries has been the location of FAO headquarters. In the first six years of its existence, from 1945 to 1951, FAO was hosted in Washington.[4] During this period, as we shall see below, the situation and issues of the forestry sector in the United States were somewhat impinging on the work of the organization. This became much less the case after 1951, when FAO headquarters moved definitely to Rome,[5] where its mother organization, the International Institute of Agriculture, had been based before World War II. This move increased the synergy of headquarters staff with the FAO European Forestry Working Group and the UNECE in Geneva and, more generally, of FAO with the European member countries and their institutions and experts.

Until 1945: The Origins of Forestry in FAO

The International Institute of Agriculture (IIA), the predecessor of FAO, was created in Rome in 1905 as an intergovernmental organization. Its main object was, in the words of its promoter, David Lubin, an American businessman and agriculturalist, "to defend the interests of farmers against industrial or commercial trusts and cartels, which impose on them their tariffs, prices and their economic and financial conditions."[6] The institute survived World War I and was active until the early 1940s, with ties to the League of Nations. It organized the two first World Forestry Congresses, in Rome in 1926 and in Budapest in 1936. In accordance with a resolution of the Budapest congress, the International Forestry Center (Centre International de Sylviculture, CIS) was established within IIA but headquartered in Berlin, with a predominantly European membership. CIS published the forestry review *Intersylva* and amassed a large amount of data that were eventually transferred to FAO headquarters in Rome.

In 1932, when the Great Depression threatened the forest interests of the Northern Hemisphere with collapse, the League of Nations' Economic Committee convened an international conference of timber experts in Geneva. Shortly afterward, the International Timber Committee (Comité International du Bois, CIB) was created, based in Vienna, with 15 European countries and the United States and Canada as members. Its main tasks were to collect and disseminate international timber statistics and facts dealing with timber supply and demand, to coordinate technical research, to collate and publish information on wood utilization, and to promote trade in timber. CIB also acted as secretariat to the European Timber Exporters Convention, whose members, all European (Austria, Czechoslovakia, Finland, Latvia, Poland, Romania, Sweden, the Soviet Union, and Yugoslavia), in 1935 signed an agreement fixing yearly quotas for their exports of sawn softwood. In the same year, at the initiative of France, the International Commission for Wood Utilization (Commission Internationale d'Utilisation du Bois, CIUB) was set up to act as a world clearinghouse for information on wood technology.

When preparations for the establishment of FAO began at the UN Conference on Food and Agriculture at Hot Springs, Virginia, in 1943, and the mandate of the new organization was initially considered, there was no mention of forestry except for a general reference in one recommendation for conserving land and water resources and, to that end, protecting forests and afforesting unprotected watersheds.

It was finally decided to include forestry within the mandate of FAO but only after a protracted, and most revealing, process, involving mostly U.S. personalities. It started after the Hot Springs Conference, with a meeting of a small group of U.S. foresters who decided to pursue the matter further: Tom Gill, who in 1950 would create the International Society of Tropical Foresters; Lyle Watts, chief of the U.S. Forest Service; Henry Graves, long-time dean of the Yale School of Forestry and a former chief of the Forest Service who had succeeded Gifford Pinchot in 1910; plus members of the national Lumber Manufacturers' Association. The U.S. secretary of Agriculture was reluctant, but Dean Acheson, then assistant secretary of State, was of a different opinion, as was President Franklin D. Roosevelt. The Interim Commission on Food and Agriculture, which had been charged to follow up on the recommendations of the Hot Springs Conference and draft a constitution for FAO, set up in early 1944 the Technical Committee on Forestry and Primary Forest Products, whose recommendations were later included in its report. The committee was chaired by Henry Graves, and among the 11 other members were three U.S. citizens—namely Lyle Watts, Tom Gill, and Walter Lowdermilk, a soil conservationist—in addition to one Canadian and European foresters. The committee stressed that FAO must go beyond "freedom from want of food" and that forest resources should be included "because there were close relationships between forestry and agriculture and because forestry could make an important contribution to an expanding world economy." Finally, the FAO founding Conference of Quebec in October 1945 endorsed the committee's recommendations and decided to include forestry in FAO's mandate.

1945–1951: The First Years in Washington

At the Quebec Conference, FAO had a constituency of 39 countries, of which 15 were from Latin America, ten from (western) Europe, and only four from Middle East, three from Asia (China-Taiwan, India, and the Philippines); only one, Liberia, was African (almost all the other African countries were still colonies or protectorates of France, the United Kingdom, Portugal, and Spain). At the end of the period, the countries numbered 69, of which 20 were from (western) Europe, 19 from Latin America, and 15 from Asia.

In the first year following the founding of FAO, a small Forestry and Forest Products Division was formed as one of the four substantive units. There were only seven officers at the start, with Marcel Leloup, formerly director-general of Waters and Forests in France, as director, a position he would hold until the beginning of 1959. In 1947, the division counted 17 officers in Washington, plus three who formed the Geneva-based European Working Group. Three officers of the 20 were U.S. foresters (including the chief of the Forestry Branch, B.B. Show), and 13 were European (of whom five were French, including Marcel Leloup, Tony François, and René Fontaine in the Geneva group), with Egon Glesinger, an Austrian citizen, chief of the Forest Products Branch, who had been secretary-general of CIB until the early 1940s (he would succeed Leloup as head of the division in 1959).

Two years later, the staff in Washington was reduced from 17 to 13, but the European Working Group added two more officers, and two regional groups had been formed with two foresters each, based in Rio de Janeiro for Latin American, and in Bangkok for the Far East. As for the nationalities of the staff, there was one less from the United States (but one more Canadian) and four more Europeans.

The work of the division was guided by the Standing Advisory Committee on Forestry and Forest Products, which had taken over from the Technical Committee. It was chaired by Lyle Watts, then chief of the U.S. Forest Service, and the vice chairman was Bernard Dufay, director-general of Waters and Forests in France. Among the 11 other members of the committee, there were three other U.S. citizens (E.I. Kotok for research, Tom Gill for the "unexploited forests," and H. Mark on chemical wood industries) and five Europeans. Two years after, the committee has expanded to 22, with the same four U.S. members but three more Europeans, four more representatives of developing countries (Latin America and Asia), and one each from Australia and New Zealand.

European foresters thus dominated the staff and the committee (it can be noted also that the composition of both entities was largely Caucasian). This "European bias" was less marked in terms of the issues addressed, however. If we look at the contents of *Unasylva* issues during this period, we observe a certain balance between subjects of concern to European countries, in particular how to meet their huge reconstruction needs for building material; those of priority for the United States, particularly soil conservation and development of the wood and pulp industries; and those concerning the rest of the world, in particular, the presentation of the results of the first World Forest Inventory and articles on forestry in tropical and Mediterranean countries and regions. It is symptomatic of that period that articles in this last category were written mostly by European tropical foresters, such as André Aubreville[7] of France and H.G. Champion of the United Kingdom, and to a lesser extent by U.S. contributors, and not by nationals of these countries, even those that had long been independent, like the Latin American nations. We must await the second half of 1949 to see articles by Indian foresters; by 1953, nationals from only two other developing countries, Brazil and Pakistan, had been published in the journal.

During this period, the two most important meetings in the field of conservation and forestry held by FAO, or with its strong involvement, are a good illustration of the balance in the "conceptual influence" of both sides of the Atlantic. The first event, organized by FAO in 1947 in Marianske Lazne, Czechoslovakia, was the International Timber Conference, which sought to address the huge wood deficit in Europe compounded by the decrease of supplies from traditional exporters to Europe (particularly the United States and the Soviet Union). The second event reflected a major U.S. concern at that time: the UN Scientific Conference on the Conservation and Utilization of Resources was held at Lake Success, New York, in 1949 in response to a proposal by President Truman to the UN Economic and Social Council, echoing a similar recommendation made at the turn of the century to President Theodore Roosevelt by Gifford Pinchot. One of the six technical sections of that conference was devoted to forestry; two others dealt with land resources and wildlife.

We find a similar balance of influence between Europe and United States in the themes of the statutory bodies established in these early years: the International Poplar Commission, created in 1947, the International Chestnut Commission,[8] and the Mechanical Wood Technology and Wood Chemistry technical panels.

However, other elements demonstrate that European forestry and foresters already had, as a whole, a certain degree of prominence at FAO, or at least that European countries gave more importance than the United States to FAO's handling of forest issues. This was probably due, in part, to FAO's assumption in 1946 of the functions (and property) of both CIS and CIB, institutes that had been based in Europe and whose work had been heavily influenced by European foresters. Among these elements, two illustrate clearly this early European bias:

- the establishment in 1947 of the FAO European Forestry Commission, which preceded the North American one by 11 years; and
- the conversion in 1948 of the international European forestry association Silva Mediterranea, the oldest one after IUFRO, into an FAO committee, which became instrumental in developing forestry cooperation in the Mediterranean basin and the Middle East, with the southern European countries, particularly France, Italy, and later Spain, as main actors.

From the Early 1950s to the Early 1970s

Up to the early 1970s, the international scene in conservation and forestry, as in other sectors, was still simple, but it would become more and more complicated. FAO was the only worldwide intergovernmental institution with a normative program in forestry. The other intergovernmental actors with some interest in forestry, either in normative or in field activities, were few and cooperated within the limits of their respective mandates. They included UNESCO and its incipient Man and the Biosphere (MAB) program, coordinated by the Division of Ecological Sciences, and the multilateral funding agencies, such as the UN Development Programme and the World Bank, which relied on FAO for policy and technical advice and assistance in this field. To a certain extent the same applied to the international conservation organizations, mainly the International Union for the Conservation of Nature and Natural Resources (IUCN, now the World Conservation Union) and the World Wildlife Fund (WWF, now the World-Wide Fund for Nature), as well as the few national ones (mostly American) with international interest in forest conservation. Forestry was not yet on the international political agenda, nor had it become a pet subject of the media. The views of competent national and international institutions were respected, if not agreed on, and not every activist or journalist was yet a self-proclaimed specialist. In this overall context, FAO was naturally recognized as the lead international agency, and could work peacefully and efficiently.

That is why the FAO foresters of the period who are still living consider this the golden age of international forestry at the organization. However, when we look back now, 40 to 50 years afterward, we realize the drawbacks of such a situation—drawbacks compounded by the limitations of the forestry profession in those days. One such limitation, which has been harped on over and over since then, almost ad nauseam and in a sometimes unfair and excessively dichotomous way, was that the profession was basically inward looking and not open to society and the public at large. This was aggravated by the reputation foresters had for behaving in an authoritative, "top-down," and punitive manner vis-à-vis or even against people, rather than in a participatory, "bottom-up," and supportive way. This accusation was not entirely unfounded, and much progress has been made since by the profession to correct its social approaches and its image.

The second limitation of the profession during this period, again seen through the eyes of today, is that it gave insufficient priority to the environmental dimension of forest management. Here, too, the picture is not strictly black and white. Foresters contend that they are the first true ecologists, who long ago coined the concept if not of sustainability, at least of "sustaining," even if it was under the restrictive label of "sustained yield." And most if not all FAO foresters of that period had environmental concerns in mind, as demonstrated in many ways. To mention just three, they developed a strong nature and wildlife conservation program at headquarters and the regional offices and in the field, with many projects for wildlife and protected area management in developing countries; they cooperated closely with UNESCO's Division of Biological Sciences in the design of the MAB program, and with IUCN and WWF, eventually leading to the World Conservation Strategy; and they were deeply involved in the preparations for the 1972 Stockholm Conference and in the creation of the UN Environment Programme.

However, it was also during that period that FAO, under the leadership of British economist Jack Westoby,[9] forcefully and consistently promoted forest-based industries as a powerful means for tackling the problems of economic and social underdevelopment, perhaps without giving due consideration to the attendant land-use and sustainability aspects. Bilateral development agencies active in forestry also worked according to the same paradigm in the 1960s and early 1970s.

This period saw a regular increase in the number of professional forestry staff of European origin, at both headquarters and the regional offices, as well as a growing number of field projects. Belgian, British, Dutch, and French foresters with tropical or Mediterranean experience joined the organization. In the case of the first three nationalities, the process was facilitated by the fact that in the many newly independent African and Asian countries, national officers gradually replaced European colonial officers, who, not being civil servants of their mother country, were available to serve in international organizations in which their overseas experience was appreciated. European presence was also facilitated by the junior ("associate expert") schemes they financed, which gave their young officers an opportunity to work and in many instances later seek more permanent employment and stay with FAO. Through these schemes and their expertise in industrial forestry, the Nordic countries reinforced their presence. After Marcel Leloup left in 1959 and Egon Glesinger departed in 1963, the division was led by N.A. Osara of Finland until 1968, and by B.K. Steenberg of Sweden until 1974. Germany and Switzerland, due especially to their junior scheme, and Spain, which was opening up, succeeded in making their presence felt in both the normative and the field programs.

Conversely, there were no more than two or three American foresters at headquarters and the regional offices, and very few served in the FAO field projects. However, studies and various papers began to be entrusted to U.S. consultants and educational and research institutions, a trend that would develop rapidly in the following period, given the wealth of American expertise. And outside FAO, American conservation groups, think tanks, and universities started to influence the approach to forestry of the financing institutions based in Washington and New York, some of them (the UN Development Programme and the World Bank) funding the organization's field technical assistance activities.

There are at least two fields in which the United States had a definite influence, despite the country's small representation in the Secretariat: first, conservation in general, be it protected area and wildlife management, or soil conservation and watershed management; and second, tropical forestry. Thanks especially to foresters like Tom Gill and Frank Wadsworth, the country was instrumental in having FAO formalize and develop its normative work in the silviculture and management of tropical high forests through the Committee on Forest Development in the Tropics, which was established in the mid-1960s and would remain active until the early 1990s.

An important development during this period was the rapid growth of the field activities of the organization. Commencing in 1950 under the UN Expanded Technical Assistance Programme for economic development of underdeveloped countries, it was soon complemented in 1958 by the UN Special Fund, until both funding mechanisms merged and were consolidated into the UN Development Programme in 1965. At the end of the 1960s, FAO was running some 70 forestry and conservation field projects. The design and implementation of these projects at national and regional levels reflected to a large extent the policies adopted within the organization, and hence the relative influence of Europe and United States, not only because FAO was the executing agency but also because the policies of the recipient countries were directly inspired by the organization as well. However, with time, individual developing countries progressed in maturity and autonomy—which was, after all, the main goal—as did the UN Development Programme.

Experts in this era were mostly European, assisted by an increasing number of junior experts, a few were Canadian and Australian, and fewer still were American. There were already some experts from developing countries, mostly Latin America and India, and their total number increased significantly in the 1970s and 1980s. Industrial forest development—including forest plantations (with the attendant genetic improvement work[10]) and pulp and paper production—was the main objective of the field projects, with conservation and institutional strengthening (administration, policy and legislation, education,[11] and research) being secondary. Latin America and the Mediterranean and Middle East region were the main areas of concentration for FAO forestry and conservation field programs, with the second gradually losing importance to the benefit of tropical Asia and Africa.

Toward the end of this period, in 1970, four years after the establishment of the Fisheries Division, the Forestry and Forest Industries Division was upgraded to the departmental level. The number of FAO regular program forestry staff peaked at that time with some 60 professionals at headquarters and about ten in the regional offices, plus a few junior experts and "trust fund" officers. These figures look small in relation to the geographic and thematic extent of the FAO mandate in conservation and forestry. However, the organization constituted then, and still constitutes today, by far the largest international forestry secretariat.

From the Early 1970s to 2000

In the early 1970s, higher priority was given to the protection of the environment by industrialized countries and the international community, as indicated by the UN Stockholm Conference on the Human Environment in 1972 and the consequent creation of the UN Environment Programme. In the late 1970s and early 1980s, the world became conscious of the serious problems affecting forests, mostly deforestation and forest degradation in the tropics,[12] with the consequent loss of biological diversity, and to a lesser extent, forest dieback attributed to air pollution of industrialized countries. Finally in the 1990s, the Rio Summit and its followup addressed global concerns about sustainability, conservation of biological diversity, and climate change. Forest conservation, rather than a more comprehensive sustainable forest development approach, came higher on the international political agenda. Forests and forest management came increasingly into the limelight and onto the international agenda—without, however, a proportional increase in real political will and official development aid.

All those changes should in principle have strengthened the forestry role of FAO as the only global intergovernmental organization with a comprehensive mandate in this field, even more so because the organization is also and primarily in charge of agriculture, the overwhelming cause of deforestation in developing countries. On the contrary, and somewhat paradoxically, forestry was weakened at FAO. The organization lost its de facto monopoly in international forestry, an evolution that reflects and parallels to a large extent a similar situation at the national level in the developed countries. On a subject that has become more prominent and even fashionable, every entity—be it a developed country, a group of countries, a governmental or nongovernmental organization, a group of individuals, or even a politician seeking a "comeback"—wants to be seen as taking action, irrespective of the work already done or being accomplished by existing competent institutions. This desire has spawned a proliferation of governmental and nongovernmental initiatives and bodies, some short-lived, others longer lasting. FAO, with its global and comprehensive mandate, was asked to take the lead in "streamlining," "harmonizing" and even "coordinating" this institutional mess, sometimes by those who themselves contributed to the confusion. It may even have happened that the organization was instructed to support the development of bodies that would become its direct competitors.[13] The frustrating and Sisyphean work of coordination and reporting took a heavy toll on the limited resources of the organization, which seemed to be the only participant serious about achieving some degree of consistency.

For information and analysis concerning the world and their own region, European countries continued to rely on FAO, on the excellent cooperation between the organization and the UNECE and on the work of their joint division in Geneva within the framework of the European Forestry Commission, the Timber Committee, and their subsidiary bodies (an increasing number of countries benefited, of course, from the work of the European Union institutions). The European timber trends studies, the global forest resources assessments, and the temperate and boreal forest resources assessments were the most appreciated products. This is why European countries continued giving high priority to these studies in FAO programs. The United States maintained a similar stand but relied more on research from its own institutions and think tanks.

The support that the United States gave to FAO work on tropical forestry fell as it became clear that the subject could not be addressed on purely technical grounds. In the saga of the FAO/UNDP/World Bank/World Resources Institute[14] Tropical Forestry Action Programme (TFAP), from the World Forestry Congress of Mexico in 1985 to the mid-1990s, the United States had a limited direct involvement, contrary to the western and Nordic European countries.[15] If in 1990 the United States showed some inclination toward a legally binding instrument on tropical forestry, after the 1992 Rio Summit, unlike many European countries, it objected strongly and persistently to a forest convention that would be applicable to "all types of forests," according to the expression used in the title of the Rio Forest Principles.[16]

During this period, European countries (the Nordic countries, the Netherlands, Italy, and to a lesser extent Belgium, France, Germany, Switzerland, and the European Community) supported in a substantial way some major FAO programs, through trust funds (and associate and other trust-funded officer positions). This is the case notably of the Forest Resources Assessment Programme and the popular Forests, Trees and People one (originally Forestry for Local Community Development). The case of the latter is interesting:

although most of the financing came from Swedish and other European trust funds, its concepts[17] of design and implementation draw heavily from the work and experience of the guild of American anthropologists and sociologists.

Officers originating from European countries continued to form the main group of the Forestry Department professional staff, but there was a significant increase over the entire period in the number of U.S. foresters at headquarters, with one or two serving in the regional offices.[18] Some of these Americans had served as Peace Corps officers in the 1960s or had worked previously in U.S. and FAO field programs.[19] However, the most significant change in staff composition was elsewhere: there was a steady increase in professionals from developing countries, at both headquarters and regional offices, and in the ever-increasing (until the early 1990s) number of field projects executed by FAO in conservation and forestry.[20]

Conclusion

This review of more than 50 years of European and American influence in forestry at FAO reveals an interesting difference. Whereas most western European governments have been steadily present and interacting with the FAO system, mostly through the trust funds and junior experts schemes, American influence has essentially been felt through individuals, universities, and research groups. FAO has benefited all along from the dedication and loyalty of the relatively few U.S. Forest Service staff and other foresters who have served it, and from the very beginning when illustrious American foresters and conservationists were instrumental in finding a home for international forestry in FAO. This is in sharp contrast with the relatively discreet involvement, regarding forestry, of the United States since the 1950s in the governing structure and Secretariat of FAO, limiting its influence on the design and implementation of the organization's forest policies and programs. One probable reason is that the political and economic stakes were not considered high enough by the Department of State and the Department of Commerce to give priority to forestry issues, except when it was to oppose a legally binding instrument on forests. In fact, American influence has been much more indirect, through the use by the FAO Secretariat of the considerable and varied forestry expertise in the United States.

This review of international forestry during the last century raises some more general questions, including these two:

- Should forestry remain a mandate of FAO, or rather, should it become that of another UN agency (or program)? A similar question can be posed at the national level in most countries: should forestry be the responsibility of the ministry in charge of agriculture, or elsewhere? All possible organizational charts within the United Nations, as well as at the national level, have their comparative merits and shortcomings for forestry. It is true that foresters often wonder whether the most adequate place for forestry is in a structure dominated by agricultural interests. However, linking forestry to agriculture has at least two significant advantages: first, the interface between the two main and competing uses of the land is handled under the same roof, and thus deforestation issues can be better tackled; second, the multifaceted contribution of forestry to food security can be more easily highlighted and valorized.
- Among the alternative institutional options, one is to have, at the international level and within or outside the UN system, a separate entity dealing with forestry, the same way there exists a separate ministry of forestry in some countries, or an autonomous forestry entity disaffiliated to a single ministry. This option was promoted by some at the time of the Rio Summit but did not go a long way, probably because it was found at the least not feasible, if not inappropriate.

A final remark, inspired by this review: browsing through the documentation from the beginning of the 20th century, one cannot help feeling a sense of repetition, of a certain stuttering in the problems found, in the diagnoses made, and in the solutions proposed. Though many will argue that new words and concepts are needed to translate new issues and that new approaches must be designed, we realize that often we deal with the same, basic, lasting problems requiring fundamentally the same sort of solutions.

References

Food and Agriculture Organization of the United Nations. Forestry home page: http://www.fao.org/forestry/index.jsp.

Philips, R.W. 1981. *FAO: Its Origins, Formation and Evolution, 1945–1981.* Rome: FAO.

Unasylva, an international journal of forestry and forest industries, 1947–2000 (issues 1 through 199). Rome: FAO. http://www.fao.org/forestry/unasylva.

Notes

1. As early as 1990, the FAO Forestry Department was instructed not to work and make any proposal on this issue, despite the comprehensive and unique knowledge and expertise it could bring to bear on the subject and the neutral forum it constituted. A determined minority of member countries had decided that the matter should not be raised at all, and later, a majority of them wanted it discussed at the level of the UN Economic and Social Council, with the eventual lack of success we know.
2. After the United States, the main contributors in 2005 were Japan (19.6 percent), Germany (9.8 percent), France (6.5 percent), the United Kingdom (5.6 percent), and Italy (5.1 percent). The 25 countries of the European Union together accounted for 38.0 percent.
3. Hence, the denomination of "funds-in-trust," or trust funds.
4. The Forestry Division was first located at 1710 New Hampshire Avenue, Washington, DC, and later moved to the Longfellow Building on Connecticut Avenue.
5. The cover photograph of the *Unasylva* issue Volume 11(2) (1957), titled "Ten Years of Forestry in FAO," shows a view of Rome from the FAO terrace, and the caption reads, "…When it was decided that the permanent Headquarters of FAO should be in Rome, the Forestry Division emigrated en masse by ship from New York to Naples, which was not the least adventurous of the Division's undertaking."
6. It is interesting to note what he wrote to Gifford Pinchot in 1908, 11 years before the creation of the League of Nations: "… But humanlike you want to know my motive before you can trust me. Well, my motive is not salary, not medal, nor social scintillations, nor is it to be a Count of Sacramento. I wish to serve the dear old Uncle, Uncle Samuel, and you laugh! But how many better men have given their lives for the Uncle. But there is a higher service still, and that is for the United States of the World. And I am happy to be a humble soldier, a private, in this Army. Do you understand? And when one is such in dead earnest, the Almighty does not mind that he is an ordinary scrub and no educated diplomat…." And in 1917, nine years later, in the midst of World War I, he wrote "An International Confederation of Democracies under a Constitution"!
7. He wrote the first article of the first issue of the journal, 1(1)(1947), the title of which is, "The Disappearance of the Tropical Forests of Africa."
8. It was formally established in 1952, but talks about a need for an international society for chestnut were held already in 1950. In this period, chestnut trees suffered severely from the blight on both sides of the Atlantic. However, for this and other reasons, the economic significance of these species declined rapidly in the 1950s, and the commission did not survive long.
9. He was chief of the Forest Economics Branch from 1958 to 1963 and then became deputy director (of the Division/Department) until he retired in 1973.
10. The FAO Panel of Experts on Forest Gene Resources was established in 1968 and has since advised the organization in its important work of coordination in this field.
11. The FAO Advisory Committee on Forest Education was set up in the 1960s and met every second year until the early 1990s.

12. The beginning of that period of worldwide concern about tropical deforestation was marked, among other things, by the U.S. Strategy Conference on Tropical Deforestation, organized jointly by the Department of State and the U.S. Agency for International Development in Washington in June 1978.
13. For instance, the expertise and institutional framework of FAO were used, together with those of UNCTAD, to prepare the International Tropical Timber Agreement and the establishment of the International Tropical Timber Organization.
14. The World Resources Institute is an American NGO based in Washington that is active in environmental affairs.
15. Many of the commonsense concepts of TFAP were resuscitated in the late 1990s within the framework of the lengthy and protracted forestry followup to the Rio Summit, under the paradigm of national forest programs, applicable to all countries, whether tropical or not.
16. Up to a point of opposing, at a certain time, any effort to harmonize the various ecoregional processes of criteria and indicators of sustainable forest management.
17. One such concept is "forest community."
18. Few, however, stay for longer than five years, for various reasons, including limitations on the leave of absence given to Forest Service employees, the difficulty spouses encounter in finding jobs in Rome, and preference for the American social environment and way of life.
19. One of them, David Harcharik, was head of the Forestry Department from 1995 to 2000.
20. From 1974 to 1994, the Forestry Department was headed by nationals from Latin America and the Caribbean.

ABSTRACT

Has the international dialogue on forests made European and American silvicultural practices and forest policy more similar? An analysis of the debate over sustainable forest management and its implementation leads to the conclusion that concepts and terminology have converged: the idea of sustainability and the criteria and indicators that define it arise from the same logic. On both sides of the Atlantic, people speak this same language. As this language is translated into modes of governance and forestry activities, however, some differences appear. American forestry, for instance, is strongly characterized by the concept of "wilderness," which makes the conservation of natural resources a major public objective. In Europe, the forest is more a constructed resource built by man over history, with forestry and agricultural activities that directly affected the landscape and forestry practices. Another difference involves the concept of participation, which takes the form of consultation in the European context of representative democracy, and communicative action in America, where deliberative democracy is a norm. Different meanings of the concept of development on the two sides of the Atlantic lead to opposite solutions and tools for reducing emissions of greenhouse gases. Given the different ecological and political contexts, the same concepts may be expressed differently, with as an expected result an active public dialogue that benefits forestry.

CHAPTER 18

The International Dialogue on Forests: Convergences and Divergences Between Europe and the United States

Gérard Buttoud

The international dialogue on forests has been focused on the so-called sustainable management of forests. Almost self-evident as formalized, this concept, which was initially articulated in international forestry spheres to promote more responsible harvesting of tropical forests, has emerged more rich and complex than it appeared at the beginning. It rapidly required forest specialists and experts to redefine the norms to be used as guidelines for action, at least those to be used for justifying their management decisions. The logic determining sustainable management is indeed new and different.

The referents are not merely technical, such as specific management strategies aimed at sustainable yield of timber production, or pure silvicultural activities, or sectoral public decisions; rather, forest management needs to evolve into a broader context, where the forest is just one of the components of environment, land use, and social and economic development. Sustainability is to be evaluated according to all its components—ecological, economic, and social.

What effectively contributes to sustainability in those three fields of reference is not fully agreed upon by the various specialists of each discipline. Not even specialists in ecology can agree on what makes an ecosystem stable; economists do not assign the same value to life; and sociologists do not assess the social logics in the same way. No one model of sustainable forest management exists, since forested land varies so much from to place.

In addition, methods of forest management that take into consideration ecological, economic, and social sustainability can only be a compromise among those different elements, which frequently work in opposition. Many tools and practices that promote ecological sustainability make it difficult to achieve objectives that could contribute to social and economic sustainability. There is a need to find an empirical balance.

Thus, whether we can reach such an integration through purely scientific expertise or by communication between stakeholders remains a basic question. Should we optimize a combination of variables, or negotiate between different positions? Both scientific and management experience clearly shows that a mixed model (as in the terminology employed by Buttoud and Samyn 1999) may be useful in a process of promoting reasonable management of the resource. It is clear that management style may strongly differ depending upon the local situation, as appreciated through its ecological, economic, and social components.

To clarify the definition of sustainable forest management, the identification of criteria (principles for action), indicators of results (performance), and means (system) by which attainment of those framework conditions can be verified will provide the needed tools for managers and auditors. In this regard, certain basic principles are commonly accepted at the international level.

The participation of the public and stakeholders is universally considered essential at all steps in the formulation and implementation of decisions, given that in today's world, technical knowledge is readily available throughout society and channels of information allow citizens and the public authorities to connect with and react to each other.

The same is true for professional expertise, which is being redefined to encourage a better integration among ecological, economic, and social aspects in a context where intersectoral links (with agriculture, environment, and land use) directly determine the management choices. Decision making is no longer restricted to its technical components. New norms of action tend to define the forester more as a moderator or auditor than as a simple manager. Finally, a decision is understood as a process that needs to be adapted to changes in the context and be evaluated from outside.

Indicators may vary from one situation to another, depending upon the local conditions but also upon the viewpoint for evaluating management. Is the issue measuring the effects of an action on the dynamics of the ecosystem, or appreciating the means and tools as a goal to improve the situation of the related forest?

The changes brought about by the international dialogue on forests in terms of governance and forest management are fundamental, and all these changes have of course affected the way forests are managed in both Europe and North America.

Convergences between Europe and America

Although the terms of the dialogue are the same on both sides of the Atlantic, even greater convergences have appeared following the international conferences on the future of the world's forests. The general acceptance of English as an international common language has led to using the same words for similar things, even abstractions, and despite discussions of interpretation, use of common terminology has helped bring divergent positions closer together, especially when forests in developing countries are under discussion.

More conceptually, the Bruntland report (published in New York and signed by a European) gave rise to the idea that as far as sustainable development was concerned, the two concepts of development and conservation were neither opposite (as held during the 1960s) nor could they be reconciled through specific practices (as promoted in the 1980s); rather, they were convergent: one could not be achieved without the other. This new vision has not only helped align producers' and environmentalists' positions, but also contributed to a convergence in philosophies of forestry action in Europe and North America, leading on both sides of the Atlantic to important revisions in management strategies.

That sustainable forest management, although difficult to define in a few words, needs to be evaluated in a rigorous manner using scientifically recognized norms also promotes the convergence of forestry approaches in Europe and America. The development of such rationalist tools as criteria and indicators, which is still far from being complete, can only contribute to a common vision of the goal of sustainable forest management, at least in the long run and in the abstract.

Both the wording and the broad concepts of sustainability make this convergence very clear, and it brings to mind the dialogue on what was forest management at Pinchot's time, when the notion of sustainable yield of timber production was current.

Divergences between Europe and America

Those evident convergences should not hide the important differences between European and American ways of considering forests, forestry, and forest policy. First, in America, the discussion about sustainability basically deals with the conservation of the forests, whereas in Europe, everything is related to what should be stated in the management plan.

As perceived by foresters but also by the public—mainly in the United States, somewhat less in Canada—sustainable management of forests is a set of methods that permit large areas of land to be maintained with as little influence as possible from human activities. Without leading to total protection, the definition of areas of considerable size, in which few economic activities are carried out, is still part of the cultural basis for the American discussion on what is a forest. Of course the American civilization has been built faster, in no more than two centuries, through a more complete alteration of nature and with greater use of technology than in Europe. The same process has taken place in Europe, but in a slower and more progressive way over the course of the last millennium. In America, the concentration of technologies on the most accessible land has left large territories where nature can be now be preserved. Thus it is not surprising that in the most forested states of United States, the international dialogue has promoted the concept of "ecosystem management," which meets more difficulty in Europe.

Indeed, in Europe the situation is the opposite: a historical connection between agriculture and forestry has led to forest lands that have the character of a resource built by man. In aiming for sustainable forestry, the first reaction in Europe is to plan a management regime based on a rationalist vision for the future, whereas in United States, the specialists find conservation the easiest arguments to develop when facing the public.

This historical divergence, which has become a cultural gap, today defines even the divergent visions of France and the United States in promoting forestry development in developing countries: French cooperation in Africa, and elsewhere as well, focuses on tools for building up the resource through management plans, while American environmental NGOs seek to promote systematic ecosystem management of the fauna and flora, considered as natural resources.

A second difference may be found in the importance given to the most promising concepts, such as participation—a notion that forms the basis of the sustainable forest management approach. In practice, the concept of participation ranges from consultation procedures in Europe in a context of representative democracy, to communicative action in America, where deliberative democracy is the norm. The participatory processes developed in forestry decision making were born in the United States with the environmental critics of the 1960s, and they have spread with the international debate on tropical deforestation and the related criticism of forestry in developing countries, where state forest management is being replaced by new modes of governance built on the participation of the main stakeholders. The major funding agencies started to promote such kinds of participation all over the world, including countries such as France, where it was not a tradition.

Certain that their conventional knowledge was superior to any ideas coming from the public and other stakeholders, French foresters were not in favor of this new concept, which could constrain their own expertise and challenge their technical authority; but based on the experience of consultative administration, which had been strong in France since the 1960s as far as timber production was concerned, they finally accepted the idea of participation through public opinion surveys, with no reconsideration of the technical aspects of the decision-making processes.

In France, the situation indeed is far from American-style communicative action, which operates in the framework of deliberative democracy, where the various stakeholders negotiate in a rather open way, without concealing their interests. When speaking about participation, then, the same wording is used on both sides of the Atlantic, but with different meanings and representations.

Last but not least, the analysis of the contribution of forest stands to carbon storage, for climate change mitigation, evidently does not result in the same decisions, policies, and statements in Europe and in the United States. In Europe, conventional forest managers were and still are reluctant to embrace the idea that there is an ongoing process of climate change, and that forest production may have a significant role in slowing its effects. Strong reservations also remain concerning the role of the market in rights to pollute (e.g., cap-and-trade emissions credits). But in the debate with environmentalists, most producers have understood that they can gain, both materially (through possible remuneration of the storage service) and symbolically (through a lessening of conflict with environmentalists), from new policies promoting carbon storage.

In the United States, despite public campaigns and intense political lobbying, the government has not yet implemented directives resulting from the Kyoto Protocol, perhaps because the country has not appreciated the international reports describing America's high level of emissions, which show the United States to be a bad actor, but also because a compromise was socially and politically difficult on so fundamental a cultural challenge as reconsidering consumption at the national level.

It may be possible that in the future, the main topic of discussion in international conferences will not be the forest as such, but global balance at the planetary level. With no strategic change on this topic in the next decade, the cultural differences between France and United States may become more apparent than now, when the conferees speak directly to the issue of forest development as such. In different ecological, economic, political, social, and cultural contexts, the same discursive concepts can be used in different practical ways, with a major objective of increasing, through a critical dialogue, the rigor of public action toward forestry development. Today, those differences appear as stimulating elements, encouraging a unique and general but abstract and nonoperative speech on what makes forest management sustainable. They also result from the impossibility, in the ongoing economic process of globalization, of canceling all the historical and cultural differences that give structure to society and ultimately a concrete meaning to general concepts.

References

Brédif, H., and P. Boudinot. 2001. *Quelles forêts pour demain? Eléments de stratégie pour une approche rénovée du développement durable.* Paris: L'Harmattan.

Buttoud, G. 1998. *Les politiques forestières.* Paris Presses Universitaires de France. « Que Sais-je ? » n° 3335.

———. 2001. *La forêt, un espace aux utilités multiples.* Paris: La Documentation Française.

Buttoud, G., and J.M. Samyn. 1999. *Politique et planification forestières.* Berne: Intercooperation.

Buttoud, G., et al. (eds.). 2004. The evaluation of forest policies and programmes. *EFI Proceedings* 52.

Germain, R.H., D. Floyd, and S. Stehman. 2001. Participants' perception of the USDA Forest Service's public participation process. *Forest Policy and Economics* 3(3–4): 113–24.

Gestion(La) durable des forêts tempérées. 1996. *Revue Forestière Française* special edition.

Humphreys, D. 2001. Forest negotiations at the U.N.: Explaining co-operation and discord. *Forest Policy and Economics* 3(3–4): 125–36.

Humphreys D. (ed.). 2004. *Forests for the future: National forest programmes in Europe.* Brussels: COST Action E19, doc. EUR 21364.

International Labour Organisation. 2000. *Public participation in forestry in Europe and North America.* Geneva: ILO.

———. 2003. *Raising awareness of forests and forestry: Building bridges between people, forests and forestry.* Geneva: ILO.

Laroussinie, O., and J.C. Bergonzini. 1999. Pour une nouvelle définition de l'aménagement forestier en tant que discipline d'ingénieur. *Revue Forestière Française*, n° special: 117–24.

Mayer, P. 2000. Forest policy in Europe: Achievements of the MCPFE and challenges ahead. *Forest Policy and Economics* 1(2): 177–86.

ABSTRACT

Conservation may be seen as the wise management of human use of natural resources. Since the 1992 United Nations Conference on Economic Development (the Rio "Earth Summit"), increasing pressure on the world's forests has prompted international efforts to manage them sustainably for the full range of goods and services they can provide and to certify good management when it occurs. The difficulty of making policy with numerous stakeholders has been alleviated by the development of participatory processes for forest planning, management, and benefits. Research has increasingly involved cooperation between governments, between research institutions within countries, and between companies. In forest-related education also, cooperation is intensifying. Interdisciplinary cooperation in policy making, forest management, research, and education will be increasingly necessary, given the coming increases in human populations and livestock, changes in climate and other environmental factors, growth in demand for energy and water, and changes in forest managerial methods and processing techniques for forest-based products. Experiences in both developed and developing countries provide valuable lessons for Europe and North America.

CHAPTER 19

The Role of Forest Conservation in Meeting Global Challenges of the 21st Century: The Necessity for International, Multisectoral Cooperation

Jeffery Burley

Mr. Chairman, distinguished participants, ladies and gentlemen, it is a great privilege and pleasure for me to be asked to address this important and historic meeting. I can lay claim to several reasons for being here; firstly, I spent four years in the Greeley Laboratory of the School of Forestry and Environmental Studies at Yale University, named after one of the major figures in American forestry; secondly, I earned my first professional dollar 30 meters up a southern pine in Mississippi while working as a student forest geneticist for the U.S. Forest Service with Francois Mergen; thirdly, I have the honor to hold honorary membership of the Society of American Foresters; and fourthly, I have long been associated with the School (subsequently Department and Institute) of Forestry at Oxford University (OFI), the institution that was formed in the same year as the Forest Service and that has taught a century of forest managers and researchers who have served throughout the British Empire and Commonwealth and recently in many other countries, too. That background and my own subsequent professional experience in many countries will hopefully allow me to present my views on some historical developments and current issues that have relevance to the links between Europe and North America.

In overview, my presentation will cover the following topics: a broad description of the world's forest types, their locations and changes; a review of the concepts of sustainable development and forest management; an outline of the contributions that forests and trees can make to sustainable development; a summary of current trends in forest policy and management; the role of research and information in achieving sustainability; the institutions that facilitate or seek sustainable forest management internationally; and a suggestion of current major issues and future challenges.

Forest Types, Locations, and Changes

Recognizing that the history of forestry in both Europe and North America has a constant thread of the conflict between forest preservation and production, I start by stating my own preference for the definition of conservation, the 1980 World Conservation Strategy, articulated by the World Conservation Union (IUCN): "Conservation is the management of human use of a resource so that it may yield the greatest sustainable benefit to present generations while maintaining its potential to meet the needs and aspirations of future generations."

In many parts of the world it is clear that the adage "use it or lose it" is becoming fundamental to maintaining appropriate forest resources. The virtually complete loss of the Brazilian Atlantic forest in the second half of the 20th century demonstrated the failure to defend natural forest with sustained management for production in the face of demand for agricultural and urban land.

However, the management that is appropriate is not constant, as there is no single type of forest. Ecologists recognize up to a hundred defined types, but for our purposes here, six are widespread and widely recognized: boreal, temperate mixed, temperate evergreen, tropical rain, tropical deciduous, and tropical dry forests. They differ in geographic location, environmental conditions, species composition, pressure from human populations, and appropriate silvicultural management.

Geographically, the world's forest area of 3.54 billion hectares was partitioned by the report *State of the World's Forests 1995* (FAO 1995) into six regions: temperate and boreal, North America, 13.2 percent; Latin America and the Caribbean, 27.5 percent; Africa, 15.1 percent; Asia and Oceania (developing countries), 14.2 percent; Asia and Oceania (developed countries), 3.0 percent; Europe, 27.0 percent.

By the time of a 2003 report (FAO 2003), because of changes in definitions, the total area of forests was estimated at 3.9 billion hectares, grouped in four types: tropical, 52 percent; subtropical, 9 percent; temperate, 13 percent; and boreal, 25 percent. The decline in forest area during the decade 1990–1999 was 9.4 million hectares, a loss of 2 percent. These areas were partitioned between "southern" developing countries (45 percent) and "northern" developed countries (55 percent) and exactly equally between tropical-subtropical and temperate-boreal regions.

Concepts of Sustainable Forest Management

There have been innumerable debates and definitions of sustainable development, but it is now generally accepted that such development occurs when the three underlying pillars or sets of factors are balanced—economic, environmental, and social. One of the commonly used definitions of sustainable forest management that embraces the ideals of most of the others is that of the 1993 Inter-Ministerial Conference on European Forests in Helsinki:

Sustainable management means the stewardship and use of forest land in a way, and at a rate, that maintains their biodiversity, productivity, regeneration capacity, vitality and their potential to fulfil now, and in the future, relevant ecological, economic, and social functions at local, national, and global levels; and that does not cause damage to other ecosystems.

Contributions of Forests and Trees to Sustainable Development

The three sets of economic products, socioeconomic benefits, and environmental amelioration factors have varying importance in different forest policies and conditions. Recently, intense attention has been devoted to valuing the nonmarketable services. However, this concern is hardly new; a Prussian, George Hartig, often considered the father of scientific silviculture, stated in 1785, "All wise forest management must...have woodlands valued...and endeavour to utilize them as much as possible, but in such a way that later generations will be able to derive at least as much benefit from them as the present generation claims for itself."

Throughout the world, timber and reconstituted wood (pulp, paper, board) are major economic products (Table 1), but in developing countries overall, more than 55 percent of all wood cut deliberately is used for firewood, charcoal, or fluid fuels. However, professional foresters have not fully appreciated the great value of forests in producing animal forage and fodder and human food, nutriceuticals, and pharmaceuticals, including from nontree and animal biodiversity within the forests. Forests also act as land banks for agriculture; whereas traditionally foresters have attempted to maintain control of forest land under forest use, there is increasing pressure for reforestation to restore degraded land into an agriculturally productive state.

Forests and trees play a major role in sociocultural-economic welfare. They provide employment and income while reducing risk in agricultural enterprise. Diet diversity and both human and animal health can be maintained and improved with forest products. Community participation in planning, managing, and benefiting from forests has been practised in the communal forests of Switzerland for centuries, but it was only in the last quarter of the last century in developing countries that joint forest management began seriously in developing countries (initially especially in India).

There is now considerable debate and demand for research on the often-quoted benefits of trees and forests on local and global environments: soil conservation and amelioration, water quality and quantity and flood control, climate and weather amelioration; shade and wind protection, site rehabilitation and restoration, and biodiversity conservation for ecosystem function and stability (*see* Calder in Marcus Wallenberg Foundation Symposium 2002, www.mwp.org).

Changes in the Forestry Sector

Although those benefits were increasingly recognized through the 20th century, a number of changes occurred that precipitated new research, management, and institutions. In the early years attention in developed countries and the colonial empires was focused on the demarcation and protection of natural forests with exploitation of only a few species for timber. The forests of Europe that were heavily devastated in two World Wars were restored by plantations as strategic reserves, largely with exotic conifers rather than indigenous hardwoods. At the same time, large plantation efforts were made in developing countries using exotic conifers and hardwoods, especially eucalypts; the original intent was to create resources for indigenous industries,

Table 1. Sales in 2003 by top-ten global paper companies (US$ billions).

International Paper USA	$25.2
Georgia-Pacific USA	20.2
Weyerhaeuser USA	19.9
StoraEnso Finland/Sweden	14.5
Kimberly-Clark USA	14.3
UPM-Kymmene Finland	14.3
SCA Sweden	11.1
Oji Japan	10.8
Nippon Unipack	10.4
Procter & Gamble	9.9
Total	**150.6**

but much of the product was eventually exported to North American, European, and Asian sawmills and pulp mills.

During the 1950s attention was paid to the silviculture and management of these plantations to maximize industrial wood volume. In developing countries natural forests were managed for export hardwoods, and collaboration took the form of shared information on silvicultural systems and allowable cuts of timber. In the 1960s attention was focused on the quality of industrial plantation timber, especially characteristics of stem form, wood density, fiber dimensions, and chemical properties that were believed to influence timber strength and paper quality. Areas of plantations and diversity of exotic species increased, especially in tropical countries, and silvicultural and breeding research were intensified. By the 1970s research and development focused on industrial pulp and paper quality, especially for plantation products and for mixed-species furnishes.

In the last quarter of the 20th century, although the industrial role of forests was still appreciated, new research and development in developing countries concentrated on the role of trees and forests in supporting agriculture and human welfare. Agroforestry systems with multipurpose species were researched intensively and both national and international agencies recognized the need for the participation of individuals and communities in forest planning, management, and benefits.

Mankind has always depended on forests for a wide variety of nonwood products, but intensive research, development, and policy decisions were made in the latter half of the 20th century on exudates, fruits and nuts, fodder, and animal products, including milk, honey, and silk. These formed the basis of work in the early 21st century to develop chemical feedstocks, nutriceuticals, and pharmaceuticals from trees and forests.

These changes within a century may be seen largely as technological and managerial, but they precipitated and reflected changes in policies and institutions throughout the world. Dominant institutional trends included the move from governmental control to private forestry, from industrial forestry to rural development support, from centralized to outgrower systems, even for major industries, and above all, the recognition that forests have global significance and are a subject for global institutional concern.

At the start of the third millennium, emerging issues included biodiversity conservation and the use of indigenous species rather than exotics; environmental quality and changes; carbon sequestration and trading; renewable energy sources; deforestation, desertification, restoration, and rehabilitation; commoditization, trade, incentives, corruption, and conflict; food security, poverty alleviation, and human health; policy reform, professional status, education, and public and political support; and the globalization of forests and forestry.

International Institutions

Professional foresters have recognized the many benefits of forests for centuries but, at the governmental level, several international forestry initiatives have been established since World War II. Collaboration in its widest sense is reflected in three major institutional drivers—intergovernmental processes, intergovernmental organizations and market-driven certification, all seeking in different ways the attainment of sustainable forest management but using similar terminologies for the definition of criteria and indicators (e.g., Maini 1993):

- *objective:* an object of action; an end as a cause of action;
- *principle:* a fundamental law or rule as a guide to action; a rule of conduct; a fundamental motive or reason for action, especially one consciously recognized and followed;
- *criterion:* a distinguishing characteristic by which a thing can be judged;
- *guideline:* a statement that leads to or directs a course of action;
- *indicator:* any variable that can be measured in relation to a specific criterion;

- *monitoring:* measurement of indicators with well-defined and commonly agreed-upon methodology and periodicity; and
- *standard:* a measurable parameter established for use as a rule or basis for comparison in measuring or judging quantity, quality, value, capacity, or other characteristics; anything recognized as appropriate by common consent, by approved custom, or by those most competent to decide.

The Food and Agriculture Organization (FAO) of the United Nations has since the 1950s been the leading UN agency with a mandate for forest resource assessment, management, and improvement. It hosted the Tropical Forestry Action Plan that led to national forestry programs, and it produced many guidelines for good forest management. The International Tropical Timber Organization (ITTO) since 1982 has supported good management and efficient use of forest products; it produced the first detailed specifications for sustainably managed forests. Since the UN Conference on Environment and Development (UNCED, the Rio "Earth Summit") in 1992, serious attempts to agree upon sets of criteria and indicators led to several regional processes for different forest types, including the Montréal Process (boreal), the Helsinki or Pan-European Process (temperate Europe), Lepaterique (Central America), Dry Zone Africa Process, African Timber Organization Process, Tarapoto Proposal (Amazon), Dry Forest Asia Process, and the Near East Process.

Those processes were developing specifically for sustainable forest management, and intergovernmental conventions were established that had direct relevance to forests—the Conventions on Biodiversity (CBD), Climate Change (UNFCCC), Desertification (UNCD), and Trade in Endangered Species (CITES). The current UN Forest Forum (UNFF) of the UN Commission on Sustainable Development at its final meeting in 2005 will decide on the future international governance instrument for forestry (which may or may not take the form of a legally binding convention). The leading organizations that constitute the Collaborative Partnership on Forests include FAO, ITTO, the World Bank, the Global Environmental Facility, the UN Development Programme, the UN Environment Programme, UNFCCC, CBD, the World Conservation Union, the International Union of Forest Research Organizations, and the Center for International Forest Research; they work in support of developing an international instrument. Despite the long processes of intergovernmental arguments over the nature and structure of such conventions, they clearly reflect collaborative intent.

While intergovernmental organizations and processes were seeking agreement on criteria and indicators of sustainable forest management, market-driven processes have also developed. Many of these are nationally based but two are international—the Forest Stewardship Council and the Pan-European Certification Scheme. These set standards for certification of forest management schemes (i.e., they are certifiers of certifiers) that encourage individual state or private forest owners to manage forests well and thereby obtain enhanced market niche, share, or value. By 2004 more than 150 million hectares had been certified as sustainably managed.

Other examples of collaboration include international political and economic organizations, such as the North American Free Trade Agreement, the European Economic Community, and the World Trade Agreement, all of which have direct and indirect effects on forest management. Still others include trade federations (e.g., the Timber Trades Federation and the International Federation of Building and Wood Workers, which seek well-managed forests as a basis of their livelihoods) and a range of joint ventures between governments and commercial organizations (e.g., ENSIS, the cooperative venture between Australia and New Zealand, or the various cooperative research centers in Australia and the United Kingdom, or the many networks of projects among institutions in countries of the European Union).

Collaboration in a managerial sense is exemplified by the government-sponsored Canadian Model Forest Programme, in which many stakeholders in a given forest share the planning, management, and benefits of the forest for multiple purposes. At least one model forest was identified in each province, and these became the models for the International Model Forest Network, with 34 forests in 14 countries—from Argentina to Canada, Cameroon to Sweden, and the Philippines to China and Russia. These are intended to set examples for wise management in other forests and countries. The Tropenbos program, sponsored by the Netherlands government and supported by several Dutch universities, conducts research and management in six tropical forests (on three continents) that are intensively described, quantified, and managed.

Specific problems have arisen in intersectoral cooperation that are not obvious in single-disciplinary work. Researchers and managers in the biophysical subjects commonly have different political masters and stakeholders from socioeconomic researchers. They have different training, background, language, and understanding of a problem. Biophysical scientists understand the difference between holistic and reductionist science, but this is more difficult when socioeconomic approaches are included. A problem that is common to all is the growing role of civil society and industry as well as governments in planning resource management and determining the problems for which research is required.

Research

It must be recognized that research is heterogeneous (Table 2) and part of a continuum from the identification of a problem through basic (pure or reductionist) research to strategic (applied or holistic) research to development and application (Burley 1999). In relation to sustainable forest management, research is required to develop technologies that may range from natural forest inventory, valuation, management, and silvicultural methods through plantation and agroforestry techniques to classical tree breeding and molecular biotechnology. These biophysical researches must be integrated with socioeconomic and cultural research to develop acceptable and appropriate technology systems.

It is apparent that the policy of many governments and industrial companies is to encourage near-market research to be financed by the producer and the market; research and development are becoming ever closer to the customer, particularly for timber and paper products. However, both industrial and forestry research seek general economic development, increased efficiency and competitiveness, improved quality at less cost, and enhanced environmental performance.

Several of the activities described earlier include small amounts of research for each selected site, and the shared results help all collaborators. However, for more than a century, research per se has been an outstanding example of international cooperation. IUFRO was formed in 1892 by the national forest research institutions in three European countries. It has subsequently expanded to include approximately 700 member organizations in 112 countries with some 15,000 individual scientists. These collaborate voluntarily in 280 research units (divisions, research groups, working parties, and task forces), each comprising 10 to 300 members working on a common topic. Activities and products include scientific meetings with proceedings, books

Table 2. The heterogeneity of research. Adapted with permission of the Commonwealth Forestry Association from Burley (1999).

Components	Design	Analysis	Information	Judgment
Steps of RDA	Pure research	Strategic research	Development	Adaptation (application, commercialization)
Benefits of trees and forests	Wood products	Nonwood products	Environmental	Social
Geographic locations	Global	Regional	National	Local
Stakeholders	International agencies and populations	National agencies and populations	Industry	Scientists
Research providers	Advanced research institute or university (international or national)	Less advanced institute or university (national)	Local field station or individual land owner	Company research unit
Sources of finance	International agencies	National government agencies	Nongovernmental organizations, national and international, (charities foundations, benefactions)	Industry
Levels of competition	Political (country or state/province)	Commercial (company)	Institutional (research institutes and universities)	Scientists (individuals or groups)
Levels of collaboration	Twinning	Multiple twinning	Networks	Cooperative centers and institutes
Types of network	Collegiate voluntary	Invitational without finance	Institutional with finance	Catalytic

and journal publications, training workshops and courses, and manuals of standard methods and common terminologies to clarify technical jargon and facilitate translation in international policy fora.

Virtually all researchable subjects are covered by existing IUFRO units or can easily be covered by new ones. Increasingly, their research is proactive and interdisciplinary in support of emerging policy issues. Moreover, they are able to support researchers in developing countries, who often work in isolation and have poor access to resources of skill, manpower, facilities, finance, information, opportunities for collaboration, and channels for dissemination of their results. A more restricted example is the European Forest Institute, based in Joensuu, Finland, with 136 member organizations in 37 countries undertaking cooperative research projects.

As we enter the second century of the U.S. Forest Service, there are a number of challenges for forest science that must be answered (and for which the Forest Service will provide outstanding support): uncertainty about the stability of climate and the extent and location of change; the impact of forests on soil structure, chemistry, and retention; the quantity, flow, and quality of water; area and quality of forests; land-use systems to preserve landscape functions; links and antagonism between holistic and reductionist research; and interdisciplinary methods. The principal environmental challenges to sustainable forest management include changes in natural environmental features of climate and diseases; manmade environmental factors, including pollution; managerial environmental factors, such as additives (fertilizers, pesticides, and possibly genetically modified trees); and technologies of processing and use of forest products.

Levels of Competition

Competition is the direct opposite of collaboration and is the striving or vying of two entities to gain the same objective or object. Humans have always competed with other organisms and with each other for mates, land, food, possessions, wealth, athletic and academic superiority, reputation, ecological niche, and total survival. In the context of this conference, several of these factors are relevant at four major levels (Table 3).

The *political level* includes countries and states or provinces within countries; here competition is occasionally for land (for physical and economic security), frequently for possessions (particularly raw materials), and always for economic wealth. The attainment of these relies occasionally

Table 3. Objectives of entering into competition. Adapted with permission of the Commonwealth Forestry Association from Burley (1999).

Objectives	Level			
	Political (country, state, province)	Commercial (company)	Institutional (research institute, university)	Individual (scientist)
Influence	Political	Market share and expansion, comparative advantage	Reputation	Fame (breakthroughs, prizes, peers)
Finance	Economy	Profit	Support	Fortune (income, intellectual property rights)
Security	Physical and economic	Know-how and income	Staff and facilities	Employment and promotion

on military strength, usually on existing economic status, and often on some form of political influence. Depending on the political system and the country, the leaders of countries may seek these benefits for personal wealth or national economic development.

It is widely accepted that the political philosophy of privatization and *commercial* competition increases efficiency while reducing waste, provided that social or environmental damage is not caused by cost-saving technologies. National and multinational companies seek expansion and increased market share while maximizing economic return to shareholders. The increased know-how derived from research maintains or improves the comparative advantage of the company.

At the *institutional level*, universities and research institutes compete to enhance their scientific reputations, to maintain or increase their income from governmental or nongovernmental sources, and to maintain their staff and facilities, often linked to training and education. Universities are increasingly assessed by their research and teaching quality in comparison with others in the country or discipline, and their governmental support depends upon their scientific rating (often judged principally by their output of peer-reviewed publications in high-quality journals rather than by their influence on development through agency reports, publications in applied science journals, and staff roles in political processes). This process does not encourage interinstitutional or interdisciplinary cooperation at a time when such activities are recognized as essential.

Individual competition among scientists is an inherent characteristic for some but is becoming common as resources decline. Scientists require continued employment and promotion, and most review or appraisal systems take account of scientific output as judged by publication. Scientists also compete (sometimes unknowingly) for fame through prizes and other recognitions of scientific progress or breakthroughs (e.g., the Marcus Wallenberg Prize). Occasionally such breakthroughs are rewarded financially by the patenting of the intellectual property rights and the sharing of profits by scientists and/or their employers.

Competition at all but the political level sharpens intellectual and practical focus but has the disadvantage of making the competitors operate in a vacuum of secrecy and discouraging productive interactions. Moreover, all such competition increases the difficulty of intersectoral collaboration.

Benefits of Collaboration

Seven major sets of benefits arise from collaboration (Table 4), most of them occurring at all four levels identified above (political, commercial, institutional, and individual). For major projects the *costs, risks, and capital equipment* may be shared; not every collaborating organization requires a mass spectrometer, a DNA sequencer, or a high-level meteorological tower, provided that adequate access at cost may be assured to all the partners. Material may also be shared with lower overall costs; examples include the classic joint collections and sharing of seed for international provenance trials that for 50 years have been the hallmark of the Commonwealth Scientific and Industrial Research Organization (Canberra, Australia), the Centre Technique Forestier Tropical (Montpellier, France), the Danish Tree Seed Centre (Humlebaek, Denmark), FAO, OFI, and many IUFRO working parties. An outstanding example is the precompetitive sharing of knowledge and genetically improved material resulting from some cooperative breeding programs. Among these, the North Carolina State-Industry Tree Improvement Cooperative deals with both hardwoods and conifers of the southern United States; its programs are jointly funded by many companies, and university staff and students conduct rigorous research and breeding strategies together with company employees. At the international level, the

Table 4. Benefits of collaboration. Adapted with permission of the Commonwealth Forestry Association from Burley (1999).

Benefits	Level			
	Political (country, state, province)	Commercial (company)	Institutional (research institute, university)	Individual (scientist)
Share risks and costs	■	■	■	
Share capital equipment	■	■	■	■
Share material	■	■		■
Share land or laboratories	■	■	■	■
Create critical mass of staff		■	■	
Stimulate interdisciplinarity		■	■	■
Share information and reduce duplication	■	■	■	■

Central American Cooperative on Research deals with Central American species and has members in many countries in the Americas and Africa.

Land and laboratories may also be shared, particularly where expensive or space-demanding experiments are concerned or where information is sought on the interactions between experimental treatments and site. Commercial companies have for many years collaborated in creating or supporting high-technology research institutions (e.g., the several national pulp and paper research institutes in Canada, Germany, Sweden, the United Kingdom, and the United States). Collaboration at this level is commonly on topics that are seen as precompetitive; currently, these include enhancing energy efficiency, improving environmental controls, reducing pollution, developing technologies for genetic engineering, and propagating improved plant material. However, gene patenting may well prejudice such collaboration.

Collaboration may be seen essentially as a means of obtaining economies of scale by bringing together relatively few resources from many sources to address widespread problems. However, such economies are achieved most effectively when the collaboration is applied to narrowly defined topics; diseconomies may result from major collaborative expenditures on diffuse topics or all-embracing institutional programs. Thus the value of multiple twinning lies in the ability of any one organization to collaborate with a number of others each on a different topic; strategic alliances should be subject-specific and may be temporary.

As the costs of research increase and government funding in particular decreases, a major advantage of collaboration is to obtain the *critical mass of staff* necessary to obtain the intellectual interactions that address problems adequately. Individual institutions are being forced to reduce staff and consequently to specialize on narrower topics, but many problems require *interdisciplinary approaches*, particularly when dealing with public goods and services rather than commercial products. It is particularly necessary in rural development research to encourage collaboration between the "hard" and "soft" sciences so that the socioeconomic needs and impacts of technological change can be established.

Underlying all those benefits is the sharing of information and expertise, thus reducing unnecessary duplication of efforts. Scientists, including graduate research students, have traditionally published their results in journals as soon as possible to achieve recognition for the first discovery. Once in the public domain, the information can be capitalized by anyone. A vast amount of information has been published but is often not known or available, particularly in developing countries; libraries such as OFI's Library of Deposit for world literature may hold the resource, but it is of use only if users can access it. Abstract services such as CABI's forestry abstracts and forest products abstracts or the Pulp and Paper Institute's pulp and paper abstracts make such information more widely available, particularly since electronic forms have become used. Yet, for specific topics, collaboration between active researchers is the most rapid way of distributing current progress—although we have to recognize the inhibitors of collaboration at the same four levels of political, commercial, institutional, and individual interest (Table 5).

As we celebrate 100 years of outstanding national and international service by the U.S. Forest Service, we must recognize the emerging challenges that

Table 5. Inhibitors of collaboration. Adapted with permission of the Commonwealth Forestry Association from Burley (1999).

	Level			
Objectives	**Political** (country, state, province)	**Commercial** (company)	**Institutional** (research institute, university)	**Individual** (scientist)
Threats	To security	To advantage	To advantage	To personal fame and ambition
Traditional behavior	Political animosities	Commercial secrecy	Unidisciplinary approach	Academic secrecy
Perceptions	Public's nationalistic suspicion	Industrial espionage	Lack of trust in partners	Lack of trust in partners
Legal		Antitrust legislation (monopolies); intellectual property rights	Intellectual property rights	Intellectual property rights
Financial		Sharing benefits	Cost of collaboration exceeds benefits	

face forestry and research in the current century: conservation of biodiversity and management of forest resources in the face of increasing human demands and uncertain climatic instability; need for clean air, water, and soil; adequate quantities and flows of water; stability of land and soil; area and quality of forest; land-use systems to preserve ecosystem functions; equity among a wide range of stakeholders with legal interest in forests; interdisciplinary problem solving; and integration from pure and applied research to development and application of technologies that conserve our forests for all the multiple benefits that they confer.

Conclusions

Throughout the century since the origin of the U.S. Forest Service, the world has seen dramatic growth in factors and dependencies relating to forests and trees: the populations of humans and their domestic animals; the recognition of the diversity of goods and services that can be derived from forests and trees; and the importance of participation by all stakeholders, not just governments or industries, in the planning, management, and use of forest resources. These have generated a growing awareness of the need for interdisciplinary research and development to solve problems of human and environmental welfare, and also the necessity of increasing the links between research, development, policy-making institutions, the media, and the public. In turn, these will continue to require well-trained professionals able to comprehend the concepts of many disciplines and communicate their research, management, and policy findings to a wide audience.

References

Burley, J. 1999. Collaboration versus competition in forestry research and development. Invited paper, New Zealand Forest Research Institute 50th Jubilee International Forest Research Conference. Rotorua, New Zealand, April 3–4, 1997. *International Forestry Review* 1(4): 207–14.

FAO. 1995. *State of the world's forests 1995.* Rome: Food and Agriculture Organization.

———. 2003. *State of the world's forests 2003.* Rome: Food and Agriculture Organization.

Maini, J.S. 1993. Sustainable development of forests: A systematic approach to defining criteria, guidelines and indicators. In *Technical report of the CSCE Seminar of Experts on Sustainable Development of Boreal and Temperate Forests.* Montreal, September 27–October 1.

ABSTRACT

Education has always been important in the development of forestry professionals, both in Europe and in America. In the United States, forestry university education effectively began after Gifford Pinchot's return from schooling in Europe. American university programs and curricula spread during the early part of the 20th century and again after World War II. The environmental movement of the 1960s brought challenges to forestry. Recent years have witnessed declines in forestry enrollment as U.S. students have entered other natural resource disciplines. In Europe, formal forestry education began during the late 18th century, and the French national forestry school at Nancy opened in 1825. Required courses today sound strikingly similar to those of the Nancy curriculum during Pinchot's time, in 1889. However, the content of the courses has evolved. American forestry schools seem to have been more flexible than the French school, especially in recent years, in altering their curricula in response to changing social conditions. Globalization should provide impetus for both sides of the Atlantic to work together to share innovations in forestry education.

CHAPTER 20

The Evolution of Forestry Education in the United States and Europe: Meeting the Challenge of Sustainable Forestry

J. E. de Steiguer, Patrice Harou, and Terry L. Sharik

It is only fitting that an international symposium that pays homage to Gifford Pinchot should consider the topic of forestry education in America and Europe. Indeed, it was the pursuit of forestry education that caused Pinchot to travel from America to Europe in 1889. There he spent time as a student at the French forestry school in Nancy, he sought the counsel of Europe's best forestry professors, and he traveled also to Switzerland and Germany to observe forestry in practice. Pinchot's stay at the French forestry school was only six months, and the entire European sojourn lasted just one year. However brief, his time in Europe was important because it would be the only formal forestry education he would receive before returning to America to begin his illustrious career as a forester and politician (de Steiguer 1994).

This chapter examines the evolution of forestry education both in the United States and in Europe and considers whether forestry education is currently meeting the challenge of sustainable forestry. In the process, we attempt to explore the importance of forestry education to the profession, the origins of forestry education, how the formation of foresters has changed over the decades, and even centuries, and lastly, just how well forestry education has adapted to the challenges of sustainable forestry in the 21st century. The two main parts—forestry education in the United States and forestry education in Europe—are, in turn, divided into subsections addressing those topics. First, however, it is important to discuss the importance of education to the professional.

Education and the Professional

The *Encyclopedia of Education* (1971) states that all professions share at least six characteristics: 1) they perform an essential social function; 2) a lengthy period of education is required to enter the profession; 3) practitioners are service-oriented (i.e., altruistic); 4) there is official recognition of professional status by the government; 5) the nature of the service rendered makes the clients incapable of appraising it; and 6) there are standards of competence.

The important criterion to note here, of course, is the second, the requirement for a lengthy period of education. The *Encyclopedia of Education* argues that a professional requires expert knowledge that can be gained most usually through formal education. Often the quality and suitability of this education is certified by some external authority, such as the government or a professional society. Furthermore, expert knowledge is distinct from a skill, which may be learned on the job; expert knowledge is not commonly possessed by the layperson.

Foresters, likewise, have long recognized the importance of education to the professional. For instance, the Society of American Foresters (1958) defined a forester, in part, as "a person who has been professionally educated in forestry." The professional forester, as opposed to the forest technician, it has been said, is well-grounded in the biological, physical, and social sciences; has a thorough grasp of principles and practices; understands not only how things happen but causes and effects as well; and gives full consideration to all factors that impinge upon forest management issues (Dana and Johnson 1962). This breadth and depth of knowledge are gained by the professional, in large part, through formal forestry education.

Without question, formal education, often at colleges, universities, or specialized schools, has been an essential part of the making of a professional forester. This has been true in both Europe and the Americas. Both regions have a rich tradition in forestry education. That of Europe, of course, is much older than that of America. But as we shall see, historically there has been a linkage between the two regions that continues even today, as evidenced by this conference and this chapter.

Forestry Education in the United States

Pinchot in Europe. In his autobiography, *Breaking New Ground*, Pinchot (1947) explains how he came to be America's first forester. He says that the idea was first placed into his head in 1885, by his father. Indeed, Pinchot's father raised this possibility with him before the son went to Yale University as an undergraduate. Gifford Pinchot had always had an interest in the out-of-doors. However, he was now divided between forestry, medicine, and the ministry as possible professions. He would need a few years as a student at Yale to make up his mind in favor of the first.

During the winter of his senior year at Yale (1889–1890), Pinchot traveled to Washington, DC. There he met with U.S. government forestry officials, such as Dr. George B. Loring and Dr. Bernard E. Fernow, who were not encouraging in his pursuit of forestry as a career. Nevertheless, young Pinchot would "stick to his guns" at his father's urging and become a forester. His dedication to pursue forestry was such that when he spoke at his Yale commencement, the topic of his talk was the "importance of Forestry to the United States."

After some arrangements and a side trip to the Universal Exposition in Paris, Gifford Pinchot enrolled on November 13, 1889, in l'École Nationale Forestière (later, l'École Nationale du Génie Rural des Eaux et des Forêt) in Nancy, France, to study forestry. His education (described in greater detail below) consisted of hour-and-a-half-long lectures covering three general areas of study: silviculture, forest organization, and forest law. There were also field excursions to the Forêt de Hayes, the Forêt Communale de Vandoeuvres, and forests in the Alps and the Vosges Mountains. Pinchot also traveled to Switzerland, to the Sihlwald, and to German forests in Saxony, Prussia, and the Black Forest region. After 13 months in Europe, Pinchot returned to the United States with the thought of organizing a national forest administration, which he of course did with the support of Teddy Roosevelt. However, a continued interest in forestry education would always be a part of Pinchot's plan to establish the practice of forestry in America.

The first half-century. Dana and Johnson (1962) have provided an excellent history of U.S. forestry education from its inception to about 1960. According to these authors, forestry education in the United States really began in 1898. Prior to that year, there were no degree offerings—only a few individual courses at universities that dealt with various forestry issues. Furthermore, with rare exceptions, none of the early teachers had formal forestry training.

The first university-level forestry education program conferring a degree was at New York State College at Cornell University, where instruction started on September 22, 1898. It was headed by B.E. Fernow, a graduate of the Forest Academy at Münden in Prussia. The course covered four years and led to a bachelor of forestry degree. Two years were devoted to basic subjects and the second two years were for technical subjects.

In the same year, the Biltmore School in North Carolina was founded and operated by Carl A. Schenck, also a German. This was essentially a technical school dominated by Schenck, who was later assisted by a small faculty. It led to a bachelor of forestry degree and then, after one more year, to a forest engineer degree. The school never had accredited standing but nevertheless turned out many leading professional foresters.

In 1900, Henry S. Graves, a Yale graduate who had studied forestry in Germany and France, assumed the directorship of the newly founded Yale School of Forestry. Only men with a bachelor's degree were admitted to the program, which covered two years and led to a master of forestry degree.

Perhaps the most unusual aspect of these very early years of forestry education was that the schools were established before there was an appreciable demand for their graduates. By 1910, for example, about 25 states had some sort of forestry instructional unit (not all of which were college programs). This supply-demand imbalance would change somewhat, however. By about 1911, there was an expansion in the demand for graduates due, in large part, to federal legislation creating the U.S. Forest Service and state legislation establishing state forestry agencies as well.

In 1909, Gifford Pinchot, to better understand the evolving forestry education situation, called for the First Conference on Forestry Education. Significant results from the conference included recognition of the need for different levels of forestry training (e.g., ranger training versus professional education). Interest was expressed in formulating a standard body of instruction for foresters, an idea that would later lead to curriculum accreditation. Also, there was a belief that foresters should be educated as logical thinkers with analytical capabilities rather than just as individuals trained to memorize facts.

By 1911, 20 schools of forestry were operating. All but two were four-year programs or longer. Because the U.S. Forest Service was the principal employer of forestry school graduates, the requirements of the civil service system largely dictated the curriculum content of the schools.

Aside from the increasing number of forestry schools, the most significant feature of the period prior to 1930 was the trend toward diversification of the programs of study. Two areas of diversification that are particularly noteworthy were forest engineering and wood products. In these areas, the demand for trained graduates easily absorbed the supply. However, in general, there was concern over uncontrolled diversification and a feeling that some uniformity among programs was desirable.

One suggested element of commonality was the number of semester hours required for a four-year college degree in forestry. A report on forestry education by Graves and Guise (1932), prepared with a $30,000 grant from the Carnegie Corporation dealt in part with this topic. The authors recommended that a four-year forestry degree consist of 128 semester hours of instruction plus ten weeks of summer fieldwork. Of the total time, 44 percent was to be devoted to general education, 23 percent to pretechnical subjects, and 33 percent to technical subjects. It is interesting to note that this recommendation mirrors to a large extent the modern bachelor's degree requirements.

To further advance degree uniformity and standards, Chapman (1935) recommended that a system of accreditation be developed to rate the forestry schools. The suggested system of weights was as follows: departmental status, 6 percent; faculty or provisions for instruction, 16 percent; personnel or faculty, 24 percent; financial support, 17 percent; equipment, 12.5 percent; field instruction, 12.5 percent; and "historical" (institutional criteria) and alumni, 12.5 percent. The executive council of the Society of American Foresters was to be the rating body, and a grade of 70 percent or better was required for accreditation. In 1936, accreditation was made a permanent and continuing part of the Society of American Foresters activities.

The establishment in 1933 of the program for federal Emergency Conservation Work (later the Civilian Conservation Corps) led to an almost immediate boom in forestry school enrollment (e.g., undergraduate enrollment increased 69 percent in 1934) and also an increase in the number of four-year forestry schools. As surprising as it may seem, however, this expansion was not viewed favorably by all. To some in the profession, the expansion of new schools and programs was merely an attempt to capitalize on new demand created by temporary government programs. For example, the Society of American Forests' Division of Education at its 1936 annual meeting issued a statement deploring the establishment of new schools except where the need could be demonstrated and the standards for professional training could be met.

In 1936, just a few years prior to World War II, the enrollment of forestry students in U.S. colleges and universities was 3,888 underclassmen, 1,518 upperclassmen, and 159 graduate students.

Postwar to 1960. World War II, of course, decreased forestry schools enrollments tremendously as GIs headed off to war. However, in 1945, with the end of the war and the availability of government assistance for university education, enrollments skyrocketed. Undergraduate forestry enrollment, in fact, jumped from 571 students in 1944 to 7,010 in 1946. The proportional gains in graduate enrollment were similar.

The result was that the forestry schools were overwhelmed. Staffs, facilities, and budgets were not prepared for such unexpected loads. However, within a few years the surged subsided. During the remainder of the 1950s, the forestry schools experienced a steady but less dramatic increase in enrollments.

The number of U.S. forestry schools also increased following World War II. Most of these new schools, however, were not accredited by the Society of American Foresters. For example, by 1959 there were 16 unaccredited forestry schools, 12 of which were established after 1945. Also, in 1959 there were 27 accredited schools, compared with 23 in 1945.

A noteworthy trend of this postwar era was the growing interest in forestry and the forest as a provider of outputs other than just timber. An increasingly mobile and affluent American public was now looking to the forests as source of recreation, wildlife management, and watershed management. Multiple-use forestry, the harmonious and coordinated management of the various forest resources without impairment of the land, was clearly becoming an interest of the public. Developments such as this began to raise questions about the adequacy of the American forester's education: should the forester be educated as a timber manager, or more broadly, as a forest land manager?

In addition to questions regarding the type of education a forester should receive, the profession began to deal with a host of other issues that would continue to characterize the struggle of the profession to meet the social changes of postwar America. There were, for example, concerns over formal statements of professional ethics and registration and licensing of foresters. Discussions about topics such as these became increasingly common and intense as forestry moved into the future. And forestry school enrollments continued to increase. By 1960, the enrollments in U.S. forestry schools totaled 8,399 undergraduates and 916 graduate students.

Older American foresters recall the 1945–1960 period with a great deal of fondness. It was a time of professional growth and prosperity, a time when forestry enjoyed a great deal of public favor. Forestry education likewise experienced growth and prosperity. However, there were rumblings of change. The public was now pressing for multiple-use outputs from the forest, including aesthetics, while at the same time asking foresters to manage the land in an environmentally responsible manner. As we shall see, the public interest in forest management would become more intense in the 1960s, and this, in turn, would have impacts on forestry education.

The environmental era. In September 1962, Rachel Carson published *Silent Spring*, a book that, in effect, ushered in the modern environmental era. By January 1963, it had become a runaway bestseller with more than a half-million copies in print, and publication in 16 countries outside the United States was soon underway (de Steiguer 1997). The immediate message in *Silent Spring* was that pesticides posed a threat to both humans and the natural environment. However, the larger theme was that humans needed to be much better stewards of the environment. Many events of the past 40 years stand as evidence of the truly enormous social impact of *Silent Spring*. Passage of the National Environmental Policy Act, creation of the U.S. Environmental Protection Agency, and Earth Day are but a few examples.

Silent Spring, likewise, affected the U.S. system of public education. Almost immediately, universities saw a rapid increase in environment-related course offerings and degrees (de Steiguer 1997). The era of environmental concern led to the creation of entirely new departments and schools of environmental study to accommodate the growing public interest. Indeed, the modern environmental era must be regarded as one of the major events of the 20th century to shape public education in the United States.

Likewise, U.S. forestry and forestry education came under increasing pressure to examine their basic philosophies and practices. It can be forcefully argued, however, that American forestry education was slow to change its ways even when faced with intense public pressure. A 1979 study of the impact of the environmental era on North American forestry education indicated that changes in forestry curricula were not dramatic, certainly not like the changes experienced by education in general. Data from the study show that the typical American or Canadian forestry school added only about one environmental course to its undergraduate degree requirements (de Steiguer and Merrifield 1979).

The decades following the release of *Silent Spring* up to the present have brought many challenges to the forestry profession and forestry education. Furthermore, traditional forestry interests have often struggled to hold their own against the tide of changing public opinion. Perhaps the most telling statistic that bears witness to the recent struggle between commodity and environmental forestry interests was the huge decline in the level of timber harvests on the U.S. national forests—a drop of 44 percent between 1991 and 1996 (Gorte 1998).

Without question, the environmental era has brought serious challenges to traditional U.S. forestry and forestry education. These challenges will be evident in the forestry enrollment data presented in the following subsection.

Current trends in forestry enrollments. Data recently collected by Sharik (2005) analyzed undergraduate enrollments in forestry and related natural resource disciplines at a representative sample of National Association of Professional Forestry Schools and Colleges (NAPFSC) institutions, by region of the country, from 1980 through 2003. The data indicate that enrollment levels dropped steadily to about 60 percent of 1980 levels by the end of the decade, rose sharply to the same level as in 1980 by the mid-1990s, then declined, and are now about back to where they were in the 1980s. Sharik (2005) also indicated that the greatest changes in enrollment levels over this period occurred in forestry, compared with other natural resource fields. In fact, unlike matriculation in forestry (and to a lesser degree, fisheries and wildlife), enrollments in parks, recreation, and tourism programs and the broad degrees, such as natural resource management, environmental studies, and applied ecology, showed a steady increase over most of this period.

Since those data were gathered in 2003, natural resource academicians and employers of the graduates of these programs have held several meetings to determine the reasons for the observed downturn in enrollments. The issue is seen as critical in light of projections of an impending demand for these graduates, given accelerated rates of retirements in the government agencies in particular (Colker and Day 2004). As a result, no less than a dozen explanations have been advanced (Sharik 2005) and are summarized below. Our comments will be confined to forestry, given the focus of this chapter.

Some have argued that prospective students have turned away from forestry as a possible career because of considerable uncertainty about jobs, and that what jobs do exist pay less than those in other professions. Others see an increasing disconnect between young people and nature and natural resources, in turn caused by increased urbanization, as a major reason for the downturn. They argue that this disconnect has a particularly adverse effect on minorities, who are disproportionately urban and moreover see no relevancy of forestry to their local neighborhoods. Still others relate the downturn to the nature of the forestry curriculum, including its lack of flexibility and heavy requirements in the "hard sciences" compared with other degree programs in natural resources and the environment, many of which reside outside colleges of forestry and related natural resources. Corollary to this argument is the longer period of time beyond a bachelor's degree that foresters need to obtain a terminal degree, compared with business, law, and the health professions, which also pay better.

Others say that forestry is associated with exploitive extraction of natural resources from forest ecosystems and seen as contributing to the degradation of these ecosystems. Still others argue that there is a lack of national intellectual leadership and charisma in forestry issues, and that better marketing, communications, and education would attract more prospective students to forestry. Finally, there are those who have attributed the recent downturn in enrollments to decreasing enrollments in colleges and universities in general in the United States. However, available statistics indicate that general enrollments have been increasing since as least as far back as 1987 and are projected to increase well into the future (NCES 2004).

Those various reasons, all advanced to explain the recent downturn in forestry enrollments, came from academicians and employers of forestry graduates. Students were not consulted, even though they are clearly the ones who decide whether to matriculate in an undergraduate forestry program. Sharik (2005) therefore conducted a survey of students attending the Society of American Foresters' National Convention in October 2004. The survey included responses from more than 60 mostly upperclass (junior-senior) students matriculated in accredited forestry degree programs at 20 U.S. institutions of higher learning. When asked what factors led to their decision to enter a forestry or related natural resource degree program, the overwhelmingly most cited reason was a love of the outdoors or nature. When asked what factors may have caused them to hesitate, the most cited reasons were low wages and lack of jobs. Although we do not have sufficient information to confirm or reject these students' perceptions, it is perceptions more than reality that drive the decision-making process of choosing a major. However, knowing the degree to which reality coincides with perceptions would help determine what strategies might be most preferred in reversing the current downturn in enrollments, assuming this is desirable given the demand for forestry graduates in the marketplace.

Although no definitive data are available on a national level, anecdotal information suggests that the 1980 enrollment levels in forestry were about half those in the mid-1970s, when enrollments reached a historical high. This peak and the sharp increase in enrollment leading up to it were attributed to a heightened interest in the environment. The downturn in enrollments following this peak may indeed have been associated with the explosion of environmental sciences and environmental studies degree programs outside colleges of forestry and related natural resources. Thus, the maximum enrollment levels we have experienced since 1980 may represent a ceiling that is not likely to be exceeded in the foreseeable future. This is only a problem, of course, if the demand for foresters outstrips the supply, and we simply do not know enough about the demand side to answer this question.

Forestry Education in Europe

Forestry higher education began in France soon after the commencement of the 19th century and thus was well institutionalized by the 1889–1890 academic year, when Pinchot was a student. The forestry education Pinchot received at l'École Nationale Forestière at Nancy was a product of almost two centuries of institutionalized practical forestry know-how on the part of Europeans; the school itself had been established in 1825.

At that time, forestry education in Nancy differed from that of central Europe in that the French insisted that silviculture imitate nature, favor natural regeneration, and take into consideration the phytosociological aspects of the forest. Instruction in forestry economics was also provided by Professor Boppe, one of Pinchot's principal professors at Nancy. The general educational approach of the forestry school at Nancy was an important reason why Professor Dietrich Brandis of Bonn—also the first forester of India—recommended the French school to Pinchot. Brandis made similar recommendations to foresters training for the British colonial service.

Even though individual new courses were occasionally developed, the overall forestry curriculum did not change much in Europe from its inception until the early 1980s. Thus, change came much later than in America. Sustainable forest management, and more recently forest global issues, has been added to forestry curricula in both the United States and the European Union. Today, the concept of sustainable forestry development seems universal on both sides of the Atlantic.

The remainder of this section will review briefly the European forestry schools and curricula both at the time of Pinchot and now, and envisage how they might further evolve on both sides of the Atlantic.

First European foresters, naturalists and forestry schools. Silvicultural knowledge was mostly derived from practical, applied experience until Duhamel du Monceau (1700–1782) published *De l'exploitation des bois* in 1764. When his book was translated into German and other languages, his European peers recognized him as a pioneer in science applied to forestry. Other contemporaries who were also recognized as important contributors to forestry science were Reaumur (1683–1757), a physician, and Buffon (1707–1788), as reported by Guinier (1952). The first practicing forester who worked closely with Duhamel du Monceau was Varenne de Fenille (1731–1794). He worked on determining the proper forest rotation age needed to maximize the volume of wood at harvest and on the thinning regimes, a rare investigation of the time. Advancement of forestry in France was delayed by the French Revolution, when Varenne de Fenille was executed. Lorentz, who was largely self-taught in matters of forestry, studied in the forests of France and Germany, as the latter was then under French administration implemented by the army of Napoleon. The foresters of Germany and France managed to stay amicable during these difficult times. Parade, a Lorentz protégé, studied in Tharandt (Saxony) and benefited from his eight years of field forestry experience with two great German foresters, authors, and teachers—Hartig and Cotta. These two emphasized forest management and the use of mathematics. As directors of the school at Nancy, Lorentz (1825–1838), followed by de Salomon (1830–1838) and then Parade (1838–1864), taught forestry science as it was instructed in central Europe but kept alive the French tradition of natural regeneration. Indeed, a famous quote of Parade's was *Imiter la nature, hâter son oeuvre* ("Imitate nature, hasten its work"). This tradition was maintained by Nanquette and Putton, who were on the faculty at Nancy when Pinchot entered. Boppe, Pinchot's professor of silviculture, became the next director of the school.

When Pinchot arrived in 1889, Boppe had just completed his *Traité de Sylviculture*, a forestry text that emphasized the newly defined concepts of ecology and phytosociology. This text was practical, concise, and well appreciated by Pinchot, who admired also the intellect of the author. During field exercises in the mountains, Pinchot became acquainted with the management of the *forêt jardinée*, or uneven-aged management. There he learned about the Liocourt Coefficient, developed by the Lorrain Francois Henri de Lallement de Liocourt (1860–1928), a disciple of Gurnaud (1825–1898). Gurnaud's theory is relevant even today in the practice of forest sustainability and biodiversity concepts (Bruciamacchie and de Turckheim, in press). During this excursion, Pinchot saw first-hand the erosion prevention techniques of Demontzey (1831–1898), a native of the nearby Vosges Mountains region, who had been educated at the forestry school at Nancy. Many other professionals influenced the development of forestry as a science in France (Peyron 1999) as well as in Europe (see bibliography in Annex 1) and thus the education of Pinchot.

The development of forestry education in Europe parallels the career of these scholars. Forestry courses began in various universities in central Europe. Private schools also appeared, but many were short-lived. The two best-known schools at the time of Pinchot were at Tharandt, founded by Cotta in 1811, and at Nancy. A forestry school program had already been planned for France as early as 1814 by Baudrillart (1774–1832). He argued for the immediate creation of a school because *la science était dans les livres, et la routine aveugle administrait les forêts* ("Science was in the books but blind routine managed the forests"). The proposed forestry curriculum was included later in his *Dictionnaire raisonné et historique des Eaux et Forêts*, published in 1823–1825. These two European schools were well established by the beginning of the 19th century. Their curriculum set the standard for forestry education in Europe at the time and, furthermore, led by espousing a new philosophy of forest management.

The forestry curriculum at the time of Pinchot. When Pinchot entered the French forestry school at Nancy, he was the last student who would be accepted in the school without having passed a preparatory year of mathematics and sciences, which meant two years at the French national agriculture school in Paris. The academic program for the year 1889–1890 found in the school archives (see Annex 2) included six modules:

1. forestry sciences comprising eight courses (75 lectures, each 90 minutes): silviculture, wood technology, dendrometry, forest economics, forest management, forest valuation, statistics, and forest history; plus all the field exercises between May 1 and July 15 and two field trips;
2. applied natural sciences (75 lectures): botany, tree pathology, lithology (soil science), geology, and applied zoology and entomology;
3. applied mathematics (50 lectures): topography, leveling, constructions, roads and river and erosion control;
4. forestry legislation (50 lectures): administrative law, civil law, penal law, mountains law, and hunting law;
5. military academy (50 lectures); and
6. German language (30 lectures).

This curriculum, with three years of science and mathematics, resulted in academically well-grounded citizens and foresters. Pinchot chose, with the advice of Putton, the director at the time, three courses that would be most useful for the objectives that he had set for himself: silviculture, forest economics, and forest law. Pinchot also discussed his year of study and the courses he should take with the German forester Brandis, former inspector general of forests for the government of India, who had chosen the French forestry school at Nancy as the place to train future British foresters leaving for India.[1] More than 120 British students would be educated at Nancy. G.F. Pearson, British official and a former colonial forester stationed in Nancy, made sure that the students did not enjoy the local night-life (*la noce*, to use Pinchot's term) to the detriment of their academic work. One of the main reasons for Pinchot's choice of the school at Nancy was the collegial nature of the school. The faculty judiciously integrated into the curriculum practical forestry experience from both Germany and France. This turned out to be a prudent economical model both for the British colonies in India and for America.

The content of the courses Pinchot took at Nancy can be seen in the books of the three main professors who taught him: Professors Boppe for silviculture; Huffel, who arrived in 1889 to teach forest economics; and Guyot, who taught forest law. Their books (Boppe 1889; Huffel 1910, 1919, 1926; Guyot 1898, 1920) were well structured, provided current knowledge, and contributed to the well-grounded education of the students. Pinchot wondered in his notes about the usefulness of the Napoleonic code[2] for America, but in his 1947 autobiography, *Breaking New Ground*, he would write that the study of forest law later proved to be most beneficial in his career. Pinchot also appreciated the field experiences. The professors were particularly pleased to have more time for field exercises when the courses in physics, chemistry, political economy, and agriculture were shifted to the first two years of education, which occurred in Paris. The practical field experiences are maintained even today as an important feature of forestry education in Nancy.

The forestry curriculum today. The forestry curriculum at Nancy has not changed dramatically since the time of Pinchot. The restructuring of courses into modules and time spent on each reveal striking similarities with the 1889–1890 curricula. Of course, the technical content of the courses has evolved with the passage of years. Environmental science, for instance, is more than just phytosociology; it now involves sophisticated descriptions of habitat for conservation and management, as evidenced by the new encyclopedic works on these subjects by Rameau et al. (2000). The evolution of ecology in the forestry curricula of Europe has been well reviewed by Michel Dupuy (2005). Modern decision-making sciences are mainstream today in many teaching modules and include the participation of stakeholders. The teaching modules at Nancy today are as follows:

1. environmental sciences;
2. engineering;

3. forest management methods;
4. forest product chain;
5. economic and social sciences;
6. humanities;
7. electives;
8. fieldtrips; and
9. internships.

Current forestry curricula elsewhere in Europe have features similar to those at Nancy. In comparison, Gritten (2004) summarizes the curricula of six forestry schools in four main modules. The four modules are defined by reference to the main components of sustainable development—environmental, social, and economic—in addition to those in traditional technical forestry. These modules have also been defined in relation to the International Union of Forestry Research Organizations (IUFRO) science divisions as follows:

1. ecology and biology (IUFRO 1, 4);
2. practical forestry (IUFRO 2, 3, 5);
3. management and economics (IUFRO 6, 7, 8); and
4. social and cultural (IUFRO 9).

An examination of six European forestry schools was conducted to determine the relative emphasis that each places on four areas of study (Harou 2004): ecology and biology, practical forestry, management and economics, and social and cultural studies. The six schools were Wageningen, Warsaw, Boku, Freiberg, Joensuu, and Lleida. The data show, for example, that most schools still emphasize ecology and biology, practical forestry, and management and economics over social and cultural studies. Wageningen, however, has placed greater emphasis in its forestry curriculum on social and cultural issues, as opposed to practical forestry—not surprising given that the Netherlands has the highest population density in Europe. Britain was not part of the data set, but similar trends have been observed there. The fast adopters of curriculum broadening do not seem to have reversed the trend of lower forestry applications and enrollments in Europe so far (Miller 2003). Forestry enrollments, as in the United States, seem to be in decline throughout.

Conclusions

Forestry schools in the United States and in Europe have demonstrated both differences and similarities. The differences arise because the regions, of course, are different in terms of their demographics, geography, and culture. Similarities arise out of the common need to address the same commodity-based forest management issues, as well as forest sustainability and global issues. These are concerns that are common to both sides of the Atlantic.

Forestry education began earlier in Europe than in America. In fact, European forestry education predates the establishment of the United States as a nation. Once underway, however, U.S. forestry schools seem to have been more flexible in altering curricula in response to changing social conditions, as in the evolution of curricula away from traditional commodity forestry more toward natural resources and environmental studies. By the early 1980s some U.S. forestry schools had already started to change their focus, as well as even their school names, using words like "natural resources" and "environmental." This American trend did not reach Europe, for the most part, until the beginning of the 21st century. One might question, however, whether even the American schools have changed enough in this regard.

Another difference in forestry education between the two continents has to do with the differing geographic scale of forest management concerns. U.S. forest management courses have been oriented more toward extensive forestry than those in Europe. The U.S. Forest Service, for example, conducts forest planning on an extensive scale, and this broad-scale planning approach has been taught in most American forestry curricula. The Resources Planning Act of 1973 and the National Forest Management Act of 1969 necessitate planning on such an enormous scale that at first, large econometric and linear programming models were used. Extensive planning saw little application in the intensively managed forests of Europe. Interestingly, the use of ecosystem planning models is now declining in the United States, although management plans continue to address forests at a much broader scale than in Europe.

Today, the concept of sustainable forest development is current on both sides of the Atlantic, and it is clearly one factor that unites modern forestry and foresters on the two continents. Sustainable forestry requires a modification of traditional forest management education. The new sustainability concept relies more on local participation of stakeholders in the development of management plans and thus may be at odds with traditional commodity forestry. This presents a challenge. Even with sustainable forest development, within each region, there are differences. The concept of sustainability is being more actively implemented in the United States than in Europe, as is apparent from Connaughton's (2001) article in an issue of the *Journal of Forestry* devoted to the subject of sustainability.

Some observers believe that with the increasing globalization of forestry, it should be the profession's duty to make sustainable development the centerpiece of the forestry curriculum of the future (Shirley 1966; Nair 2004). This would ensure that graduates are well prepared to resolve global forestry issues using innovative policies and technologies, to address climate change and biodiversity through international institutions, and to manage forests sustainably using ecocertification (Lele 2002; Harou 2004). In the future, to work in only one part of the globe will not suffice, as greenhouse gases and global warming have demonstrated. A united effort will be required.

Today's revolution in information technology should enable forestry institutions to share their curricula worldwide and engage with stakeholders wherever they may be. A renewed commitment to an American-European partnership in forestry education should help us address the new global challenges. This global education should extend to everyone and ensure the active participation of all citizens, as suggested by James Speth, dean of Yale University's Pinchot School of Forestry and Environmental Studies, in his recent book *Red Sky at Morning* (Speth 2004). In short, it is important for the European Union and the United States to contribute to knowledge for all in forestry and the environment.

References

Baudrillart, J. 1823–1825. *Dictionnaire général raisonné et historique des Eaux et Forêts*, 2 volumes. Arthus Bertrand.

Boppe, L. 1889. *Traité de sylviculture*. Berger-Levrault, ed. Paris.

Bruciamacchie, M., and B. de Turckheim. In press. *La Futaie irrégulière*. Edition Edisud.

Carson, R. 1963. *Silent Spring*. Boston: Houghton Mifflin.

Chapman, H.H. 1935. Professional forestry schools report: Giving the comparative status of those institutions that offered instruction in professional forestry for the school year 1934–35. Washington, DC: Society of American Foresters.

Colker, R.M., and R.D. Day (eds.). 2004. Federal natural resources agencies confront an aging workforce and challenges to their futures. *Renewable Natural Resources Journal* 21(4): 6–29.

Connaughton, K. 2001. Sustainability—The key forest policy issue of the new millennium? *Journal of Forestry* 99(2): 7.

Dana, S.T., and E.W. Johnson. 1962. *Forestry education in America: Today and tomorrow.* Washington, DC: Society of American Foresters.

de Steiguer, J.E. 1997. *The age of environmentalism.* New York: McGraw-Hill.

de Steiguer, J.E., and Merrifield. 1979. The impact of the environmental era on forestry education in North America. *Unasylva* 31(123): 21–25.

Dupuy, M. 2005. *Un siècle d'écologie forestière en Europe.* ENGREF.

Duhamel du Monceau. 1760. *Des semis et plantations des arbres et de leur culture.* Guerin et Delatour.

Encyclopedia of education. 1971. New York: Macmillan.

Gorte, R.W. 1998. *Federal timber harvests: Implications for U.S. timber supply.* Congressional Research Service, Natural Resources Policy Division. March 10. 98–233 ENR. Distributed by the National Library for the Environment.

Graves, H.S., and C.H. Guise. 1932. *Forest education.* New Haven, CT: Yale University Press.

Gritten, D. 2004. Comparative analysis of forestry curriculum of six European Universities. In *ICT in higher forestry education in Europe*, edited by L. Tahvanainen and P. Pelkonen. Silva Publications, Joensuu University Press.

Guinier, Ph. 1952. Preface. In *Sylviculture*, by Henri Perrin. TI. École Nationale des Eaux et Forêts Nancy.

Guyot, Ch. 1898. *L'enseignement forestier en France: l'École de Nancy.* Nancy. Crepin-Leblond.

Guyot, Ch. 1908–1912. *Droit forestier.* L. Laveur (ed.). TI-III. Paris.

Harou, P. 2004. The new international context for forest policies—Case of the revised World Bank forestry strategy. In *International frameworks in forest politics*, edited by M. Baldus and R.A. Fuckner. Proceedings of a seminar organized by the International Forestry Students Association, Freiburg, Germany, June 25–28, 2003. Verlag Isle.

Huffel, G. 1910, 1919, 1926. *Economie forestière.* T I-III . Paris.

Lele, U. 2002. The way ahead. In *Managing a global resource—Challenges of forest conservation and development*, edited by U. Lele. WB series on Evaluation and Development, V5. London: Transactions Publisher.

Miller, H. 2003. Trends in forestry education in GB and Germany. Mimeo. Review for FAO.

Nair, C.T.S. 2004. What future for forestry education? *Unasylva* 55(2004/1): 3–9.

National Center for Education Statistics (NCES). 2004. Undergraduate enrollment in degree-granting institutions, with alternative projections: Fall 1987 to fall 2012. http://nces.ed.gov/pubs2002/proj2012/figure_25.asp.

Peyron, J.-L. 1999. L'École Forestière au XIX dans son contexte historique. Miméo. Nancy: ENGREF. March.

Pinchot, G. 1946. Quelques souvenirs de Gifford Pinchot sur l'École Forestiere de Nancy. Revue du bois et de ses applications. 6: 3–5.

Pinchot, G. 1947. *Breaking new ground*. Washington, DC: Island Press.

Rameau, J.-C., C. Gauberville, and N. Drapier. 2000. *Gestion forestière et diversité biologique*, vol. 3. ENGREF, ONF, IDF.

Sharik, T.L. 2005. Trends in undergraduate enrollments in natural resources at NAPFSC Institutions, 1980–2003. NAPFSC General Assembly and SAF National Convention, Edmonton, Alberta, October 2004. http://www.napfsc.org/NAPFSC_trends%20revised.pdf.

Shirley, H.1966. Forestry education in a changing world. In *Proceedings of the Sixth World Forestry Congress, Madrid*. Rome: FAO.

Society of American Foresters. 1958. *Forest terminology*. Washington, DC: Society of American Foresters.

Speth, J.G. 2004. *Red sky at morning*. New Haven, CT: Yale University Press.

World Bank. 2002. *Constructing knowledge societies: New challenges for tertiary education*. Washington, DC: World Bank.

Annex 1. Forest History Society Bibliography on European Forestry Education

Alaoui, M. Y. 1997. 25 ans d'enseignement à l'École nationale forestière d'ingénieurs de Salé (Maroc) (25 years of teaching at the National Forest Engineering School in Salé, Morocco). *Revue forestière française* 49(4): 379–87. Forestry education at the school since its establishment in 1968. Text in French.

Blais, R. 1990. Contribution à l'histoire de l'administration forestière: digression sur un témoignage concernant l'école forestière secondaire de Villers-Cotterêts (Aisne) en 1874. *Revue forestière française* 42(1): 77–86. Development of forestry education in France, 1790–1884.

Boulaine, J., and C. Feller. 1985. Louis Grandeau (1834–1911), professeur á l'école forestière. *Revue forestière française* 37(6): 449–55. The life and work of the agronomist at the forestry school at Nancy.

Broda, J. 2000. Die Geschichte der Forstwirtschaft als Lehrfach im Forststudium (Forest history as an academic subject at the Department of Forestry). *News of Forest History: Forest History in Poland* (29): 50–54. Forest history education at Warsaw University, Lvov Polytechnic, and Poznan University in Poland since the 1950s. Text in German.

Brown, J.C. 1877. *The schools of forestry in Europe: a plea for the creation of a school of forestry in connection with the arboretum at Edinburgh*. Edinburgh: Oliver and Boyd, 1877. Survey of forestry schools and of curricula covered in European forestry education during the nineteenth century.

Brown, J.C. (comp.). 1887. *Schools of forestry in Germany, with addenda relative to a desiderated British National School of Forestry*. Edinburgh: Oliver and Boyd.

Corvol, A., and C.D. de la Boissonny (eds.). 1992. *Enseigner et apprendre la forêt: XIXe–XXe siècles*. Papers presented at a colloquium held October 8–10, 1990, at Nancy,

France. Paris: L'Harmattan. Forestry and forestry education in France in the 19th and 20th centuries.

de Steiguer, J.E. 1994. The French National Forestry School. *Journal of Forestry* 92(2): 18–20. The French school in the 19th and 20th centuries.

Grzywacz, A. 2000. Polnischer Forstverein, seine Geschichte und Tätigkeit (The Polish Forest Society: Its history and direction of activities). *News of Forest History: Forest History in Poland* (29): 16–24. History of the Polish Forest Society, a professional organization for foresters that publishes research, maintains a library collection, produces the forestry journal titled *Sylwan*, and assists in creating forestry education curriculum, in the 19th and 20th centuries. Text in German.

Hafner, F. 1984. The study of forestry at the university level in Austria. *Allgemeine Forstzeitung* 95 (November): 319, 321. Forestry education at the Agricultural University of Vienna. Text in German.

Harrison, A. 1980. Sic vos non vobis. *Quarterly Journal of Forestry* 74 (July): 158–64. Some history of the Dean School of Forestry in England, 1904–1954.

Jacamon, M. 1991. Henri Gaussen (1891–1981): Les forestiers se souviennent. *Revue forestière française* 43(4): 333–37. Life and career of this French forestry educator.

Killian, H. (ed.). 1990. Forest history in Italy. *News of Forest History* 13/14 (October): 1–30. Articles on forestry education and museums in Italy since the 19th century. Also, brief reviews of recent works in the field. Text in German, English, and French.

Matthews, J. 1991. Creating a forestry tradition. *Quarterly Journal of Forestry* 85 (April): 83–89. Includes growth of professional forestry in Great Britain in the 20th century.

Miller, H. 2002. The state and future of professional education in forestry. *Quarterly Journal of Forestry* 96(3): S3–S6. Changing requirements in the workplace, curriculum design, and problems in the provision of education and training at the University of Aberdeen (Scotland) Department of Forestry in the 20th century.

News of Forest History. 1992. Forest history in Denmark. *News of Forest History* 15 (March): 1–26. Articles on forest policy and forestry education in Denmark. Summaries in French and German.

News of Forest History. 1999. Forest History in Romania. *News of Forest History* 28 (September): 1–76. Articles on climate change, forest vegetation, forest utilization, forestry education, forest ownership, and forest management in this southeastern European nation (and its previous, autonomous principalities prior to unification in 1861) from the medieval era to the present.

Pardé, J. 1990. Il y a 100 ans: Le traité de sylviculture de L. Boppe. *Revue forestière française* 42(3): 370–72. Excerpts from Boppe's textbook on forest management, written in 1889. Text in French.

Rondeux, J. 1997. *La forêt et les hommes: Arrêt sur images 1900–1930.* Gembloux [Belgium]: Presses agronomiques de Gembloux, 1997. Chiefly photographs, covering the early history of the forestry school of the Sciences Agronomiques de Gembloux, Belgium, including forestry education, forest utilization, plantation forestry, forest products, and silviculture. Text in French.

Stebbing, E.P. 1906. *Notes on a visit to some European schools of forestry.* Forest Bulletin 5. Calcutta: Office of the Superintendent of Government Printing. The similarities and differences between zoological collections and methods of tuition at major European forestry schools of the early 20th century, with illustrations. Stebbing (1870–1960) compiled notes for this work while visiting academic institutions in Russia, Prussia, Germany, France, and Austria in 1904.

Steen, H.K. 2001. *The conservation diaries of Gifford Pinchot.* Durham, NC: Forest History Society.

Viney, R. 1962. L'oeuvre forestière du second Empire. *Revue forestière française* 14 (June): 532–43. Forestry in France during the 19th century.

—Compiled by Cheryl Oakes, Librarian and Archivist
Forest History Society, Duke University, January 2005

Annex 2. Courses at the Nancy Forestry School, 1889–1890

Enseignement. — D'après le nouveau programme chaque promotion doit recevoir 330 leçons ainsi distribuées :

Sciences forestières (Sylviculture, technologie, dendrométrie, Économie forestière, aménagement, Estimations, Statistique forestière, histoire de la science forestière) 1 heure 1/2 _____ 75

Sciences naturelles appliquées aux forêts (Botanique, Maladies des arbres, Lithologie, géologie de la France, Zoologie et entomologie appliquée) 1 heure 1/2 _____ 75

Mathématiques appliquées (topographie, nivellement, constructions, routes, torrents) 1 heure 1/2 _____ 50

Législation forestière (Droit administratif et civil appliqué aux forêts, droit pénal forestier, législation de la chasse et des montagnes) 1 h. 1/2 _____ 50

Art militaire. 1 heure 1/4 _____ 50

Langue allemande. 1 heure _____ 30

Total égal _____ 330

Notes

1. In the first edition of Mowgli, 1893, Rudyard Kipling talked about the dune fixation method following the Nancy methods. Another part of Kipling book mentioned the student songs, 'chansons paillardes,' learned in Nancy!
2. Gifford Pinchot's grand father, Amos, was a Colonel in the Napoleon army when 19 years of age. He had to escape from France at the time of the Restoration allegedly for having attempted with others to plan the escape of Napoleon from St. Hélène. He escaped to the U.S. through the UK using a fisherman boat (Pinchot, 1946).

EPILOGUE

Catalysts for Positive Change

Dennis Le Master, V. Alaric Sample, Franz Schmithüsen, Dominique Danguy des Deserts, and Patrice Harou

The preceding chapters, based on papers given during the two-part Pinchot Colloquium, examined how Europe and the United States have learned and continue to learn from each other in the allocation of forest resources in their respective geopolitical regions. As the first U.S. born—and European trained—professional forester, Gifford Pinchot gained knowledge and experience at the École Nationale Forestière in Nancy, France in 1889 and in the Sihlwald of Zurich, Switzerland in 1890. Pinchot came to Nancy under the advice of a German forester, Sir Dietrich Brandis, who had been in charge of forestry in the British colonies. The philosophy at the French national forestry school at Nancy was to work with Nature rather than to dominate it, foreshadowing the ecology-based approaches to forestry so familiar to later generations. Pinchot instinctively sensed that this approach would be more appropriate for America. He adapted this knowledge and experience to the unique conditions in the United States, and he later brought them to bear as chief of the U.S. Forest Service. That Pinchot would go to Europe to study forestry is less remarkable than what he learned and applied once he was in a position to do so.

The ties between Europe and the United States—or more inclusively North America—are well known. They are historical, cultural, institutional, economic and political. Their forests are also comparable, mainly temperate and boreal. Indeed, Europe, North America, and Siberia share a virtually intact belt of boreal and temperate forests that encircles the Northern Hemisphere. Forestry scientists, administrators, and practitioners from both Europe and North America have learned from one another during the 20th Century. Further, their ideas about multiple-use, sustained-yield forest management and landscape conservation have been converging because:

- for the most part, both deal with temperate and boreal forests;
- both have democratic forms of government and market economies (with some countries still in the process of moving toward full democracy in Eastern Europe);
- both share about the same proportion of forest land area because Europeans have invested in and expanded their forest capital, while the United States drew its forest capital down during the 19th century and the first half of the 20th century;
- both participate in global markets for forest products and engage in international forest development and conservation projects;
- both Europeans and North Americans make complex, multiple market and non-market demands upon their forests; and
- both have large forestry research and educational establishments and both engage in frequent mutual exchanges of scientific and educational information.

It is questionable, however, whether Europe and United States will ever fully converge in multiple-use, sustained-yield forest management for at least three reasons.

1. Forests play a significantly greater and more positive role in European culture than they do in U.S. culture where they were historically regarded as a barrier, something to be overcome.
2. Private property rights tend to be overdrawn in the U.S. as compared to Europe, which often obscures the public character of many important forest uses.
3. The scale of forestry is smaller in Europe, where forests are developed and managed by over 40 nation states—countries that are separate political and cultural entities—while in the U.S. forests are developed and managed in 50 states that *are not* wholly separate political and unique cultural entities.

What is it that Europeans and North Americans will likely learn from each other—ecologically, economically, and socially—over the next several decades? First, much knowledge will be developed and exchanged among European and North American scientists in ecology, genetics, and global climate change, and this knowledge will have important implications for forestry practices, particularly as they relate to biological diversity, resource sustainability and cultural values of landscape. There will also be much progress in managing forests as a renewable resource for wood production and bio-energy. Hence, a convergence of scientific knowledge in forestry is likely between Europe and North America.

Second, Europeans will likely learn that stronger and sustained economic growth will require them to increase the flexibility of their markets as well as to decrease the level of regulations, making them more responsive to market forces for wood products but also for environmental goods and services. In so doing, they will likely experiment with new and different forest tenure systems. On the other hand, the United States, whose markets are significantly less regulated, will probably learn the social costs of forest fragmentation are too high to allow private forest ownerships to be transferred at the whim of market forces, as can be seen in the current sell-off of forest industry timber lands. That a stable, productive forest land base is essential for the benefits yielded by forests, including biodiversity conservation, water supply and quality, climate change mitigation, sustainable wood production, and sustainable development of rural communities. The United States will also experiment with new and different forest tenure systems and associations. Hence, a convergence of economic institutions is likely in Europe and North America.

Third, the social demands upon European and North American forests will continue unabated, requiring changes in allocation of forest resources. Public participation in forest land decision making will increase. As a result, new institutions will be developed in both geopolitical regions. Some will be successful, and some will not, and the lessons learned will and should be freely exchanged. Careful analysis and evaluation of private and public decision-making will be essential since a tendency to over react in either direction is probable. Like Pinchot suggested over one hundred years ago, a basic question that must be answered is whether the remedy is appropriate for the problem. A convergence will occur on remedies for problems that are comparable.

The introduction to this volume included a quote from Gifford Pinchot's book, *The Fight for Conservation* (1911): "Our responsibility [as forestry professionals] to the Nation is to be more than just stewards of the land. We must be constant catalysts for positive change." And change in Pinchot's vision knew no bounds. Well ahead of his time, Pinchot had a global perspective on the importance of sustainable natural resource development and environmental justice to achieving world peace: "The conservation of natural resources is the key to the safety of the American people, and all the people of the world, for all time to come.... It is a foundation of permanent peace among the nations, and the most important foundation of all" (*Breaking New Ground*, 1947). Today our responsibility in forestry is global, beyond just Europe and North America. International cooperation, collaboration, and participation are essential to meeting that responsibility of being catalysts of positive change world wide for the protection, management, and sustainable use and development of forest resources.

It is this legacy of Gifford Pinchot that we have sought to continue, through the commemoration of the centennial of the U.S. Forest Service, and through the strengthened linkages of international cooperation that will endure long afterwards. More than at any time in human history, we now recognize the interdependency of all peoples and all nations in ensuring the sustainability of life on Earth. Conserving and sustainably managing the world's forests is a key component in achieving that goal. We wish to express our gratitude to those who contributed to this effort, and we invite others to join with forestry professionals around the world in working toward the vision of a just and sustainable society in the 21st century.

APPENDIX

Toward Sustainable Forest Management: The Divergence and Reconvergence of European and American Forestry

Vers un Aménagement Durable des Forêts: La Divergence et Reconvergence de la Foresterie Européenne et Américaine

A Colloquium in Commemoration of the Centennial of the United States Forest Service

European Colloquium

École Nationale du Génie Rural des Eaux et des Forêts
Palais des Congrès
Nancy, France
March 7–8, 2005

Monday, March 7, 2005

8:30 am	**Accueil des paticipants /** *Welcome* Palais des Congrès – Nancy, France
9:30 am	**Allocution d'ouverture /** *Opening* Gérer la forêt des deux côtés de l'Atlantique / *Managing the forest on both sides of the Atlantic* **Dominique Danguy des Déserts**, retired, former director, École Nationale du Génie Rural des Eaux et des Forêts, Nancy, France **V. Alaric Sample**, president, Pinchot Institute for Conservation, Washington, D.C.
10:00 am	**Contexte historique /** *Historical context* Moderator: **Patrice Harou**, visiting professor, École Nationale du Génie Rural des Eaux et des Forêts (ENGREF), Nancy, France

- Les acquis de la sylviculture européenne lors de la formation de Gifford Pinchot à Nancy / *European silviculture knowledge during Pinchot training in Nancy*
  Jean-Luc Peyron, director of ECOFOR, Paris, France
  Marie-Jeanne Lionnet, Head Librarian, ENGREF, Nancy, France

- Gifford Pinchot et ses contemporains: création du cadre institutionel, légal et politique sur lequel la foresterie américaine c'est développée, 1900–1950 / *Gifford Pinchot and his contemporaries: Creating the institutional, legal, and policy framework on which American forestry developed, 1900–1950*

Char Miller, professor of History, Trinity University, San Antonio, Texas (presented by Steven Anderson, president and CEO, Forest History Society, Durham, North Carolina)

- La politique forestière en Amérique comme une nation en développement; démocratie Jeffersonienne, la domestication de la nature sauvage et la montée du mouvement de Conservation, 1800–1900 / *Forest policy in America as a developing nation: Jeffersonian democracy, the taming of the American wilderness, and the rise of the Conservation Movement, 1800–1900*
Michael Williams, emeritus professor in geography at the University of Oxford, Oxford, England

- Le contexte économique et social de l'émergence du rendement forestier soutenu en Europe: la magna carta, le déclin du féodalisme, et le développement d'une base légale du droit de propriété sous le régime de la « Common Law » anglaise et du Code Napoléonien / *The social and economic context for emergence of sustained-yield forestry in Europe: The Magna Carta and the decline of feudalism, and the development of a legal basis for land rights under English Common Law and the Napoleonic Code*
David Adams, retired professor, Department of Forestry, North Carolina State University, Raleigh, North Carolina

2:00 pm   Les évolutions forestières / *Forestry evolutions*
Moderator: Jean Paul Lanly, retired, former director, Nature Conservation, Forests, and Wood Section of the Council of Rural Development, Waters, and Forests (Conseil Général du GREF) of the Ministry of Agriculture

- Sciences et forêts: acquis, évolutions et défis / *Science and the forest: Achievements, evolution, and challenges*
Yves Birot, retired, former director of the INRA Forest Research Department, Nancy, France
François Houllier, scientific director for Plant and Plant Products, INRA, Paris, France

- Sylviculture: conversion et reconversion / *Silviculture: conversion and reconversion*
  Heinrich Spiecker, director of the Institute for Forest Growth, Albert-Ludwigs-University, Freiburg; faculty of Forest and Environmental Sciences, Frieberg, Germany

- L'homme, la nature et la forêt: les grands débats d'idées en cours / *Man, nature and forest: Today's important issue*
  Christian Barthod, deputy director for Natural Areas, Ministry of Ecology and Sustainable Development, Paris, France

- L'avenir du bois dans nos sociétés en évolution / *The future of wood in our evolving societies*
  Michel Vernois, scientific director, Centre Technique du Bois et de l'Ameublement, Paris, France

- Le débat international sur les forêts: nouveaux concepts et nouvelles politiques / *The international dialogue on forests: New concepts and policies*
  Gérard Buttoud, professor and director of the Laboratory of Forest Policy at the French Institute of Forestry, Agricultural, and Environmental Engineering (ENGREF), Nancy, France

5:30 pm	**Présentation de *The Greatest Good* centennial film et cérémonie à la mémoire de Gifford Pinchot /** *Presentation of* The Greatest Good *centennial film and ceremony commemorating Gifford Pinchot* **Dale Bosworth**, Chief, U.S. Forest Service, Washington, D.C.
8:30 pm	**Recital de piano a l'Hotel de Ville /** *Piano concert at the Hotel de Ville* **Yakov Alvaz**

Tuesday, March 8, 2005

8:15 am   Foresterie et société / *Forestry and society*
Moderator: **Jean-Luc Peyron**, director of ECOFOR, Paris, France

- Le consensus émergeant sur les principes de la gestion forestière durable aux États-Unis: objectifs communs de conservation pour le 21ème siècle / *The emerging consensus on general principles of sustainable forest management in the US: Common goals for the next century of conservation*
  V. Alaric Sample, president, Pinchot Institute for Conservation, Washington, D.C.

- Education forestière des deux côtés de l'Atlantique / *Forestry education on both sides of the Atlantic*
  Joseph-Edward de Steiguer
  Patrice Harou, visiting professor, École Nationale du Génie Rural des Eaux et des Forêts, Nancy, France
  Terry L. Sharik, head, Department of Environment and Society, Utah State University, Logan, Utah

- Le rôle de la conservation des forêts pour répondre aux défis environnementaux mondiaux du 21éme siècle: le nécessité d'une coopération internationale multisectorielle / *The role of forest conservation in meeting the global environmental challenges of the 21st century: The necessity of international, multi-sectoral cooperation*
  Jeff Burley, former director of the Oxford Forestry Institute, Oxford, England

10:30 am   Table Ronde. Quel futur pour la foresterie? Responsabilités communes E.U.-U.S. au niveau domestic et mondial et opportunités pour renforcer leur coopération et coordination / *Roundtable discussion. What future for forestry? Common E.U.-U.S. responsibilities at the domestic and global level and opportunities for reinforced cooperation and coordination*

Moderator: **Franz Schmitüsen**, professor, Swiss Federal Institute of Technology (ETH), Zurich, Switzerland

Dale Bosworth, directeur du Service Forestier des États-Unis / *Chief, U.S. Forest Service*
Madame Claire Hubert, sous-directrice de la forêt et du bois au ministère chargé de l'agriculture / *deputy director of Forestry, Ministry of Agriculture*
Gilbert Rodts, représentant du directeur général de l'Office National des Forêts / *representative of the French National Office of Forests*
V. Alaric Sample, président du Pinchot Institute / *president, Pinchot Institute*
Joss Crochet, représentant de la forêt privée européenne / *president of the European Private Forest Owners Association*

12:30 pm **Clôture du congrès /** *Closing speech*
Hôtel de Ville, Place Stanislas, réception par André Rossinot, Maire de Nancy

U.S. Colloquium

Grey Towers National Historic Landmark
Milford, Pennsylvania
June 27–28, 2005

Monday, June 27, 2005

8:30 am **Welcome**
David Tenny, deputy under Secretary,
Natural Resources and Environment,
United States Department of Agriculture
Sally Collins, Associate Chief, U.S. Forest Service

Introduction
V. Alaric Sample, president, Pinchot Institute for Conservation, Washington, D.C.
Dominique Danguy Des Déserts, retired, former director, École Nationale du Génie Rural des Eaux et des Forêts, Nancy, France

9:00 am **Common roots of forestry and forest science /** *Les sources communes de la foresterie et de la science forestière*

- The social and economic context for emergence of sustained-yield forestry and development of a legal and policy framework / *Le contexte economique et social de l'émergence du rendement forestier soutenu en Europe et le développement d'une base légale et d'un cadre politique*
David Adams, retired professor, Department of Forestry, North Carolina State University, Raleigh, North Carolina

- European silviculture knowledge during Pinchot training in Nancy / *Les acquis de la sylviculture européenne lors de la formation de Gifford Pinchot à Nancy*
Marie-Jeanne Lionnet, head librarian, ENGREF, Nancy, France

- The evolution of forest management in Europe / *L'évolution de l'aménagement forestier en Europe*
  Heinrich Spiecker, director of the Institute for Forest Growth, Albert-Ludwigs-University, Freiburg, Faculty of Forest and Environmental Sciences, Frieberg, Germany

10:30 am   **Divergence in forestry in the U.S. and Europe in the 20th century** / ***Divergence en foresterie entre les États Unis et l'Europe***

- Evolution of forestry education in both Europe and the U.S. / *Evolution de l'éducation Forestière en Europe et en Amérique*
  Patrice Harou, visiting professor, École Nationale du Génie Rural des Eaux et des Forêts (ENGREF), Nancy, France

- Man, nature and forests: the evolution in social attitudes toward forestry and nature in the E.U. / *L'homme, la nature et la forêt: L'évolution des attitudes sociales vis à vis de de la foresterie et de la nature dans l'Union Européenn*
  Christian Barthod, deputy director of Natural Areas, French National Forest Health Department

1:30 pm   **Divergence in forestry in the U.S. and Europe in the 20th century** / ***Divergence en foresterie entre les États Unis et l'Europe (continued)***

- Gifford Pinchot and John Muir: The diverging (or converging?) paths of two icons of the American conservation movement / *Gifford Pinchot et John Muir: les chemins divergents (ou convergents?) de deux icônes du mouvement de conservation américain*
  John Perlin, Professor, Department of Physics, University of California, Santa Barbara, California, author of *A Forest Journey: The Story of Wood and Civilization*

- European and U.S. influence on forest policy at UN FAO, 1945–2000 / *L'influence européenne et américaine sur la politique forestière aux Nations Unies*

Jean Paul Lanly, retired, former director, Nature Conservation, Forests and Wood Section of the Council of Rural Development, Waters, and Forests (Conseil Général du GREF) of the Ministry of Agriculture

- Back to the future: Approaches to U.S. forestry at the turn of the (20th and) 21st centuries / *Retour sur le future: les approches de la foresterie américaine au tournant du 21ème siècle*
  Paul Hirt, associate professor of History at Arizona State University

- New challenges in forestry in Germany / *Nouveaux défis de la foresterie en Allemagne*
  Konstantin von Teuffel, director of the Forest Research Institute of Baden-Württemberg, Freiburg, Germany

- European forests: Heritage of the past and options for the future / *Forêts européennes: Héritage du passé et des options à l'avenir*
  Franz Schmithüsen, professor, Swiss Federal Institute of Technology (ETH), Zurich, Switzerland

4:00 pm   **Signing of agreement between ENGREF and Yale University School of Forestry & Environmental Studies /** *La signature d'un accord entre ENGREF et l'Université de Yale, École des Forêts et Etudes Environnementales*
Cyrille Van Effenterre, director, ENGREF;
Gordan Geballe, Associate Dean, Yale University

Tuesday, June 28, 2005

Reconvergence toward common goals and objectives in sustainable forestry in the 21st century / *Reconvergence vers des buts et des objectifs communs de la foresterie durable du 21ème siècle*

- The emerging consensus on general principles of sustainable forest management in the US: Common goals for the next century of conservation / *Le consensus émergeant sur les principes de la gestion forestière durable aux États-Unis: Objectifs communs de conservation pour le 21ème siècle*
  V. Alaric Sample, president, Pinchot Institute for Conservation, Washington, D.C.

- France's National Forest Plan and implications for biodiversity conservation / *Le plan forestier national français et ses implications pour la conservation de la biodiversite*
  Jean-Jacques Bénézit, deputy director, International Relations, Ministry of Agriculture and Fisheries, Paris, France
  Cyrille Van Effenterre, director, École Nationale du Génie Rural des Eaux et des Forêts, Nancy, France

- The continuing evolution in social, economic and political values relating to forestry in Europe and the US: Implications for forest policy / *L'évolution continue des valeurs sociales, économiques et politiques liée à la forêt en Europe et en Amérique: Implications pour la politique forestière*
  Dennis Le Master, professor emeritus of Purdue University, West Lafayette, Indiana
  Franz Schmithüsen, professor, Swiss Federal Institute of Technology (ETH), Zurich, Switzerland

10:30 am   **Roundtable discussion.** What future for forestry? Common E.U.-U.S. responsibilities at the domestic and global level and opportunities for reinforced cooperation and coordination / *Quel futur pour la foresterie? Responsabilité commune E.U.-U.S. au niveau domestic et mondial et opportunités pour une coopération et coordination renforcée*
Moderator: **Dennis Le Master**, professor emeritus of Purdue University, West Lafayette, Indiana

Jean-Jacques Bénézit, deputy director, International Relations, Ministry of Agriculture & Fisheries, Paris, France
Sally Collins, associate chief, U.S. Forest Service
Jim Grace, state forester, Pennsylvania Bureau of Forestry
Roland Burrus, president, French National Association of Private Forests
Robert Flies, head of Forests Unit, European Commission, Brussels, Belgium

11:45 am   **Summary and next steps /** *Résumé et suivi*
V. Alaric Sample, president, Pinchot Institute for Conservation, Washington, D.C.

12:00 pm   **Closing remarks /** *Remarques de cloture*
Peter Pinchot, director of Milford Experimental Forest, Milford, Pennsylvania, and grandson of Gifford Pinchot

Contributions to U.S. Forest Service centennial time capsule / *Contribution à la capsule temps du centenaire du service forestier américain*

CONTRIBUTORS

David A. Adams received his degrees from North Carolina State University in wildlife management and plant ecology. He was North Carolina's first chief park naturalist and also served as curator of the North Carolina Museum of Natural History, commissioner of Commercial and Sports Fisheries, and assistant secretary of the Department of Natural Resources and Community Development. He held federal positions as a member of the President's Commission on Marine Sciences, Engineering, and Resources and as a staff member on the National Council on Marine Resources and Engineering Development in the Executive Office of the President. In the private sector, Adams founded Coastal Zone Resources Corporation, an environmental consulting company. At North Carolina State University he was a professor in the Division of Multidisciplinary Studies and the Department of Forestry, where he taught environmental science, environmental impact assessment, and renewable resource policy and wrote *Renewable Resource Policy: The Legal-Institutional Foundations*. Adams is currently retired.

Dr. Steven Anderson is President and CEO of the Forest History Society in Durham, NC, an international organization that aims to preserve and interpret the documents of forest and conservation history and bring a historical context to environmental decision-making. He brings to that position more than 25 years experience as a university professor, teacher, and administrator. He has worked for Native American tribes in Alaska, and worked throughout the southern United States helping industry establish research studies in forest fertilization and management; and spent ten years on the faculty of Oklahoma State University as program leader for their Extension Forestry, Wildlife, and Aquaculture Program. Dr. Anderson received a Bachelor of Science in Forest Management in 1977 from Rutgers University in New Jersey; a Masters in 1979 in Forest Soils from the University of Washington, and a Ph.D. in Forest Economics from North Carolina State University in 1987. He has authored over one hundred articles on various aspects of forestry and forest history. Most recently he co-edited two books in the IUFRO Research Series entitled *Forest History: International Studies on Socio-Economic and Forest Ecosystem Change* and *Methods and Approaches* in *Forest History*. Dr. Anderson currently edits the *Forest History Today* magazine as well as administers the production of the quarterly journal *Environmental History*.

Christian Barthod established France's national Forest Health Department, which he managed for seven years. He was appointed deputy director in charge of forestry in 1995, and deputy director in charge of natural areas (including national parks and reserves) in 2002. He directed France's international forest negotiations between 1988 and 1995. He previously worked in private forests of the Massif Central and held two jobs in the forest research sector. Barthod is the coauthor of four books on forests and wood and is the author of more than 120 articles and other publications on forest protection and forest policy, among other topics. He also drafted two laws, the Forest Law of 2001 and the National Parks Law of 2006.

Jean-Jacques Bénézit, born in 1957, is a senior engineer, graduate of the École Nationale du Génie Rural, des Eaux et des Forêts (French Institute of Forestry, Agricultural and Environmental Engineering). He first held several management positions in Regional Directorates of the Office National des Forêts (ONF) (1982–1991 Alsace, Lorraine, and Normandy). He then was appointed head of Development and International Department at the ONF General Directorate in Paris (1991–1996). Subsequently, he was posted as agricultural counsellor to the French Embassy in London (1996–2001) and eventually became deputy director, International Relations at the Ministry of Agriculture and Fisheries in Paris (2001–2006). He is now counsellor at the Delegation of France to the Organization for Economic Cooperation and Development (OECD), in charge of agriculture, fisheries, environment, and sustainable development portfolio.

Yves Birot graduated from Institut National Agronomique in Paris and the School of Forestry at Nancy. He joined the Centre Technique Forestier Tropical in West Africa as a researcher in plantation forestry and silviculture. Back in France, he joined INRA, where he specialized in forest genetics applied to conifers improvement and breeding, holding positions in Nancy and Orleans. He became head of the Forest Research Unit at Avignon, before being deputy director and director of the INRA Forest Research Department (1989–1998). Before retiring in 2002, he carried out several missions for

INRA: European Scientific Cooperation, scientific assessments, expanding forest research capacity. Birot has been member or chair of many boards or scientific advisory boards of national (ONF, ECOFOR, CIRAD-Forêt, Cemagref, ENGREF) and international institutions (Marcus Wallenberg Prize Selection Committee, COST, European Forest Institute, CIFOR). He is currently the chairman of the Scientific Council of the European Forest-based Technology Platform.

Dale N. Bosworth, a second generation forester and Forest Service employee, was born in Altadena, California and raised on ranger station compounds. He received his B.S. in forestry from the University of Idaho, in 1966 and began working for the agency on the St. Joe National Forest, in Idaho. Just prior to becoming the 15th Chief of the U.S. Forest Service in 2001, Dale served as the regional forester for the Northern Region, where he developed a program of stewardship contracting, involving citizens in planning and land management activities. Dale has had a breadth of experience on the ground, and in several leadership positions in the agency, including forest supervisor of the Wasatch-Cache National Forest, regional forester for the Intermountain Region and deputy regional forester for the Pacific Southwest Region.

Jeffery Burley, CBE, took an honors degree in forestry at Oxford and an M.F. and Ph.D. at Yale. He headed the Forest Genetics Research Laboratory of the Agricultural Research Council of Central Africa for four years before returning to Oxford University to become a senior research officer, lecturer, and ultimately director of the Oxford Forestry Institute and professor of forestry. From 1995 to 2000 he was the president of the International Union of Forestry Research Organizations. For many years he was director-at-large of the International Society of Tropical Foresters. In 1991 he was presented with the CBE (Commander of the British Empire) by the Queen for services to international forestry. He is an honorary member of the Society of American Foresters and honorary fellow of the UK Institute of Chartered Foresters. Since 1994 he has been a member (and since 1998 the chairman) of the committee that selects the winner of the annual Marcus Wallenberg Prize for forestry and forest products research.

Gérard Buttoud is professor and director of the Laboratory of Forest Policy, at the French Institute of Forestry, Agricultural and Environmental

Engineering (ENGREF) in Nancy, France. From 1983 to 1995, he acted as the director of the INRA (National Institute of Agricultural Research) Laboratory of Forest Economics, also in Nancy. His present field of research is forest policy analysis, with focus on formulation and evaluation of national forest policies and programs, and changes in modes of governance at world level. Buttoud chairs the scientific board of the European Observatory of Mountain Forests (EOMF) and was a member of the scientific advisory board of the European Forest Institute (EFI) from 1999 to 2004. He is the author of seven books and more than 160 papers published in national and international journals.

Dominique Danguy des Déserts is retired chief engineer of the French Institute of Forestry and Environmental Engineering (ENGREF), which is now under AgroParisTech. Since 2005, he has been a member of Conseil General (Central Committee) in the French Department for Agriculture and Fisheries where he is in charge of various missions, especially on forest issues. From 1996 to 2005 he was in charge of the forest training center in Nancy, France, where he managed the "formation des ingénieurs forestiers" (a master's degree) and developed international relationships in forestry. From 1972 to 1996 he occupied several positions in the Office National des Forêts (Provence and Vosges/Lorraine) and in agriculture and forest policy (Central Mountains, Brittany, Poitou). He is a 1970 graduate of the Institut National Agronomique in Paris and a 1971 post-graduate of the École Nationale du Genie Rural des Eaux et des Fôrets (ENGREF) in Nancy with a specialization in forestry.

J.E. "Ed" de Steiguer is professor of natural resource economics and policy in the School of Natural Resources at the University of Arizona. Previously, he was a research economist and policy analyst with the USDA Forest Service. In 1992 he was associate professor at l'École Nationale du Génie Rural des Eaux et des Forêt in Nancy, France, where he lectured in forest economics. De Steiguer holds a bachelor's degree in economics from Lamar University, an M.F. from Stephen F. Austin State University and a Ph.D. in forestry from Texas A&M University. In 2001, he was named a Udall Fellow by the Udall Center for Studies of Public Policy and the Institute for Studies of Planet Earth. He is the author of more than 100 books and articles dealing with forestry and the environment. His teaching and research interests focus on the economic requirements of policies and laws related to planning on the U.S. national forests.

Patrice Harou is visiting professor, L'École Nationale du Génie Rural des Eaux et des Forêt, Nancy, France. He earned a degree in agricultural engineering at the Catholic University of Louvain, Belgium, and his Ph.D. in natural resources economics at the University of Minnesota. Harou has worked for the UN Food and Agriculture Organization in Honduras and Brazil, and as director of the forestry program at the University of Massachusetts, Amherst. He has researched the forestry incentives systems in E.U. countries at Albert-Ludwigs University, Freiburg, Germany, and consulted for international organizations, banks, and firms. In 1990, Harou joined the World Bank, where he synthesized research and trained professionals and practitioners of client countries in environmental economics and policy. Today he conducts assessments for the bank's Independent Evaluation Group. He has written more than 100 publications on natural and forestry resources and environmental economics and contributed to several books on these subjects.

Paul Hirt, who holds degrees from the University of Arizona in philosophy, Asian studies, and U.S. history, is associate professor of history at Arizona State University; he previously taught for eleven years at Washington State University. Hirt is a specialist in the history of the American West, environmental studies, and public lands policy. He is the author of *A Conspiracy of Optimism: Management of the National Forests since World War Two* (Nebraska, 1994), and the editor of two anthologies on the environmental history of the Pacific Northwest: *Terra Pacifica: People and Place in the Northwest States and Western Canada* (Washington State University, 1998), and Northwest Lands, *Northwest Peoples: Readings in Environmental History* (University of Washington, 1999). He is currently writing a book on the history of electric power in the Pacific Northwest. Besides his academic work, Hirt has been active with a variety of NGOs promoting biodiversity and wildlands conservation for the past 25 years.

François Houllier graduated from École Polytechnique and the Nancy School of Forestry and has a Ph.D. in forest biometrics from the University of Lyon. At the French National Forest Inventory Service, he contributed to the development of information systems for assessing and predicting timber resources. At ENGREF, he taught forest biometrics and established a research group focused on modeling forest dynamics and predicting timber quality as the result of growth processes. He was director of the French Institute of Pondicherry (India), a multidisciplinary research center in human, social, and ecological sciences, where he worked on tropical forest biodiversity and dynamics. He was then director of the Plant Modeling Research Unit in Montpellier. As head of the Forest Department at INRA, he created the Department of Forest, Grassland and Freshwater Ecology before becoming scientific director in charge of plants and plant products. Since 2001 he has been a member of the Board of the European Forest Institute, which he has been chairing since 2004.

Jean-Paul Lanly graduated from École Polytechnique (an advanced institute in mathematics, physics, and chemistry) and the Nancy School of Forestry and has a Ph.D. from the University of Toulouse. At the Centre Technique Forestier Tropical, he designed and supervised large-scale forest surveys and management research activities in tropical Africa and Latin America. At the UN Food and Agriculture Organization, he served in several capacities before becoming director of the Forest Resources Division and had responsibilities in global forestry programs such as the Forest Resources Assessment programme, the Tropical Forestry Action Plan, and the forestry component of the 1992 Rio "Earth Summit." Back in France, Lanly led the Nature Conservation, Forests and Wood Section of the Council of Rural Development, Waters and Forests (Conseil Général du GREF) of the Ministry of Agriculture. Since his retirement from the civil service in 2003, he remains active in forestry and related fields and is a member of the Academy of Agriculture (forestry section).

Dennis C. Le Master is professor emeritus of Purdue University. He was professor and head of the Department of Forestry and Natural Resources, Purdue University, from 1988 through 2004. Before that, he was professor and chair, Department of Forestry and Range Management, Washington State University (1979–1988). He served as staff consultant for the Subcommittee on Forests, U.S. House of Representatives, for the 95th Congress (1977–1978). Earlier, he was director of resource policy for the Society of American Foresters. Le Master received his bachelor's, master's, and doctoral degrees from Washington State University. He received his Ph.D. in economics in 1974. His teaching and research interests are in resource economics and policy. He has authored or coauthored more than 90 scholarly publications and is currently a senior fellow at the Pinchot Institute for Conservation and was a member of its board of directors for eight years, serving as chair from 2000 to 2002.

François Le Tacon is emeritus director of research at l'Institut National de la Recherche Agronomique (INRA), France, and has been president of the Forest Research Center of Nancy for ten years. From 1980 to 2002, he was permanent adviser at the International Foundation for Science, Stockholm, in the field of forestry. He received his Ph.D. in soil science from the University of Nancy and prepared a science doctoral thesis in forest biology at the National Polytechnic Institute of Lorraine. His research has focused on forest ecology and forest tree nutrition. For the past 30 years he has conducted studies on ectomycorrhizae, and he is now researching the functional diversity of ectomycorrhizal fungi, molecular ecology, and fungal phylogeny. Le Tacon has published more than 100 scientific papers in international journals in the fields of forest ecology and mycorrhizae.

Ambassador Jean-David Levitte presented his credentials to President Bush on December 9, 2002. Born in 1946 in the south of France, Ambassador Levitte earned a law degree and is a graduate of Sciences-Po (the renowned Institute for Political Science in Paris) and of the National School of Oriental Languages, where he studied Chinese and Indonesian. Ambassador Levitte has had a distinguished and outstanding career in the French foreign service, serving on the staff of two French Presidents and holding various senior positions in the French foreign service. Mr. Levitte was assigned to his first position in the United States as Second Counselor at the Permanent Mission of France to the United Nations in New York. Upon returning to Paris, Mr. Levitte was appointed Deputy Assistant Secretary in the African Bureau. He was then assigned as Deputy Chief of Staff to the Foreign Minister, a position he held from 1986 to 1988. In 1988, he was designated to his first position as Ambassador and served as the French Permanent Representative to the United Nations Office in Geneva from 1988 to 1990. Returning to Paris in 1990, he held senior positions in the French Foreign Ministry, first as Assistant Secretary for Asia and then as Undersecretary for Cultural and Scientific Cooperation.

After the presidential elections in 1995, President Chirac asked Ambassador Levitte to be his Senior Diplomatic Adviser. He served in that position from 1995 to 2000. President Chirac appointed him as French Permanent Representative to the United Nations in 2000, his most recent position before being appointed French Ambassador to the United States.

Marie-Jeanne Lionnet, now retired, has an archivist's degree from the School of Librarians and Archivists of Paris. She worked for two years as an archivist in the French Institute for Agronomic Research (INRA) in Versailles. From 1969 to 2005 she was the head librarian at the École Nationale du Génie Rural des Eaux et des Forêts (ENGREF) in Nancy. Because ENGREF has played a major role in the evolution of forestry at the national and international levels, the library has regularly expanded its holdings with writings from all over the world and become an important resource for forest education and training. Communication technologies have recently allowed the further expansion of its collections. Lionnet has contributed to the new edition of the trilingual version of *Decimal Forest Classification*, published by IUFRO. She has also been involved in the publication of several books on French forest history. Gifford Pinchot and the exchange between two continents are her new field of research.

Francis Martin received a Ph.D. in plant physiology from the University of Nancy and a science doctoral thesis in plant molecular biology from the Paris XI University. He has worked at the University of Nancy, University of California Los Angeles, the CSIRO Division of Forestry in Perth. Today he is head of the forest microbial ecology program and leader of the Center of Excellence for Eco-Genomics at the Forestry Research Center of l'Institute National de la Recherche Agronomique in Nancy. He also coordinates the research program on the functional genomics of poplar at INRA and is a member of the International Mycorrhizal Genomes Steering Committee. His current research interests include the interactions between trees, ectomycorrhizal fungi and rhizospheric bacteria; the physiological mechanisms that allow the integration of symbiosis in the biology of the tree and the mycobiont; and eco-genomics for understanding the role of fungal and bacterial communities in nutrient cycling in forest ecosystems.

Char Miller is a member of the History Department and director of Urban Studies at Trinity University in San Antonio. He is a senior fellow at the Pinchot Institute for Conservation, a member of the board of directors of the Forest History Society, and an associate editor for the *Journal of Forestry* and *Environmental History*. Miller is author of *Gifford Pinchot and the Making of Modern Environmentalism*, coauthor of *The Greatest Good: 100 Years of Forestry in America* (1999, 2004), editor of *The Atlas of U.S. and Canadian Environmental History*, and author of *Deep in the Heart of San Antonio: Land and Life in South Texas* and *Ground Work: Conservation in American Culture* (2006). In 2004–2005, as centennial lecturer for the USDA Forest Service, he gave more than 70 talks across the United States, and was a consultant on the documentary, *The Greatest Good: A Forest Service Centennial Film*.

John Perlin is author of *A Forest Journey: The Story of Wood and Civilization*, which recounts wood's major role in the culture, demographics, economy, politics, and technology of the great civilizations of Sumer, Assyria, Egypt, China, Knossos, Mycenae, Greece, Rome, Europe, and North America. Harvard University Press has chosen *A Forest Journey* as one of its "One-Hundred Great Books." After five printings at Harvard University Press, a second edition has been published by Countryman Press. Perlin's latest work, *From Space to Earth: The Story of Solar Electricity*, covers the development of solar cells and their applications. He was recently the principal investigator for the National Renewable Energy Laboratory and the National Center for Photovoltaics' gathering of exhibit and historical material for the 50th anniversary of the crystalline silicon solar cell by Bell Laboratories. Currently, with Nobel Laureates Walter Kohn and Alan Heeger, Perlin is producing a film, *Power from the Sun: A Century since Einstein's Photon/Fifty Years of Modern Photovoltaics*.

Jean-Luc Peyron is currently director of ECOFOR, a French public body and a federal platform in charge of research coordination in the field of forest ecosystems. Located in Paris, ECOFOR comprises nine scientific, statistical, and professional organizations and manages research programs on functioning, dynamics, and sustainable management of forest ecosystems, in both temperate and tropical environments, and also works on information systems for knowledge and management. Peyron helps edit the *French Forest Review* and is very active in the IUFRO unit on managerial economics and accounting. He has an engineering degree in quantitative sciences, an engineering degree in forest, agricultural, and environmental sciences, and a Ph.D. degree in economics. In his doctoral thesis, he proposed a forest economic accounting system based on physical and monetary stocks and flows, intended to better integrate forestry into national accounts.

Jean Pinon is a plant pathologist and director of research at INRA-Université Henri Poincaré Nancy I in Champenoux, France. His main subjects of interest are wheat rust in Canada (electron microcopy); poplar diseases, including pathogen variability and host resistance; Dutch elm disease and the selection of resistant clones; and oak wilt and the testing of European oaks in the United States for their susceptibility. He has been head of the Forest Pathology unit in Nancy since 1995 and head of Forest Pathology in INRA since 2001. He is also the former secretary of the working group on disease of the International Poplar Commission.

V. Alaric Sample has served as president of the Pinchot Institute for Conservation in Washington, DC, since 1995. He is a fellow of the Society of American Foresters and a research affiliate on the faculty at the Yale School of Forestry and Environmental Studies. Sample earned an MBA and an MF as well as a doctorate in resource policy and economics from Yale University. His professional experience spans public, private, and nonprofit organizations, including the U.S. Forest Service, Champion International, the Wilderness Society, and the Prince of Thurn und Taxis in Bavaria. Sample has served on numerous national task forces and commissions, including the task force on biodiversity on private lands for the president's Commission on Environmental Quality, and the National Commission on Science for Sustainable Forestry. He is the author of numerous research papers and books on national and international forest policy. His most recent book is *Forest Conservation Policy*, with Antony Cheng, published in 2004.

Franz Schmithüsen is professor emeritus and former chair of Forest Policy and Forest Economics at the Swiss Federal Institute of Technology (ETH) in Zurich. He studied forestry and business economics in Freiburg, Vancouver, and Zurich and graduated from the University of Freiburg and ETH Zurich. His research and teaching focus on national and international forestry development, landscape conservation, and comparative analysis of renewable resource policies and environmental legislation. As author and coauthor he has published numerous monographs, textbooks, and reports on forest policy and law in Europe, North America, Africa, Asia, and Latin America. He is honorary member of the International Union of Forest Research Organizations (IUFRO) and foreign member of the Academy for Agriculture in France, and was invited to give the Pinchot Distinguished Lecture 2003. In recognition of his multidisciplinary approach toward forest, landscape, and society, he has received honorary doctorates from the Aristotle University in Thessaloniki, Greece, and from the Czech Agriculture University in Prague.

Terry L. Sharik is Professor of Forest Ecology in the Department of Environment and Society in the College of Natural Resources, Utah State University, Logan Utah. He served as Head of the Department from 2002–2007. He earned his B.S. in Forestry and Wildlife Management from West Virginia University in 1964 and a Master of Forestry in Forest Recreation in 1966 and a Ph.D. in Forest Botany in 1970 from the University of Michigan. Prior to arriving at Utah State University, Sharik served as Assistant Professor of Biology at Oberlin College from 1971–1973, Assistant Professor of Forest Biology at Virginia Polytechnic Institute from 1975–1982, and as Associate Professor and Professor of Forest Ecology at Michigan Technological University from 1986 to 1993.

Heinrich Spiecker is director of the Institute for Forest Growth, Albert-Ludwigs-University Freiburg, Faculty of Forest and Environmental Sciences. He received an MSc in the economics of forest management at University of California–Berkeley, a Ph.D. in forest management, and a habilitation (postdoctoral lecture qualification) in forest growth and silviculture at Freiburg. Spiecker worked as guest professor at the University of Curitiba (Brazil) and for several years with the forest administration in southwestern Germany. His research interest is forest growth and forest management. He is involved in international teaching programs and coordinates international research projects analyzing environmental changes and their impacts on forests and wood production. He is a fellow of the European Forest Institute, where he chaired the Scientific Advisory Board. He is an associate of the University Wisconsin at Stevens Point and adjunct professor at the University of Toronto. He is president of the Association for Tree-Ring-Research and a general board member of IUFRO.

Cyrille Van Effenterre is president of Paris Tech (Paris Institute of Technology) which includes ten of Paris's well known French engineering schools including Agro Paris Tech. He was previously, and since 2000, director of l'École Nationale du Génie Rural des Eaux et des Forêts (ENGREF), the French national school for water, forest, and environmental sciences which has now be integrated into AgroParisTech, the main agriculture school in France. From 1998 to 2000, he was the director of rural land and forest at the French ministry of agriculture, where he supervised the writing and implementation of various laws and policies regarding forestry and the environment. In the early stages of his career, he held positions dealing with soil conservation, natural hazards in mountains, forest fires, the environment, and land planning in the French Alps, and was appointed as a minister's technical adviser in 1991. Van Effenterre holds two engineering degrees from École Polytechnique and from ENGREF.

Michel Vernois has served as scientific director for the Technical Center for Wood and Furniture in Paris since 1996. He has headed research and development programs in the field of wood, wood composites, and furniture for the industrial sector and been a member of several research committees at the national and European level. He was previously director of research and development and quality control at a major company in the wood panel industry, and also a research engineer in the pulp and paper industry. Vernois holds a master of science degree and a Ph.D. in organic chemistry from the University of Strasbourg and a master of applied science from the University of Toronto. He has been also active in polymer chemistry at the Research Center for Macromolecules in Strasbourg and contributed to several patents and publications in this field.

Konstantin von Teuffel is director of the Forest Research Institute of Baden-Württemberg, Freiburg. He is also a member of the board of the European Forest Institute, coeditor of *Annals of Forest Science*, and coordinator of a IUFRO research group. He graduated from Göttingen and undertook professional training at the State Forest Administration of Baden-Württemberg. He conducted forest management planning at Forstamt

Ochsenhausen and subsequently served as head of the State Forest District in Ulm, Baden-Württemberg; head of the Division of Forest Policy at the State Forest Service Regional Office in Tübingen; and head of the Division of Silviculture, Forest Management Planning, Hunting and Forest Research at the Ministry of Rural Space, Stuttgart. Von Teuffel has managed several international research and development projects on the E.U. level, including conservation of the ecological diversity of floodplain-forests of the Upper Rhine Valley, harvesting and salvage in storm-damaged forests, nutrient fluxes in soils on permanent plots, and forest decline inventory.

Michael Williams is emeritus professor in geography at the University of Oxford and a fellow of Oriel College. He previously taught at Adelaide, Australia, and at the universities of Wisconsin (Madison), Chicago, and California (Los Angeles). His research interests are initial settlement and landscape evolution in Britain, Australia, and the United States; and global land-use and land-cover transformation, particularly forest clearing. He is the author of *Americans and Their Forests* (1989) and, most recently, *Deforesting the Earth: From Prehistory to Global Crisis* (Chicago University Press, 2003, 2006). He is currently writing a biography of Carl Sauer, the American geographer and polymath. Williams has been editor of *The Transactions of the Institute of British Geographers* and joint editor of *Progress in Human Geography* and *Global Environmental Change*, and is currently on the editorial board of *Environmental History* and Environment and History. He was elected fellow of the British Academy and fellow of the Forest History Society, Inc. (Durham, NC) in 1989.